GARDEN LIBRARY
PLANTING FIELDS ARBORETUM

D1435755

GARDEN LIBRARY
PLANTING FIELDS ARBORETUM

The Garden Plants of China

The Garden Plants of China

Peter Valder

TIMBER PRESS
Portland, Oregon

Pre-title pages Cultivars of *Rhododendron* × *pulchrum*
in the Yi Yuan, Suzhou

Frontispiece *Nelumbo nucifera*, the lotus

This page A terracotta plaque in the Forbidden City,
Beijing, showing lotuses, bullrushes, and ducks

Contents pages Weeping willows on the Bai Causeway,
West Lake, Hangzhou

First published in North America in 1999 by
Timber Press, Inc.
The Haseltine Building
133 S.W. Second Avenue, Suite 450
Portland, Oregon 97204, USA

Text and photographs © Peter Valder 1999

This book is copyright under the Berne Convention.
No reproduction without permission.
All rights reserved.

ISBN 0-88192-470-9

A catalog record for this book is available from
the Library of Congress

Editor Gillian Gillett
Designer Stan Lamond, Lamond Art & Design

Printed in Hong Kong

ACKNOWLEDGMENTS

Were it not for the kindness and support of a great many people it would not have been possible for this book to have appeared. I am particularly indebted to Gilbert Teague of Florilegium for having agreed to be the principal publisher and for his enthusiasm and encouragement throughout. My particular thanks go also to Allan McNeish, who has accompanied me on several visits to China, exhibited astonishing patience, made endless investigations on the Internet on my behalf, read the manuscript, and made many helpful suggestions. Likewise I am very appreciative of the advice given by Richard Clough, who also read the manuscript and, with his knowledge of Chinese matters, provided much assistance. And I am most grateful to Gillian Gillett for editing the text and drawing my attention to various discrepancies, and to Stan Lamond for the expertise he brought to the design of the book.

Professor Zhang Zhiming of the Beijing Botanical Garden, Chinese Academy of Sciences, has provided transliterations of the Chinese names of all the plants included, along with their meanings where appropriate. He has also been most helpful in other ways, as has Professor Zou Xiuwen, who showed me the chrysanthemum collection at the Beijing Botanical Garden and gave me a copy of her book about the lotus in China. Also I am indebted to the many people in China who guided me to things I wished to see, in particular to Tan Jie in Beijing, Bush Jiang in Yangzhou, and Joy Dong in Ningbo. I am most grateful, too, to Terese Tse Bartholemew of the Asian Art Museum of San Francisco for the information she provided concerning botanical motifs in Chinese art.

In Sydney I have been given great assistance by Dr Wang Nianen and Ms Tang Luhua, who have provided translations and transliterations, as well as unravelling many of the complexities and confusions faced by one with little knowledge of the Chinese language. Here, too, my thanks go to all those librarians who have responded so helpfully to my requests. The efforts of Anna Hallett and Miguel Garcia at the library of the Royal Botanic Gardens Sydney have been especially appreciated, as have those of Jane Wallace Alling, who sent me copies of material from the Pennsylvania Horticultural Society. For invaluable advice about the names and identities of plants I must thank Professor David Mabberley and Dr Peter Michael of the Royal Botanic Gardens Sydney, Mr A. C. Whiteley of the Royal Horticultural Society, and Mr Tom Savige, who has also provided information about the lesser-known camellias. Others who have helped are Peter Dakers, Johan Harder, Ted and Lesley Jacob, Jim and Richard Prentice, Keith and Libby Raines, Alec and Dorothy Robertson, and Stephen Ryan, all of whom allowed me to photograph plants in their gardens, the late Dr Gerald Straley, who provided information about Chinese herbaceous peonies, and Antoinette Duncan and Margaret-Anne Mayo, who for some years have patiently overseen my travel arrangements.

Permission to reproduce illustrations (figs 5.2, 5.3) from Joseph Needham's *Science and Civilisation in China*, Vol. 6, 1986, and from Osvald Sirén's *Gardens of China*, Ronald Press, New York, 1949, (fig. 12.8), was kindly granted by Cambridge University Press and the Östasiatiska Museet, Stockholm, respectively.

For providing photographs my thanks go to the Freer Gallery of Art, Washington, DC (fig. 2.15), the Metropolitan Museum of Art, New York (figs 2.1, 7.1), the Museum of Fine Arts, Boston (figure 20.14), the National Palace Museum, Taipei (figs 2.7, 2.12, 6.8, 16.3, 18.8, 19.57, 19.73, 19.74, 20.2), the Palace Museum, Beijing (figs 2.14, 16.8, 16.9, 20.6, 20.26, 20.27), and the Royal Horticultural Society, Lindley Library (fig. 1.5). For permission to photograph plates from copies of *Curtis's Botanical Magazine*, *Edwards' Ornamental Flower Garden*, *Edwards's Botanical Register*, and the *Transactions of the Horticultural Society of London* I am indebted to the library of the Royal Botanic Gardens Sydney. The photographs reproduced as figures 2.4, 2.13, 3.2, 6.11, 7.7, 7.12, 9.6, 9.7, 13.3, 13.15, 15.11, 18.4, 19.22, 19.41, 19.96, 19.114, 20.8, 20.13, 20.15, and 20.31 were taken by Jaime Plaza, to whom I am most grateful, as I am to Richard Clough for providing the transparency used for figure 5.12. The remainder of the photographs are my own.

CONTENTS

INTRODUCTION

Nowadays it is possible once more to visit most parts of China, to repeat and extend the journeys of the famous plant explorers of the past, and to collect many beautiful plants not previously brought into cultivation or, in many cases, even known. In the light of such excitements we tend to forget that the Chinese plants which have had the greatest influence on Western gardens have come to us not from the wild but from Chinese gardens or, perhaps more correctly in many cases, from Chinese nurseries. Moreover many ornamental plants that we associate with Japan were originally brought there from China.

Over a very long period the Chinese have selected and developed an array of ornamental plants to use in the composition and adornment of their gardens. On the whole, unlike the gardeners of the Western world, they have not continuously sought novelty, preferring plants rich not only in physical beauty but in historical, literary, and symbolic associations. And the connection in China between the cultivation of ornamental plants and their representation in art, literature and decoration has long been an intimate one. In short, whether they be highly developed cultivars which can survive only with human aid or merely species

Figure 0.1 View in the Yu Yuan, Shanghai

saved from extinction or near extinction, the garden plants of China make up a very significant part of its cultural heritage. As mentioned above, they are valued not merely for their appearance but as icons of values and ideals accumulated over many centuries of appreciation. And these products of some 2000 or more years of Chinese horticultural endeavour eventually came to enrich gardens in much of the rest of the world, albeit cut off from their deep cultural roots and meaningless in symbolic terms to those who have little or no idea of their long past. Amongst the best known of these plants are peaches, peonies, chrysanthemums, camellias, gardenias, azaleas, forsythias, wisteria, and crabapples, to mention but a few. And the development of the modern repeat-flowering roses would not have occurred had the

so-called monthly roses not been brought to Europe from Chinese gardens.

China is an enormous country, stretching more than 5000 km from the tropics in the south to Siberia in the north (fig. 0.2). It is not surprising, then, to find that the garden flora changes as one goes from one part of the country to another. Even so the amount of variation encountered is much less than one might expect, as the tendency to use only the plants traditionally admired by men of letters is very strong. And many of these plants seem able to grow well under an unexpectedly wide variety of climatic conditions. In addition, frost-tender plants are kept indoors for the winter in the north and brought out again when conditions are suitable. Likewise in the south, plants which do not thrive in

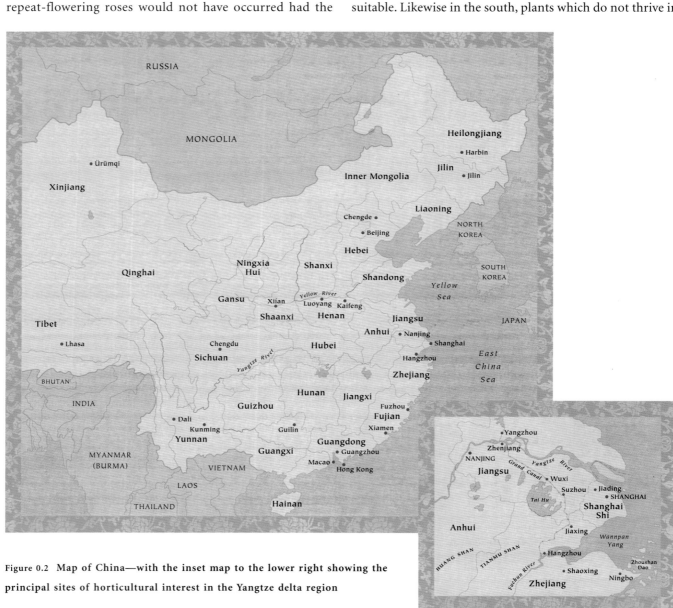

Figure 0.2 **Map of China—with the inset map to the lower right showing the principal sites of horticultural interest in the Yangtze delta region**

areas where the winters are not cold are brought annually from further north and discarded after flowering. Thus one finds tree peonies in Guangzhou (fig. 0.3) and *Hibiscus rosa-sinensis* in Beijing. It is this state of affairs which proved such a disappointment to Robert Fortune who, on visiting Tianjin and Beijing in 1861, found almost nothing he had not already collected in the south in earlier years. He was greatly surprised that the nurserymen there grew few plants which did not need protection in winter (Fortune, 1863, 1864).

As one might expect, it is in the south that one encounters the greatest number of introduced ornamental plants, many of them having long ago found their way there from India, Ceylon, and the adjoining regions of south-east Asia. These of course are chiefly tropical or subtropical species. But many other exotic plants arrived there relatively early as well, presumably because the chief ports of call for foreign ships were in Guangdong and Fujian. Plants were also introduced overland via central Asia, beginning at least 2000 years ago, and American plants began to arrive after the Portuguese reached the East and the Spanish established a galleon route across the Pacific from Mexico. It is not surprising, then, that Shen Fu (1983: trans. of original of c.1809), who thought there were no flowers he did not know, recognised only six or seven out of every ten of those he saw when he came down from Suzhou to Guangzhou somewhere around 1800.

Although much has already been written about Chinese garden plants, apart from H.L. Li's excellent books (1956a, 1959), which deal with a limited number of plants, it is either in Chinese, dealt with incidentally in works of broader scope, or scattered in various notes and articles. It has been my intention then to bring together at least some of the available information about the history, occurrence, use, and symbolism of Chinese garden plants. It is hoped this will be of interest to visitors to China, to everyone who grows these plants, and to those who simply like to read about them or admire depictions of them in Chinese art.

It would be both unrealistic and beyond my competence to attempt to include every ornamental plant that has been or is now in cultivation in China. Thus, with few exceptions, trees used in forestry have been omitted, as have the majority of the vast number of different species and cultivars now to be found in China's numerous botanical gardens, but which were not previously cultivated. And while allowances have been made for plants brought from other countries in

Figure 0.3 A potted tree peony in Guangzhou

earlier times and which became absorbed into the established Chinese garden flora, most of the annuals, perennials, bulbs, and shrubs recently introduced and now popular in public plantings have been excluded. Also, it would have been unrealistic to attempt to include every one of the many plants used for penjing but which are rarely if ever cultivated in any other manner.

On the other hand the temptation to mention some of the Chinese fruits and nuts has not been totally resisted, particularly since many of them have attracted the notice of artists and writers since ancient times and some of them occasionally appear in gardens as well as in commercial plantings. Some knowledge of these enhances the pleasure and interest to be gained by visitors to China at meal times and in the shops and markets where they are displayed.

Meat, poultry, and fish are expensive by Chinese standards and are used sparingly in most households. This, together with the fact that those adhering strictly to Buddhist tenets eat no meat, has ensured that the Chinese have become masters of vegetarian cookery, just as they are of the vegetable gardening which provides the ingredients. Thus it

Figure 0.4 A vegetable garden in front of the Xumifushou Zhi Miao, Chengde

is not surprising that the principal form of gardening in China should be vegetable gardening (fig. 0.4). Even apartment dwellers manage to have a pot or two of something on a balcony or window ledge. Hence, although this is principally a book about ornamental plants, I have included one or two Chinese vegetables that have particularly attracted my attention. Otherwise the topic is not covered, since there are already excellent books in English describing oriental vegetables (Herklots, 1972; Larkcom, 1991; Zee & Hui, 1990). Likewise the cereals and field crops, which are such a feature of the Chinese landscape, are mentioned, if at all, only in passing.

In order to place the garden plants of China in context, I have commenced this book with a survey of Chinese horticulture followed by chapters about the role of plants in Chinese culture and the introduction of Chinese garden plants to the rest of the world. The remainder of the book deals with the plants themselves. When I started writing I thought it would be best simply to arrange them in alphabetical order. This, however, would not have allowed the most esteemed plants to receive the prominence they deserve. Also it would have prevented the division of this section into chapters, a state of affairs which I felt would be uninviting to the reader. Likewise to treat the subject in a botanical manner, plant family by plant family, would have produced a similar result. Eventually I settled for a series of chapters based on a 'plants of the seasons' approach, beginning with the conifers and their allies and following on with the best-known plants more or less in the order of their blooming or use for ornament. This I felt would take into account Chinese sensibilities concerning the progression of the year. Plants not included in these chapters are to be found in those entitled 'Further Trees and Shrubs' and 'Other Herbaceous Plants'. Within chapters the entries are in alphabetical order except in Chapter 5 where, for botanical reasons, I have placed *Ginkgo* and *Cycas* before the conifers. I hope readers will not find this arrangement irrational. Those encountering difficulty in finding a particular plant are referred to the index.

It is not the aim of this book to provide detailed descriptions nor to give cultural notes. Information of this type may be found in various specialist books and in encyclopaedic works such as *The New Royal Horticultural Dictionary of Gardening* (Huxley et al., 1992) and *The Royal Horticultural Society A-Z Encylopedia of Garden Plants* (Brickell, 1996). References are cited throughout by author and date, as I have commenced doing above, the full reference in each case being listed alphabetically in the References. There is also a table of the Chinese dynasties with their dates and a short glossary of terms.

In the citing of a botanical name for each of the plants included my aim has been to make things as clear as possible for the general reader. This, however, has involved me in a certain amount of soul-searching. For a start, many Chinese garden plants have had species names assigned to them by Western botanists even though they have no known wild equivalents and are clearly of complex hybrid origin. Then, while it is desirable to be as up to date as possible, at a time when the taxonomy of many groups is being or has recently been revised, it is difficult to be right up to the minute. In fact, even if one is, this can result in the use of names with which those other than specialists are unfamiliar and which therefore can be puzzling until they become generally accepted. Hence I have tended to be conservative. Also, when it comes to plant families, on the whole I have stayed with the old names, for example Compositae, Cruciferae, Gramineae, and Leguminosae, which are familiar, have historical associations, and are descriptive, rather than opting for their modern alternatives Asteraceae, Brassicaceae, Poaceae, and Fabaceae, which lack these attributes.

As happens elsewhere, the common names of plants in China often vary from one part of the country to another, so cannot always be taken as a reliable guide for identification. However, after the botanical name given above the line in the heading for each plant, I have added, in Chinese characters, a name or names commonly used in Chinese botanical works, followed by the pinyin transliteration of each and, where appropriate, a translation. It must be pointed out, however, that in some cases the Chinese name of a plant is no more than that, as with 'oak' and 'ash' in English. Where a plant also has a commonly used name in English this is given below the line.

With regard to the Chinese names, it should be recognised that translations and transliterations cannot bring with them all the nuances and associations which the appearance of a character can bring to the Chinese mind. Also I would like to draw attention to the fact that the words *cai* (vegetable), *cao* (grass), *hua* (flower), *mu* (tree or wood), and *shu* (tree) are frequently used in the composition of plant names. In this regard the word *cao* is often used in the names of plants which are not grasses in the Western sense, and as a result of this I have taken the liberty of translating it as 'herb' or 'plant' where this seems more appropriate.

The pinyin system of romanisation is used throughout except for the citation of authors and the titles of books and articles, where the original romanisation has been retained. However, while I have managed generally to bring myself to discard the English names of Chinese localities and use, for instance, Guangzhou instead of Canton, I have made exceptions for the Yangtze, Yellow, Pearl, Mekong and Salween Rivers, as I felt it might be confusing for readers were I to use their Chinese names. For the same reason I have retained the traditional transliterations of Cantonese pronunciations in the cases of the Fa Tee Gardens, Hong Kong, Chiang Kai-shek, and Sun Yat-sen.

It has not been feasible to illustrate every plant mentioned but, wherever possible, the illustrations used are of plants photographed in China during the visits I have been fortunate to have made. It is the pleasure and interest derived from Chinese plants and gardens during these visits that have moved me to write this book.

CHAPTER 1

CHINESE HORTICULTURE

The Chinese have been cultivating plants for several thousand years. It seems that rice was grown in the Yangtze delta region as early as 5000 BC and that various indigenous and introduced crop plants were added as time went by. Compared with Europe, Chinese farming concentrated on grain production with very little grazing. Apart from draught animals such as buffaloes, oxen, and mules, the Chinese have traditionally kept pigs and fowls, which thrive on household waste, and ducks which can be husbanded on bodies of water. Meat plays a restricted role in the diet, only pork, poultry, and fish being really popular, and until recently milk products have been almost totally absent (Needham, 1984).

Since little land or grain was required for feeding livestock, any area could support a greater density of population than its equivalent in Europe. As the population grew, new land had to be brought under cultivation. Long ago most of the country was stripped of its natural vegetation, the alternative to reckless clearing often being starvation.

Figure 1.1 **A display of chrysanthemums in Beijing**

But as the organisation of agriculture improved and towns and cities appeared, plants began to be cultivated for ornament as well, as happened almost everywhere in the world.

In spite of the upheavals caused by the repeated overthrow of one dynasty by another and the periodic accidental and deliberate destruction of collections, as a result of which much may have been lost, there has come down to us a succession of books and records stretching back to the middle of the1st millennium BC. The oldest of these incorporate material from even earlier times. A study of this literature reveals a continuous steady growth in knowledge about almost everything, including the plants. There has been no equivalent in China of the so-called Dark Ages in Europe. As a result the Chinese were world leaders in many fields, at least until the Renaissance, a state of affairs which has not received the recognition due to it. Not the least of Chinese achievements has been the domestication and cultivation of cereals, legumes, fruits, vegetables, and ornamental plants. It would be hard now to imagine the world without the edible, useful, and ornamental plants developed over thousands of years by Chinese farmers, orchardists, and gardeners.

Since I am unable to read Chinese, let alone accurately interpret ancient texts, it is principally upon the work of others that I must depend for the information given here. While no doubt further documentation concerning Chinese garden plants and horticulture remains to be brought to light, much has already been interpreted by Western writers. Early records of plants in China can be found in the writings of the Arab and European visitors quoted by Major (1854) and in the subsequent summaries of the topic by Bretschneider (1880, 1898).

According to Bretschneider (1898) Marco Polo, who visited China in the 13th century, mentioned, amongst other things, grapes, mulberries, enormous pears, and peaches, both yellow and white. However, since he is believed to have dictated the text of his book, *Il Milione*, from memory during a spell in prison in Genoa in 1298 and no authentic and original copy now exists, it would be unwise to give too much credence to what are claimed to be his observations.

Another to write about China was J. Gonzalez de Mendoza, who recorded various cereals, praised the pine nuts, apples, chestnuts, melons, and lychees, and mentioned three kinds of oranges in his *History of the Great and Mighty Kingdom of China*, which was originally published in Rome in 1585. Alvaro de Semedo, whose *Relatione della Grande*

Monarchia della Cina appeared in 1643, also noted a variety of fruits and seems to have been the first European to mention the wintersweet (*Chimonanthus praecox*). A few years later Martino Martini produced the *Novus Atlas Sinensis*, published in Munich in 1655, in which he recorded many plants, including the tree peony. He appears to have been the first Westerner to notice this striking plant. Michael Boym's *Flora Sinensis* appeared the following year, the earliest European publication devoted exclusively to Chinese plants (Bretschneider, 1880). Chinese plants were mentioned too in J. B. du Halde's *Description Géographique, Historique, Chronologique, Politique, et Physique de l'Empire de la Chine et de la Tartarie Chinoise* published in Paris in 1735 and subsequently in The Hague in 1736. Du Halde had never been to China but drew on the letters of the Jesuit missionaries for his information. Many of these letters had appeared in the *Lettres Édifiantes et Curieuses*, a fascinating source of information about China at the time, published in Paris in a series of volumes from 1702 to 1776. These contain passing mention of plants but from a gardener's point of view are most notable for Jean-Denis Attiret's letter of 1 November 1743 describing the enormous imperial garden, the Yuanmingyuan, on the outskirts of Beijing.

The *Lettres Édifiantes* were followed by the *Mémoires concernant l'Histoire, les Sciences, les Arts, les Moeurs, les Usages, etc. des Chinois par les Missionaires de Pé-kin* published from 1776 to 1814. In this series of volumes a wealth of information about Chinese plants and horticulture is recorded, particularly in the articles written by P. M. Cibot, who went to China in 1759 and died in Beijing in 1784. He has left us the earliest detailed accounts in any Western language of many Chinese ornamental plants and horticultural procedures. Judging by the detail of his descriptions he would appear to have had some botanical training. Not only this but he refers to mention of these plants in the Chinese classical literature and he comments on their history and symbolism. No doubt it was his training and devotion to the church that caused him to be less than kind when mentioning some of the superstitions influencing Chinese horticultural practice.

But as far as Western studies of Chinese botany and horticulture are concerned, it was not until the country was opened to foreigners after the second Opium War (1856–1860) that any real progress could be made. An outstanding figure in this field was Emil Bretschneider. He was physician at the Russian Legation and Ecclesiastical Mission in Beijing from 1866 to 1883. A botanist of considerable merit, in his impressive *Botanicon Sinicum* (1881, 1893, 1895) and other works (1880, 1898) he gathered together a vast amount of information, laying the foundation for further research.

I am particularly indebted, too, to the historical sketch of Chinese horticultural literature given by Li (1959). This refers as well to works in European languages and gives an excellent summary of the subject together with a bibliography. This has been followed by the monumental work of Needham (1954–96) and his co-workers, of which volume 6 part 1 (1986) contains the bulk of the material about plants and gives extensive bibliographies of works in Chinese, Japanese, and Western languages. All these authors have provided rich sources of information and are recommended to anyone wishing to enquire further into the subject. It must be pointed out, however, that very little of the Chinese horticultural literature of the past 2000 years has been translated into Western languages and so remains inaccessible to most people, including of course myself. It is a matter of regret then that the material being prepared by Georges Métailié for incorporation in Needham (1986) has not yet seen the light of day, as it was to have included horticulture and its techniques. As a result there is as yet, as far as I know, no comprehensive overview of this fascinating subject available in any European language, although much can be gleaned from writers as separated in time as Cibot in the 18th century, Livingstone and Fortune in the 19th, and Meyer in the early part of the 20th.

Just as a survey of the surviving literature by someone capable of translating and interpreting it would undoubtedly tell us more concerning the plants known and cultivated at various times in China, so it seems probable that more would be revealed by a wide-ranging examination of Chinese art, particularly paintings, by competent horticultural botanists. On the whole, Western authors who have already provided identifications of plants mentioned in the Chinese literature or depicted in paintings have done a good job, though it is unrealistic to expect those with little experience in botany and horticulture always to have been correct. Also it must also be acknowledged that, while the majority of the plants depicted in Chinese art have, at least at some time or another, been cultivated, this cannot be taken for granted. For instance wild plants and weeds are known to have been admired and used in flower arrangements (Li, 1956a).

Since there are several distinct aspects of Chinese

horticulture upon which I have elaborated below, it seemed sensible to me to deal with the subject under a series of subheadings rather than as a single lengthy exposition. Since some hundreds of different plants are or have been grown in traditional Chinese gardens, it also seemed sensible to me to start with a few words about Chinese botanical nomenclature and the efforts which have been made to relate Chinese names to their Western equivalents.

Botanical nomenclature

In China a system of naming and classifying plants emerged in ancient times, but one of the great differences between China and Europe was that the latter had in its past history a dead language, Latin, from which scientific names could be formed, set apart permanently from the common names used by most people. In China there was never such a distinction. However, as Needham (1986) points out, the fact that the Chinese scholars were extremely conservative and sought to perpetuate the values of antiquity had the incidental effect of imparting a certain stability to the biological terminology and nomenclature. Many Chinese plant and animal names are still employed in the same sense as they were in the 1st millennium BC. All the same this is certainly not always the case and, as happens elsewhere, plants may have several names in one region, or different names in different parts of the country, or they may share names with other plants. On top of all this there is the phenomenon of *baizi*, a character wrongly written as that of another pronounced in the same manner. This is understandable in a language where most characters have the same or similar pronunciation as several others. Hence, although the identifications of plants mentioned in the ancient literature are the result of over 1700 years of deliberation by Chinese, Japanese, and Western scholars, it would be wise to recognise that uncertainties still exist.

It is to Bretschneider's (1893, 1895) monumental works that we are principally indebted for bringing together and clarifying the names of plants mentioned in the classical literature, including the pharmacopoeias. As Needham (1986) makes clear, Bretschneider's work constituted the climax as it were of a movement that had been going on since the 17th century—the translation of Chinese plant names into Western and then into modern Latin binomial nomenclature. Lists of botanical names with their Chinese equivalents are also given by Henry (1888), Faber and Macgregor (1907),

Matsumura (1915), and Chin (1989). In modern Chinese studies botanical names are given together with recognised Chinese ones.

When it comes to cultivar names the Chinese have been as imaginative as the rest of the world but, as in other aspects of Chinese endeavour, literary and historic references are more frequently encountered. And while plants are sometimes named after well-known figures from history, literature, or mythology, the 'Lady Alice Fitzwilliam' genre does not seem to have received the same support as the 'Thousand Petals of Jade' school of thought. Where the names of people are used, and they frequently have been, it is often in the context of their being the originator or introducer of a cultivar, for example the famous tree peony cultivar 'Yao Huang' ('Yao Family Yellow'). Places of origin feature too, as with 'Dali Cha' ('Dali Camellia'), while in other instances plants are given names that are straightforwardly descriptive, such as 'Gold-Edge Leaf', an orchid cultivar. Nowadays, with the importation of plants proceeding apace, various Western cultivar names are being translated into Chinese, so I fear it may not be long before the indigenous equivalents of 'Bloomin' Easy' begin to appear, if they have not done so already. But then I suppose if Mrs Archibald Little (1901) was able, late last century in Shanghai, to admire a spiky little chrysanthemum called 'The Tiger's Moustache', it is probably foolish to lose any sleep over the matter.

History of the cultivation of Chinese ornamental plants

While in the oldest writings it may be difficult to distinguish the legendary from the factual, it seems that pines may have been cultivated for ornamental, religious, and ceremonial purposes as long ago as 2000 BC. Apparently their popularity was challenged between about 1500 and 1000 BC by the Oriental Thuja (*Platycladus orientalis*) (fig. 1.2). According to Wang (1988), before the middle of the 3rd century BC bamboos had become indispensable, the sap of *Toxicodendron vernicifluum* was being used to make lacquer, and the Chinese Juniper (*Juniperus chinensis*), the Chinese Parasol Tree (*Firmiana simplex*), the Chinese Elm (*Ulmus parvifolia*), the Scholar Tree (*Sophora japonica*), *Catalpa ovata*, *Toona sinensis*, and the willows (*Salix babylonica* and *S. matsudana*) were in cultivation for various purposes. Also, amongst the food plants, chestnuts (*Castanea mollissima*), hazelnuts (*Corylus heterophylla*), the Chinese Date (*Ziziphus jujuba*),

peaches, apricots, Japanese plums (*Prunus salicina*), cherries (*P. pseudocerasus*), pears, and mandarins (*Citrus reticulata*) were being grown. *Prunus mume* (Plum or Japanese Apricot) had also been domesticated.

In relation to their beauty and to their flowering at particular times of the year, mention was made of daylilies, lilies, magnolias, orchids, osmanthus, herbaceous peonies, the trumpet creeper (*Campsis grandiflora*), daphne, hibiscus, hollyhocks, and the lotus. Whether or not these were cultivated at this time is unclear.

When the Qin state succeeded in unifying China in 221 BC, its king, Huang Di, became the first emperor. During his reign writing was simplified and standardised, the Great Wall was built and, in an attempt to destroy Confucianism and all other thought that did not support absolute political power, Confucian scholars were persecuted and there was the famous Burning of the Books in 213 BC. However, horticulture seems to have been encouraged and, according to Wang (1988), the emperor amassed a collection of rare plants in his capital in Shaanxi. He also sent an expedition to annex the southern regions, the source of such valuables as rhinoceros horn, ivory, jade, and pearls, as well as of various tropical plant products, including bananas and lychees. So it was in this period that interest in tropical and subtropical plants began to grow.

In the succeeding Han dynasty (202 BC–220 AD), mandarins, bamboos, lotuses, and gardenias were cultivated on a large scale and further plants were brought into cultivation. Emperor Wu is recorded in 138 BC as having an enormous park with over 2000 different plants, including tender species brought from the south, such as lychees and longans, for which heated houses were provided. Subsequently Yang Di, second emperor of the Sui dynasty (581–618 AD), built the largest gardens in history and searched far and wide for rare and beautiful plants (Li, 1959). Bamboo was enormously popular and tree peonies became known to gardeners (Wang, 1988). Along with all this interest came books about plants, their cultivation and propagation. Ji Han's *Nanfang Caomu Zhuang* (Account of the Plants and Trees of the Southern Regions), traditionally dated at 304 AD, is the earliest purely botanical work in the Chinese language and a notable example, as is Dai Kaizhi's *Zhu Pu* (Treatise on Bamboos), written about 460, in which he described a large number of genera and species. Of a more general nature is the *Weiwang Huamu Zhi* (Records of the Flowers and Trees

Figure 1.2 An ancient Oriental Thuja (*Platycladus orientalis*) in Zhongshan Park, Beijing

[in the Gardens] of the Prince of Wei), written somewhere between 480 and 535 (Needham, 1986).

The Tang dynasty (618–906) was one of the most powerful dynasties in the history of China and with it came a period of peace and prosperity. During this time people were able to devote more time and money to gardening and to ornamental plants. According to Schafer (1965) this development seems to have accompanied a gradual change from the appreciation of nature, compressed into the space of a garden, to the feeling for plants and birds as individual and unique aesthetic objects. Crabapples, wisteria, magnolias, and azaleas were mentioned as garden plants for the first time, and tree peonies came into prominence, particularly in the capital, Changan (now Xian), and subsequently at Luoyang. Provincial cities became noted for certain other other flowers. Chengdu in Sichuan, for instance, became famous for its plantings of *Hibiscus mutabilis*, and Quanzhou, on the coast of Fujian, for *Erythrina variegata* (Li, 1979; Wang, 1988). Ornamental plants became the subject of poems and paintings, and books about them proliferated.

During the Northern and Southern Song dynasties (960–1279) horticulture received even greater attention. Further plants appeared in cultivation, new cultivars were developed, and evergreen foliage plants began to be used for interior decoration in winter (Wang, 1988). With paper making and printing well established in China, scholars must have been aware that their interest in a particular plant or plants could be shared by thousands of people not personally known to them. It is not surprising, then, that the golden age of Chinese botanical and horticultural monography followed, dealing principally with the plants which were grown for their beauty in gardens. For instance, during the Song dynasty there appeared at least five books on chrysanthemums, four on tree peonies, three on herbaceous peonies, three on *Prunus mume*, and two on orchids (Wang, 1988). This period is also notable for the first appearance of encyclopaedic works about ornamental plants, for example Chen Jingyi's *Quanfang Beizu* (Complete Chronicle of Fragrances) of 1256. There were also books about gardens, such as Li Gefei's *Luoyang Mingyuan Ji* (Notes on the Famous Gardens of Luoyang) (Needham, 1986). The most renowned garden of the time was undoubtedly the Genyue, an enormous pleasure park built at Kaifeng by Huizong, the last emperor of the Northern Song. While smaller than the enormous imperial parks of the Han and Tang, the beauty of its landscape and

its collection of rare plants, animals, and rocks undoubtedly rivalled and probably surpassed that of any similar royal enclosure of the preceding dynasties. A special bureau, the Huashigang (Flower and Rock Network), was set up in Suzhou to search for trees, flowering plants, and rocks to be sent to this garden. As a result rarities were sent from almost every part of the empire, though the activities of the Huashigang are reported to have brought misery and suffering to the common people (Hargett, 1988–89). This extraordinary garden was destroyed in 1127 by the Jin, who overran the northern part of the empire. The survivors of the court moved south to Hangzhou where the dynasty, the Southern Song, remained in power until it was overthrown by the Mongols in 1279.

The remarkable books of the Song writers were followed by numerous others, particularly during the Ming (1368–1644) and Qing (1644–1911), so that it is possible to obtain a considerable amount of information about the plants which were grown and admired. The 30-volume *Qunfang Pu* (Assembly of Perfumes) of 1630 by Wang Xiangjin is a notable encyclopaedia, listing a wide range of ornamental plants (Needham, 1986). There are also treatises on the methods of cultivation of popular plants, which include advice on soil mixtures, fertilisers, propagation, control of diseases and pests, pruning, choice of kinds, and the times of year for carrying out various operations. One such work is Chen Haozi's *Hua Jing* (Mirror of Flowers) of 1688, which was translated into French by J. Halphen and published in 1900 in Paris as *Miroir des fleurs: guide pratique du jardinier amateur en Chine au XVIIe siècle* (Rivoire, 1928). Chen deals with the pleasures of the four seasons, lists the tasks to be carried out each month, just as our own gardening books do, and then goes on to describe more than 350 kinds of ornamental plants (Giles, 1901; Haw, 1986b). Other works deal with the dwarfing of trees, table-plant culture, and flower arrangement. Much information is also included in notebooks of all kinds, especially those dealing with the 'art of living'. Such books were most numerous in the 16th and 17th centuries, when there was much emphasis on the full enjoyment of life. And since nearly all the important ornamental plants that originated in China are cultivated in Japan, there is much to be found in the horticultural literature of that country as well.

An examination of the relevant literature by Needham (1986) suggests that, in the case of chrysanthemums, very few new varieties were recorded between about 1200 and 1600, so that enthusiasm for them during the Ming may have been mostly a matter of disseminating existing varieties. Renewed activity in breeding resumed towards the end of this period and an enormous number of new varieties appeared in the Qing. While enthusiasms for particular plants are known to wax and wane, it may be that this state of affairs applied generally to garden plants. At any rate there was certainly renewed interest in novelty and diversity during the Qing.

Books about ornamental plants and gardening have continued to appear this century in China, though on nothing like the scale that has occurred in the West. Numerous small works dealing with the traditional garden plants and methods for growing them are common, and since the Cultural Revolution there have been many larger books along Western lines, illustrated with colour photographs and dealing with topics such as orchids, rhododendrons, camellias, penjing, and the classical Chinese gardens.

Introduction of plants to China

As one might expect, not all the plants cultivated by the Chinese are indigenous. Wheat and barley were probably introduced about 1500 BC, and broad beans are thought to have arrived in Han times. But the first foreign plants for which written records exist are lucerne (alfalfa) and grapes (fig. 1.3), brought from central Asia in 126 BC (Laufer, 1967; Needham, 1954). The pomegranate is first recorded in the late 3rd century AD, followed by the chive, walnut, pistachio, pea, coriander, safflower, sesame, flax, beet, lettuce, fig, cucumber, melon, and other central-Asian, Middle-Eastern, and Mediterranean plants. It seems likely that most, if not all, of these arrived overland via the Iranian region. Several actually have names which include the character *hu*, which meant 'from the northern or western regions', beyond China's borders. Another very old route, little used until about the 8th century AD, led from India into Burma and thence to Yunnan and Sichuan. It has been suggested that some American plants may have entered China this way after their introduction to India by the Portuguese. Undoubtedly other plants, such as the pomelo, moved into Guangdong from Malaysia and Vietnam. And from the north yet others probably came by the route which was used to bring products overland from Korea and Manchuria to the point where the Great Wall meets the ocean.

Many plants also arrived by sea. From the north the shipping was largely in the hands of the Koreans. The Japanese, too, usually came via Korea, but in the 8th century, after Japan and Korea had become enemies, Japanese ships were obliged to come direct to Chinese ports. In the 9th century the Japanese merchants resorted to the better navigated Korean ships once more, the principal destination being Yangzhou, strategically placed where the Grand Canal crosses the Yangtze. They also used Chinese ships which took them further south (Schafer, 1963). However, it is not known how or when Japanese ornamental plants first reached Chinese gardens. No doubt the exchange of plants between the two countries commenced some time after the introduction of Buddhism and Chinese culture to Japan, principally from the 6th century on. And while plants such as the tree peony, chrysanthemum, peach, and *Prunus mume* may have reached Japan soon after, the Japanese plants now seen in Chinese gardens are chiefly cultivars which were probably not in existence in Japan before the Edo period (1600–1853).

But it was from the Persian Gulf, not from the north, that most of China's overseas trade came. Ships sailed via India, and sometimes Ceylon, before rounding the Malay peninsula and heading up the South China Sea. Their chief destinations

Figure 1.3 **A vineyard in Jiangsu**

Figure 1.4 **Frangipani (*Plumeria rubra* f. *acutifolia*)**

the frangipani (fig. 1.4), which was taken up by the Buddhists to such an extent that it early became known as a temple tree. And it is hard to imagine Asia without chillies, which must have arrived by one or both of these avenues.

Whatever the truth of all this may be, it is clear that from the 16th century on routes were available by which foreign plants could be brought to China from most parts of the world. That foreign residents were anxious to grow the plants with which they were familiar also had a pronounced effect. Graham (1938), writing of Beijing early this century, makes it clear that the

were North Vietnam and Guangzhou, which became a prosperous port. Two kinds of jasmine, *Jasminum officinale* and *J. sambac*, recorded as popular in Guangzhou as early as the 3rd century, are thought to have been brought by this route (Schafer, 1948).

The arrival of the Portuguese in Brazil in 1500, together with their establishment of trading posts in India from 1501 on, opened up the first direct route from the eastern coast of America to the western coast of India. The Portuguese reached Macao in 1513, and Guangzhou soon after, and in 1542 Portuguese merchants were allowed to settle at Ningbo. A second route for the introduction of American plants became possible in 1565 when the Spanish started sailing between Acapulco and the Philippines, establishing a galleon route which lasted 250 years (Merrill, 1954).

According to Merrill it is no accident that tropical American plants began to appear at this time in southern and eastern Asia. Plants which he suggests arrived via the Portuguese route are the peanut, pumpkins and squashes, popcorn, the grain-bearing species of *Amaranthus*, and the African marigold. A much larger number of American plants is believed to have arrived via the Philippines, including tobacco and

Chinese took over the new kinds brought by the foreigners and offered them for sale at the appropriate season. And in more recent times many flower, fruit and vegetable varieties not previously known in the country, particularly those suited to mass production, have been imported.

General horticultural techniques

Although full of admiration for the dexterity of the Chinese with the hoe in forming beds and trenches with 'astonishing neatness and regularity', John Livingstone (1824) was moved to remark of their horticultural practice that 'independently of that routine which has been followed with little variation, from a very high antiquity, they seem to be entirely ignorant of all the principles by which it would have been placed on a scientific foundation'. The casual visitor to Chinese gardens 170-odd years later might be forgiven for thinking that little has changed. The usual potting mix appears to be dust or silt, weeding does not seem to be given high priority, water appears to be applied in a somewhat desultory manner, and leaves are still swept up with bundles of sticks. Chinese attitudes to these matters are undoubtedly different from those exhibited by gardeners in many other countries, so perhaps

Figure 1.5 A painting of three different chrysanthemums grafted on one stock, from the Reeves Collection
of the Lindley Library of the Royal Horticultural Society

it is inappropriate to draw attention to them. The more so, perhaps, when most of the ornamental plants one sees look healthy, handsome blooms are produced, and diseases and insect pests are rarely to be noticed. Aside from this, it is clear that many of the potted plants used in public displays and for the decoration of hotel entrances and office buildings are produced on large scale by modern horticultural methods. No doubt the general care and cultivation of garden plants is, like everything else in China, in the process of modernisation.

The past achievements of Chinese horticulture and botany are, none the less, very considerable. Soil types were classified early in history, and the effectiveness of leguminous green manures and the value of organic wastes were recognised (Needham, 1986). There is much emphasis on the value of soil dredged up from the bottoms of lakes, ponds, and canals, both for spreading on cultivated land and as an ingredient of potting mixtures. Livingstone (1824) noted that fertilisers were applied close to plants rather than to the soil in general, no doubt a sensible economy. He recorded the use of lime, ash, horn, bones, feathers, crabs, soot, street sweepings, and the dung and urine of animals. As might be expected, human, poultry and animal excrement receive frequent mention in the literature, as do liquid manures made from these. A variety of other liquids have been recommended from time to time for particular plants. Tea, for instance, was said by a Song author to be good for a particular orchid (Watling, 1928).

The choice of plants for use in the garden is still greatly influenced by tradition, as is the manner in which they are used. Whether they are grown in the ground or in containers, and where they are placed in relation to each other, and to rocks, water, and buildings are all matters to be considered. Some plants in containers are left in position more or less permanently, some are removed to shelter during the winter, and some are brought out only when in flower. Likewise some are considered suitable for display in the open air, while others are used chiefly to ornament the interiors of houses and garden buildings (figs 14.11, 20.7).

A great many species were brought into cultivation, and by collecting variants from the wild, by raising seedlings and making selections, and by propagating sports, many cultivars of each of the most popular flowers were produced. As far as I can determine, deliberate cross-pollination was not carried out, but natural hybrids were no doubt introduced from the wild and others must have arisen fortuitously in gardens and nurseries. As a result of all this there are numerous garden plants which cannot be reliably assigned to any wild species. Many roses, camellias, peonies, azaleas, and cymbidiums, for example, fall into this category, as do Chinese apples, pears, peaches, and other fruit trees.

There are records in the ancient literature of the propagation of woody plants by seeds, cuttings, layers, and suckers (Needham, 1996). Grafting and budding were also widely practised, and much information about these procedures and about rootstocks and the appropriate seasons for carrying out various operations was gathered and written down.

Methods of grafting various plants have been described in the Chinese literature for at least 1000 years and, as might be expected, this continues to be the case, even in the small books produced for the general public (e.g. Wei, 1986). European descriptions of Chinese grafting methods are few, though Cibot (1778b) and Fortune (1852) give accounts of procedures used for tree peonies. Fortune (1863) observed that the Chinese were extremely fond of grafting several species or varieties onto the one stock (fig. 1.5). In Tianjin he saw 'two species of Thuja and one Juniper, all growing together on one stem'. A little later someone writing under the pseudonym Cantoniensis (1867) reported: 'Nothing is more common among Chinese gardening operations than grafting'. He or she records inarching (fig. 1.6), cleft grafting, and side grafting, and also mentions the grafting of herbaceous plants and the practice of circumposition or, as we would call it, air layering. According to Huxley et al. (1992) it is sometimes called Chinese layering, having been practised there for over 4000 years (fig. 1.7). All these techniques are still practised in China today.

While Cantoniensis said that inarching was the method most frequently adopted for fruit and other large woody trees, Livingstone (1822) earlier stated that air layering was the general method used. The procedure involved removing a ring of bark from the branch chosen, after which the exposed area was covered with clay tempered with almost any kind of earth. For most species it seems this sufficed on its own but for large branches of elm, *Michelia champaca*, and suchlike a covering of straw and/or coarse cloth was added. As far as Livingstone could ascertain no watering was carried out, the sap of the boughs providing sufficient moisture. Under these conditions roots formed and the branches could be detached after 6–12 weeks.

Figure 1.6 **Inarching of osmanthus in Yunnan**

Figure 1.7 **Air layering of a Japanese cherry in Yunnan**

Figure 1.8 **An advanced pine prepared for transplanting at the Wofo Si, Beijing**

The Chinese are adept, too, at transplanting large trees, going to considerable lengths to prepare them, to lift them with a large root-ball intact (fig. 1.8), and to support and water them, so that they quickly become established in their new positions. This skill, together with the survival of traditional techniques for the construction of rockwork and garden buildings, has enabled the restoration of many famous sites in a manner such that the casual visitor is unable to tell what is original and what is new.

Just as there is a great deal of discussion in the Chinese literature about the best sources of water for making tea, so there is concerning water for irrigation. Rain water, well water, and water from fish ponds and other sources all have their advocates. But in no field does the choice of suitable water receive greater emphasis than when it is used to fill vases for cut flowers. Also, in relation to traditional floral art, the old literature contains much information about methods of cutting and preparing flowers as well as extending their life in the vase (Li, 1956a). Advice is given about the times of day for cutting various species, and there are instructions for 'bathing flowers' to remove mud and dust, undoubtedly necessary for anything grown in the open in northern China. The importance of regularly changing the water and the value of the singeing, crushing, or immersion in hot water of stems to prevent wilting are all mentioned.

While on the subject of water it would be remiss of me, in view of their influence on horticulture, not to mention drought and flood. These two great scourges have repeatedly afflicted the Chinese over the millennia. Hence it is not

surprising to find that some of China's most remarkable achievements have been in the fields of irrigation and flood control. For instance Emperor Yu, who is said to have lived in the Xia dynasty (c. 2000–c. 1520 BC), is credited with controlling floods in the Yellow River basin by constructing levees and canals and deepening river beds. While this claim cannot be substantiated, it is clear that many such works were carried out. Nearer to our own time it is definitely known that the division of the Min River into two great feeder canals for irrigation near Chengdu was completed about 230 BC. And another remarkable achievement, though not meant primarily for irrigation, was the construction a few years later of a canal connecting the headwaters of two rivers in Guangxi, one flowing south, the other north. This made it possible to travel by water from Guangzhou to the Yangtze. The joining together of various lakes, rivers, and canals in the period 581–618 AD resulted in the formation of what became known as the Grand Canal, stretching from the lower Yangtze valley to the Yellow River. Then, with the establishment of the Yuan dynasty in the late 13th century, the canal was extended further north to the new capital, Beijing, its total length being some 1700 km. All in all the achievements of the Chinese in canal building and irrigation are astounding, their irrigated land making up one third of the world's total (Needham, 1971).

On a more mundane note, it perhaps should also be pointed out that China is not free of the mythology which afflicts horticulture around the world. 'When the orange tree crosses the Huai River it turns into the thorny limebush' is, according to Needham (1986), one of those utterances (first recorded at least as early as the the 1st millennium BC and constantly repeated with numerous variations) which is characteristic of classical Chinese literature. The thorny limebush is *Poncirus trifoliata*, widely used as a stock on which to graft citrus varieties (fig. 18.13). It seems likely that the saying arose because, when grafted oranges were taken to northern China, the scions were killed by cold and the stock grew up and replaced them. Similarly there are numerous references to the production of new cultivars by grafting. It is known that on rare occasions shoots consisting of tissue from both stock and scion may develop from the region of a graft union, giving rise to a new type of plant known as a graft chimaera. While this may have happened in China from time to time, it seems less likely that the claims that new colours and forms arose following the grafting of scions onto completely unrelated stocks, with which union is unlikely to have taken place, have any basis in fact. Cibot (1778b), for instance, records the extraordinary claims by Chinese gardeners of the successful production of new varieties of tree peonies by grafting on parsnips, radishes, ashes, hazelnuts, and the like. In spite of his scepticism about this he seems to have accepted the claim that the giant quinces which he saw had resulted from the grafting of the common quince on the orange. Similarly he seems not to have questioned assertions concerning the grafting of grapes on *Ziziphus jujuba* and apricots on oranges (Cibot, 1780b). Claims of this sort are also mentioned by Li (1959), regarding a tree peony said to have been grafted on *Toona sinensis* so that it could be viewed from an upstairs window, and by Fortune (1853a), who was told that it was a common thing for *Myrica rubra* to be grafted on the pine, as a result of which the fruit became much larger and of finer flavour. He also mentioned a juniper and a *Photinia serrulata*, growing on a tomb near Ningbo, which appeared to be joined and which the Chinese believed to be the case. This occurrence is similar to the famous tree at the Kasuga Shrine in Nara, Japan, which is said to be composed of a wisteria and species of several other genera grafted on the one stock. My careful examination of this object of veneration in 1991 did nothing to counter my scepticism. There are also claims in China that flower colours can be changed by pouring various substances around the roots of the plants (Li, 1959). For instance, saffron was said to turn white tree peonies red. Whether there is any evidence to support these assertions I do not know.

Chinese greenhouses

In the northern parts of the country the use of greenhouses seems to have been an early development. The use of heated houses for the protection of tropical plants in Changan during the Han dynasty has already been mentioned, and in the gardens of high officials shallots, onions, and madder were raised in winter in such structures during this period. Apparently this practice was not viewed favourably by everyone, as in 33 BC Shao Xinchen, who was Minister of Natural Resources Revenue, asked that it be discontinued, saying that these were untimely things which would be injurious to men and inauspicious as offerings. It seems unlikely, however, that the procedure was done away with, as a Tang poem refers to melons that were already advanced in the second month as the result of the use of heating (Tun, 1965). Later, during the Ming dynasty, heated houses were extensively used for the production of flowers near Beijing. By carefully controlling the temperature the growers managed to have plants such as *Prunus mume* in bloom for the Chinese New Year and to protect others, citrus varieties for instance, all of which were in great demand at this season. Fortune (1849b, 1852), on visiting flower shops in Shanghai in January, was surprised to find purple magnolias, tree peonies, double peaches, and *Prunus glandulosa* all in bloom. He said they were brought from Suzhou, 'the great emporium of Chinese fashion and luxury', and were produced in houses with straw stuffed into the crevices and heated with charcoal fires. Later, when visiting nurseries in Tianjin, he found that most of the plants on offer were those of the southern provinces and needed protection in winter. Hence there were in every garden a number of 'winter houses' in which the plants could be safely stowed away (Fortune, 1863).

Tun Li-ch'en (1965), writing in Beijing at the very beginning of this century, recorded that peonies, jasmines, plums, and peaches in bloom were on sale in the 12th month. He said that such plants were always called *tanghua* (hall flowers) by the florists because 'the delicate beauty of the three spring months' could all be found within one hall. He also remarked that these plants continued to be produced in spite of the frequent warnings given in the ancient literature of the natural disasters which will overtake anyone who follows at any one season the rules pertaining to another.

Also writing of Beijing in the early part of this century, Bredon and Mitrophanov (1927) record peonies, magnolias, and peaches in bloom and nandina, oranges, lemons, and

Buddha's Fingers (a cultivar of the citron) in fruit on sale at this time of year. Osbert Sitwell (1949), too, was greatly impressed during his visit to Beijing in the 1930s by this ability of the shops to offer in midwinter 'a synthesis of spring and early summer'. He mentions oranges with both flowers and fruit, dwarfed roses, apples, quinces, peaches, plums, and cherries, together with white and mauve wisterias in bloom '2–3 feet high, coiled and serpentine', peonies, begonias, jasmines, and lilacs. According to Bredon (1931) these, when bought, 'are then packed in paper-lined baskets, warmed with hand braziers and, thus snugly protected from frost which would otherwise wither them in an instant, are transported to the home of the purchaser to add to the decoration of the house or shop at this joyous season'. Plants of this type are still to be seen at New Year in many parts of the country (fig. 1.9).

In Beijing the principal places where these forced plants were displayed for sale were at the fairs held regularly in the grounds of the Huguo Si and Longfu Si. The Longfu Si (Longfu Temple) has disappeared, but the remains of the Huguo Si can still be seen to the north-east of the Forbidden City (fig. 1.10). It seems that quite large plants could be obtained from the growers who displayed their wares at this temple, as Dorothy Graham (1938) records that pomegranates were brought to her garden from this source in spring, having been cared for in the greenhouse during the cold weather. Presumably the camellias and oleanders, which she also mentions as being 'brought', also came from there.

A detailed description of the Chinese greenhouses has come down to us from Cibot (1778a), who examined them in and around Beijing in the second half of the 18th century. He said that they were like large frames, with walls of mud, half sunk in the ground and built facing south to catch

every possible ray of the sun, as was noted nearly a century later by Fortune (1863). The framework of the south-facing surface was covered, as were windows at the time, with special paper. At night straw mats were unrolled over it to conserve heat, being rolled up again in the morning to allow the sun to penetrate once more. As required heaters were introduced and even containers of boiling water to maintain humidity, a procedure also reported by Juliet Bredon (1931). Dorothy Graham (1938) was also impressed by the effectiveness of these simple greenhouses. And George Kates (1967), in describing an old temple outside Beijing in the 1930s, wrote:

Figure 1.9 (right) **Part of a display of plants forced into bloom for the New Year in Zhongshan Park, Beijing, showing a contorted flowering peach with** *Prunus triloba* **and azaleas behind**

Figure 1.10 (below) **The Huguo Si, Beijing, at which formerly there was a famous plant market**

Behind the guardian's lodging there was further a Chinese greenhouse, half sunken pit and half mud wall, the slanting frames on its sunward side covered with Korean paper. Out of this primitive contrivance, with the slightest of heating from a tiny smudge stove in winter, the guardian's assistant, a simple soul, drew numerous thriving flowers. In good weather they were transferred in tubs and pots to the tower, or else to the low platforms under it where Buddhas and Lohans had once been enthroned.

As well as their use for protecting tender plants and forcing flowers, such houses were also used occasionally to grow onions and certain other vegetables in winter, as has been mentioned above. Tun (1965) noted in his day that, while such means were feasible for the production of luxuries, they

were certainly not for that of ordinary crops, as the cost would be too high and hence they could not be grown as food for everyone. Under ordinary circumstances almost the only vegetables available in northern China at this time of year were Chinese cabbages, still to be seen in huge mounds on the pavements of Beijing in late autumn (fig. 1.11). These, together with root vegetables such as turnips, can be stored and until relatively recently they reputedly made up 80 percent of the vegetables consumed in winter in northern China. But in the last 20 years or so things have changed a great deal. Not only are vegetables transported long distances from warmer regions, but cucumbers, melons, onions of various types, other vegetables, and flowers are grown on a large scale locally in modern versions of the traditional greenhouses described above (fig. 1.12). These have appeared in huge numbers in the northern Chinese countryside, usually not sunk in the ground, but otherwise much the same, and always facing south. Instead of the Korean paper they are now covered with polythene. This combined with the straw mat procedure is most effective, even though the polythene, unlike the paper, is not biodegradable. After disintegrating it blows about, accumulating conspicuously in fields, ditches, trees, and shrubberies to the annoyance of visiting photographers and adding to the pollution of various sorts that afflicts the country. All the same it is clear that the growers of centuries gone by knew what they were doing when it came to the trapping and conservation of the sun's heat.

Ice houses

At the other end of the temperature scale the Chinese were dab hands at keeping fruits in sound condition for long periods, and thus having them on the market the year round. Fortune (1863)

Figure 1.11 (left) **A pile of cabbages on the pavement outside the actor Mei Lafang's house, Beijing**

Figure 1.12 (above left) **Chinese greenhouses on the outskirts of Beijing**

Figure 1.13 Penjing at Tiger Hill, Suzhou

described the ice houses in Tianjin, the floors of which were considerably below the level of the surrounding ground. In these apples, pears, and grapes were packed away in round tubs placed one above the other, the spaces between and above them being filled with ice. They were then left there and taken out as required.

Penjing

The Chinese art of dwarfing trees and creating miniature landscapes in containers is known as *penjing*, a word which means 'potted scenery' (figs 0.1, 1.13, 1.14 , 4.22, 9.5, 19.63). The earliest known depictions of penjing are in the wall paintings of a Tang tomb built in 706 AD. These show oval trays planted with miniature fruit trees in rocky landscapes. Penjing had undoubtedly become popular by this time, and from then on references to it occur regularly in poetry, prose, and painting (Huxley et al., 1992). This art has probably been practised in China for over 2000 years, certainly for hundreds of years before it was passed on to Japan. The Chinese word for a plant in a pot is *penzai*. The characters which this romanisation represents are pronounced 'bonsai' by the Japanese, who apply the word to miniature landscapes as well as to individual dwarfed plants.

In China the earliest surviving book on the subject appeared about 1170 AD (Huxley et al., 1992), and popular gardening books still contain sections describing methods for

the training and care of plants in containers (e.g. Hou, 1986; Wei, 1986). Few Chinese today are in a position to grow plants other than in pots, hence the cultivation of plants as penjing has remained popular.

Over the years a great many styles have arisen. Penjing may consist of one or more plants in a container, either alone or in combination with rocks. In some cases the rocks are more important in the overall design than the plants. Sometimes plants are dispensed with altogether, the scene being composed entirely of rocks, pebbles, and occasionally water. All these scenes have a strong relationship to

Figure 1.14 Large penjing at the Flower Nursery, Hangzhou

According to Livingstone (1822), who observed the production of penjing in southern China, air layering was the preliminary step in the formation of nearly all dwarf trees and shrubs. After separation of the layer 'Any exuberance of growth is first carefully repressed by cutting off the extreme points of the branches, as well as by rubbing off part of the buds and leaves'. This was followed, as it is today, by bending and contorting the branches with the aid of wires or other mechanical means. Livingstone went on to say 'It is the custom in Fo-kien [Fujian], where the best dwarfs are said to be formed, to entice ants to destroy the heartwood, into which openings are made to introduce sugar and the like'. He said that the dwarfed tree most admired was the plum (presumably *Prunus mume*) and that 'Should it be hollow, with its boughs twisted and contorted into all kinds of fantastic forms, it is so much the more esteemed; for such a dwarf tree a very large price will be cheerfully given. Of forest trees the Elm seems to be the most approved, and it is the most common; but being much more easily formed than the Plum, its value is never so considerable'.

While a great many enclosures are devoted exclusively to the display of penjing, they are also used, as are ordinary pot plants, to ornament the traditional Chinese garden and house (fig. 0.1). This use of pot plants and penjing has led in turn to the manufacture of special tables and stands on which to display them (figs 14.2, 14.6, 14.11, 20.7). An early Western record of this practice was made by Archibald Little (1888), who visited a country house outside Chongqing in April 1883, the guest hall of which looked onto 'a handsome enclosed courtyard, filled with orange, camellia, and azalea trees growing in large pots placed on stone pedestals'.

Perhaps Chinese sensibilities concerning trees in general and the appreciation of penjing have contributed to the aesthetic manner in which trees are trained and pruned. Those lining roadsides and city streets are usually excellently

landscape painting, whether they be tiny or several metres across (fig. 1.14).

Tree penjing, too, range from specimens a few centimetres tall to those two or more metres in height. They may be trained to represent ancient trees on windswept mountainsides or into curious shapes to represent concepts or to suggest objects or animals (fig. 19.56). Quite frequently the trunk and branches are trained into the shape of an auspicious Chinese character, and the more fanciful shapes may well defy interpretation by the untutored eye.

Many different kinds of plant are used for penjing, but on the whole the pines reign supreme. And, as with bonsai in Japan, various methods are employed to achieve the end result, particularly where an appearance of great age is desired. Shen Fu (1983), writing in the early 19th century, gives some detailed advice concerning the procedure and says that the proper training of a tree takes at least 30 or 40 years. Nowadays seedlings are still patiently trained, twisted, and pruned, taking many years to reach perfection. Alternatively, such an outcome may be achieved more rapidly when small, gnarled specimens are dug from inhospitable natural sites, potted, pruned, and trained. Plants dug up for this purpose may still be seen for sale in many parts of China. A somewhat more pragmatic method is that in which established plants are cut back severely to encourage the production of branches, those best fitted to the design being retained and cut back in turn.

managed and most handsome. They are almost invariably cut up or back with sensitivity, show no ugly wounds, and are trained into pleasant shapes, even when overhead wires and other obstacles have to be taken into account (fig. 19.87). In this regard municipal authorities in many other parts of the world have much to learn. Likewise in gardens great care is taken to achieve picturesque shapes. For instance Shen Fu (1983) pointed out last century that when trees are planted it does not matter if they are set in at an angle and that, in fact, if a tree is planted upright it will be difficult for it to grow into a striking shape.

Those wishing to know more about penjing are referred to more detailed treatments of the topic (e.g. Hu, 1982, 1988; Huxley et al., 1992).

Control of diseases and pests

As far as the control of plant diseases is concerned, the Chinese do not seem to have introduced any notable advances. However, good hygiene, the removal of diseased plants or plant parts, the provision of suitable conditions for growth, and the maintenance of good air circulation all seem have been recognised as useful.

While the control of pests appears to have depended principally on vigilance and manual destruction, advances were made in devising other methods to control or repel insects, in particular the early discovery that plant products could be used. According to Needham (1986), who reviews the history of natural plant pesticides in China, there is a record of the use of 'plant smoke' for the control of insects during the Han dynasty. Smoke from the burning of 'moxa', made from a species of *Artemisia*, was used to preserve stored grain in the 1st century BC, and a reference of 1085 AD records that moxa smoke was used to ward off mosquitoes, no doubt the antecedent of the modern mosquito coil. Nearer to our own time, Fortune (1853b) observed that Buddhist monks in Zhejiang used a wild *Artemisia* species, sun-dried and twisted into bands, to produce smoke which kept bees at bay while honey was removed from their hives, and noted that this was also used against mosquitoes.

In a book of 1273 AD it was stated that if trees were affected by borers these could be controlled by stuffing stems of *Daphne gengkwa* (the flowers of which were used as a fish poison) into their holes, and by the Ming many other plants were known to have insecticidal properties. For instance the bark of *Catalpa ovata*, the leaf and oil of *Aleurites fordii*, the fruits of *Melia azedarach*, and unnamed parts of *Rhododendron molle* were regarded as efficient insecticides, and an infusion of the ground leaves of *Sapium sebiferum* was used as a spray to control agricultural and garden pests. In addition to these, Needham (1986) lists various other plants used by the Chinese to control insect pests.

The most extraordinary of all their procedures was the deliberate encouragement of ants believed to control citrus pests. Needham (1986) points out that the recognition that certain insects prey on crop plants, and that some insects prey on other insects, can be traced back to very early records. References to both types of observation can be found in the *Shi Jing* (Book of Odes), a compilation of folk songs and ceremonial poems collected from the 9th to the 5th centuries BC. The earliest record of putting such an observation to use is in Ji Han's *Nanfang Caomu Zhuang* (Account of the Plants and Trees of the Southern Regions) of 304 AD (Li, 1979). This records the selling of the nests of a particular ant to orange growers, who introduced them to their orchards for the purpose of controlling damaging insects. To what extent the trees actually benefited remains a matter for argument, but at least it represents the earliest known attempt to control insect pests biologically, a procedure now attracting much attention. The practice continued until well into the 20th century before being replaced by powerful organic insecticides. However, as a result of the rising cost of chemicals and the development of resistance to them by the insect pests, the procedure has been reintroduced in some areas (Huang & Pei, 1987). The devising and introduction of arrangements which allow the ants to overwinter in the orchards has cut costs, and it is claimed that they control the insect pests as well as chemicals do. Equally important is the fact that the use of ants does not interfere with the natural enemies of minor pests such as aphids, mites, and soft scale insects.

The exploitation of insects which feed on plants

The relationships between insects and plants have been exploited in other ways in China. A notable example is the deliberate cultivation of scale insects on privet and ash trees for the production of white wax. According to Wilson (1913) this wax first came to notice in the 13th century, and subsequently its production became a considerable industry. It has a high melting point and was used chiefly for making candles and for glazing paper. But the most remarkable procedure of all has been the cultivation of the mulberry and the development of the silk industry. The domestication of silkworms and the discovery of the method of unwinding the silk filaments of the cocoons may have taken place as early as the 3rd millennium BC. This knowledge was jealously guarded for perhaps two or three thousand years before the art of raising silkworms eventually spread to Japan, India, Iran, and, ultimately, Europe.

The effect of the events of the last 150 years

Those visiting China today should be aware that the cultivation of ornamental plants has not been favoured by the political turmoil of the past 150 years. Firstly, the authority of the Qing dynasty was challenged last century by the industrialised foreign powers, whose interference in Chinese affairs can hardly be seen as admirable. As a result of the Opium Wars of 1839–42 and 1856–60, the Chinese government was forced to grant many concessions, and in Beijing the enormous imperial palace and garden, the Yuanmingyuan, and neighbouring imperial properties were destroyed by British and French troops. Also, after the Boxer uprising of 1900 there was further considerable destruction by foreign troops in and around that city.

In addition to this and earlier, there had been the Taiping Rebellion (1850–64), led by Hong Xiuquan. As an unexpected consequence of Christian efforts to save the souls of the Chinese, Hong had come to believe himself to be the younger brother of Jesus Christ who, according to Hong's view of things, was living contentedly in heaven with his wife and five children (Spence, 1996). Hong assembled a vast army of supporters but his unsuccessful attempt to overthrow the old order and establish a 'Heavenly Kingdom', with equality for everyone, resulted in the devastation of much of the country south of the Yangtze and the deaths of an estimated 20 million people. It was during this period that Robert Fortune (1855b) was moved to record his melancholy feelings on returning to Shanghai to find that nurseries where he had earlier obtained plants had been destroyed, that trees which he had admired had been felled for firewood, and to remark that 'The celebrated peach gardens near the south and west gates of the city, which at this time of year (April) used to be one sheet of bloom, have now nothing remaining except the stumps of the trees'. He went on to comment: 'The picture I have endeavoured to paint applies, unfortunately, to many other parts of the country'.

Subsequent to all these events the Qing dynasty never regained its former hold over the country and finally gave way to a republic in 1911. Members of the imperial family, officials, and private individuals could no longer afford to maintain their estates, most of which were abandoned or fell into decay, although some were granted a temporary reprieve by being leased to foreigners. Fortunately foreign visitors during the 19th and early 20th centuries were able to record for us many glimpses of the old order before it disappeared.

Following the overthrow of the Qing dynasty the country's problems remained unsolved. The old social system persisted, there were internal struggles, and the territorial ambitions of the foreign powers continued to cause distress. It is not surprising then that the failure of the revolution to change the social order and improve conditions led to the formation of the Chinese Communist Party in 1921. However, when Chiang Kai-shek set up his Nationalist Government in Nanjing in 1927, membership of the Communist Party became a crime punishable by death. On the other hand he offered little resistance to the Japanese, who seized Manchuria in 1931 and gained a base for their troops in Shanghai in 1932. In 1937 they began a campaign to seize the whole of China, and by the end of 1938 most of the eastern half of the country, including the main urban centres, was in Japanese hands and the Nationalist Government moved inland to Chongqing. Then, when the Japanese withdrew at the end of World War II, the differences between the Communists and Nationalists remained so great that civil war broke out once more. In spite of American support for the Nationalists, the Communists eventually triumphed, and on 1 October 1949 Mao Zedong proclaimed in Beijing the establishment of the People's Republic of China.

Under the new republic land was redistributed, landlords were punished, often executed, cooperative farms were set up, industrialisation was recommenced, education reformed, the written language simplified, and an attempt made to eliminate the traditional corruption and dishonesty of public life. In 1966 the Cultural Revolution, which has been interpreted as the expression of the fear of reversion to traditional or western modes of thought and behaviour, began (Hookham, 1972). The aim was to eliminate the 'four olds': old ideas, old habits, old customs, and old culture. With the aid of the Red Guards these aims were pursued with alarming vigour and the depredations of the Taipings were repeated on a nationwide scale (fig 1.15).

As has happened in most countries, the construction of elaborate gardens and the search for horticultural novelties was very much a pastime of the elite. Over a century ago Fortune (1863) observed that, in his experience of Chinese towns, there was 'a curious connection between nursery gardens and a thriving trade which produces wealth. Where the one is found the other is generally not far off'. It is hardly surprising, then, to find that, while ornamental plants were incorporated in ceremonies organised by the Party and planted in public places, the growing of plants for pleasure was otherwise frowned upon. They were regarded as signs of luxury and relics of discarded customs (Goody, 1993).

During this period commercial flower growing ceased, flower shops were smashed, the New Year ceremonies were banned, records were lost or destroyed, and gardens were neglected or razed, as were temples in the courtyards of which many fine plants grew. Botanic gardens were ploughed up, their collections damaged or destroyed, and their staff sent to work elsewhere. The Beijing Botanical Garden, for instance, became a pig farm.

The enormity of all this is brought to life by Cheng Nien (1987) in describing the looting of her house in Shanghai, when antique porcelain and works of art were destroyed and a pot of jasmine was smashed on the floor. During this period her gardener hid seedling boxes under beds when the Red Guards were about. Also her old teacher, Miss Pang, was criticised for painting pendulous branches of plum blossom, since this was taken as symbolising the downfall of the Communist Party.

In the light of all the above you might well be wondering whether there is much left for me to write about. Mercifully, however, many of the old palaces and

mansions were spared due to their being occupied by government departments or party officials, and their gardens, although neglected, still exist (fig. 1.16). Also in areas remote from the disturbances much has survived. And after Mao died in 1976, Deng Xiaoping began his policies of freeing up the economy, the New Year ceremonies were reinstated, the commercial production of flowers and ornamental plants was permitted again, botanical gardens began to be revived, and the personal freedom and material welfare of many Chinese improved. However, the gap of 26 years under Mao, preceded by the difficult war years, meant that horticultural expertise was not readily available among the young. As Goody (1993) remarks: 'Today the literati and local elites of earlier times have effectively disappeared, having been killed off, metaphorically and sometimes literally, following the events of 1949'.

Figure 1.15 (right) **A marble ornament broken from a balustrade by Red Guards at the Beitang (North Church), Beijing**

Figure 1.16 (below) **The Ke Yuan, a decaying Qing garden in Beijing**

Even so there was rapid growth in the demand for plants and flowers once the restrictions were removed, no doubt a reflection of their significance as symbols of longevity, wealth, and happiness. This worried the Party, which in 1983 attempted to forbid their use in factories and offices, which were among the main consumers at the time of the Chinese New Year (Goody, 1993). But the renewed injunctions did not last long and a garden, flower, and plant culture has re-emerged. And just as people flocked to enjoy displays of flowers centuries ago, so they continue to do at the present time. Sites famous for plum and peach blossom are thronged in spring, artists and photographers are to be seen recording the splendour of the tree peonies, exhibitions of azaleas, orchids, roses, and chrysanthemums are mounted in their seasons, and collections of penjing are on permanent show in many places and across the country. Societies devoted to orchids, roses, and other favourite plants have also begun to spring up, and books about gardens and ornamental plants are once again in the bookshops.

Happily, though perhaps many Chinese cultivars may have been lost, enough of the old Chinese ornamental horticulture remains to attract the attention of those with an interest in plants. There has been much replanting with traditional species and their cultivars in recent years. Most of the favourites mentioned both in the Chinese literature and the writings of earlier visitors to the country are still to be seen.

Nevertheless it cannot be denied that the present modernisation of China continues to take its toll of traditional gardens and their plants. The old quarters of the cities have now largely been swept away and replaced with endless blocks of apartments. Gone are most of the old houses, some of which once had gardens and most of which had courtyards with one or more trees, a few shrubs, a lotus in a tub of water, goldfish, and many plants in pots. Those with the space had frames or a service area from which potted plants were brought out for display when at their best. Others relied upon itinerant sellers who came round bringing with them the plants of the seasons. The well-to-do often had contracts with such people or with nurserymen who removed plants no longer required and replaced them with

Figure 1.17 (opposite page) **Part of a tulip display in Taiziwan Park, Hangzhou**

those currently in flower or fruit (Livingstone, 1820b). Charming glimpses of this now almost extinct horticultural world are given by Dorothy Graham (1938) and George Kates (1967) in their descriptions of Beijing in the 1930s.

The plant sellers are once more to be seen but, not surprisingly, their range of offerings now reflects the international influences that affect us all. Here and there, too, there are shops and markets devoted to plants, the accessories needed for their culture, and garden ornaments. At the large plant and flower market near the Temple of Heaven in Beijing, for instance, a great array of living and plastic vegetation is displayed together with containers, Chinese lions, and quite wonderful replicas of some of the masterpieces of European sculpture, not made of artificial stone but carved most competently from marble. Amongst the plants there are modern roses in pots, chrysanthemums, araucarias, palms, and the usual array of indoor plants, no doubt better suited to the needs of the apartment dweller than many of the old favourites.

In the public parks stupendous displays of tulips (fig. 1.17), hyacinths, and daffodils elicit admiration of the type formerly reserved for *Prunus mume*, tree peonies, orchids, and chrysanthemums. At the same time indigenous plants previously ignored by Chinese gardeners are beginning to be cultivated, not only in the botanical gardens where Western influences have been most strongly felt, but in public plantings. In Beijing, for instance, *Kolkwitzia amabilis* frequently appears, and in the south *Michelia maudiae* (fig. 8.11) is beginning to make its presence felt.

At the present time, too, florists shops are proliferating, stocking the bulbous iris, carnations, gladioli, roses, and so on to be found wherever one goes in the world. In short, along with everything else, Chinese horticulture is rapidly being 'globalised', if that is how one should put it these days. Even so, although the Chinese are now following the Japanese and Koreans in taking up with a vengeance the Western-style wedding, complete with hire cars, photographer, video, dinner suit, and an elaborate white wedding dress, traces of the old traditions can be glimpsed in the red flowers that feature prominently in the bouquets and as decorations on cars and elsewhere. It seems that still no wedding is possible without this auspicious colour and the prominent display of the 'double happiness' character.

CHAPTER 2

ORNAMENTAL PLANTS IN CHINESE CULTURE

It is clear from the ancient Chinese records that ornamental plants have been admired and no doubt cultivated at least from the first millennium BC. From early times horticulture has been promoted, both intellectually and practically, by scholars, artists, the imperial court, Buddhist and Daoist priests, and those interested in flower arranging, to say nothing of ordinary gardeners. The efforts of all these have combined to produce the most elaborate culture involving ornamental plants that the world has known, a culture in which the involvement of plants in art, literature, religion, folklore, and everyday life has been more intimate and complex than in any other civilisation (Bartholemew, 1980, 1985a, b; Bredon & Mitrophanov, 1927; Doolittle, 1866; Goody, 1993; Gulland, 1928; Koehn, 1952). Amongst the numerous authors who have recorded the involvement of plants with everyday life both Tun Li-Ch'en (1965) and

Figure 2.1 *Pear Blossoms* by Qian Xuan (c. 1280)

Figure 2.2 Princess Der Ling and the Dowager Empress Cixi descending Peony Hill at the Summer Palace in the snow, from *Two years in the Forbidden City* by Princess Der Ling, 1911. Note the elaborate floral embroidery on the gowns.

share their passion for plant collecting. For example Clarke Abel (1819), who accompanied the embassy of Lord Amherst and visited the Fa Tee Gardens in January 1817, was moved to record that 'Plants remarkable for their dazzling colours and singular forms were, however, more cultivated than those of great rarity'. Somewhat later Fortune (1850e) commented: 'The gardens of the English residents in Shanghae far excel those of the Chinese in the number of trees and shrubs they contain, and also in the neat and tasteful manner in which they are laid out and arranged'. And towards the end of the century Westland (1894) wrote:

> The Chinese florists are peculiarly conservative in the matter of stock, and it is unusual to find more than fifty genera represented in their gardens. Novelties are scouted as new-fangled innovations. It is only well-tried plants that have been hallowed by the appreciation of their forefathers, and endeared by association, that are thought worthy of careful cultivation. And yet there are thousands of native plants in the 'Flowery Land' that rival and excel the cultivated kinds; but, as the ancient philosophers who established the canons of taste in China did not proclaim their beauty in some poem, they remain unappreciated.

It may still seem surprising that the vast bulk of the rhododendrons, lilies, primulas, and other decorative plants growing wild in China, particularly in the mountainous regions of the west, were not brought into cultivation by the Chinese. But presumably these were unknown in earlier times to the literati, who lived in the centres of wealth and culture in the east. There was no opportunity, then, for them to be praised in poems, depicted in paintings, or considered suitable for gardens. Likewise, apart from those exotic plants which arrived prior to the establishment of the precepts of Chinese taste, it appears that comparatively few introduced ornamental plants have captured the imagination of Chinese gardeners or become part of the fabric of Chinese culture. Undoubtedly this is because they did not bring with them any of the symbolism or relationship to classical literature and art which had accumulated around the early favourites. As a result a very conservative tradition persisted until recent times.

That this should be so can better be understood if we look more closely at the conservatism of Chinese culture as whole. For a start it is remarkable that this culture, unlike

Princess Der Ling (1911) (fig. 2.2) are of particular interest, since they were both Manchus who lived in Beijing prior to the overthrow of the Qing dynasty and the era of tremendous change which has followed.

Because of the complexity of the involvement of ornamental plants with Chinese life and thought, I have dealt with the topic under a series of sub-headings, as I have with Chinese horticulture in the previous chapter. And because visitors to China so frequently comment on the relatively small number of different plants to be found in traditional gardens, I decided to commence by looking at this phenomenon in the light of Chinese cultural conservatism as a whole.

Chinese conservatism

Europeans visiting the country in the 19th century tended to find it puzzling, even annoying, that the Chinese did not

that of other lands, has continued in an unbroken sequence for thousands of years. And this has taken place in an enormous area, populated by a variety of different peoples, which persisted as a large empire for more than two millennia, despite many invasions and changes of dynasty. Unlike Europe, China did not evolve into a collection of competing nationalities but was held together by influences favouring political and intellectual unity. Not the least of these influences was the existence of a centralised government administered by scholar-bureaucrats and with the emperor at its head. The scholar-bureaucrats obtained their positions by succeeding in the imperial civil service examinations which tested their knowledge of the classical literature and their ability to write in the accepted literary style.

This in itself seems unlikely to have encouraged innovative thought. Also, censorship and the often fierce supression of criticism and opposition appear to have been features of Chinese administration in all periods.

In China the power and prestige of the scholar-bureaucrats, which they retained until the beginning of the present century, surpassed that of anyone else save the emperor and his close intimates. And, as Bodde (1991) points out, during a large part of Chinese imperial history, the emperor, at least outwardly, shared with them their veneration for books, their respect for the past, and their orientation toward human institutions. The nobility usually enjoyed little political power, the merchants were traditionally discriminated against, and the military remained subordinate to the civil bureaucrats except in times of dynastic crisis.

It has been argued that an important factor contributing to uniformity and conservatism has been the Chinese language. The script and its accompanying literary language gave China a cultural continuity in time and a unity in space. The characters have always retained their meanings, irrespective of their varying pronunciations in different dialects or at different historical epochs. Thus Chinese writing and literary language have been immensely powerful agents for preserving and diffusing Chinese cultural values. At the same time their effect has been strongly conservative, in that they made the entry of foreign ideas difficult, slowed down change, and discouraged cultural variation, especially within the small but dominant literate minority (Bodde, 1991).

The conservatism of the Chinese with regard to garden plants, then, should be seen, not as an isolated phenomenon, but as part of a general cultural conservatism and uniform-

ity. It is thus not surprising that Robert Fortune should have found the garden plants of Beijing to be substantially the same as those he had seen in the south, and that the Chinese should have gone to such lengths to grow the southern favourites in the north and vice versa.

The Chinese view of the universe

To look more deeply into the involvement of garden plants with Chinese culture, it is necessary to know at least a little about ancient and medieval Chinese beliefs concerning the workings of the natural world. That a person of my background and training should dare to comment at all on Chinese philosophical matters can perhaps be seen as incautious. However, I feel it is necessary to say something, albeit abbreviated and over-simplified, in order to enable readers to appreciate some of the symbolism with which the Chinese were able to endow the appearance of plants.

The ancient Chinese picture of the workings of the universe involved elements of Shamanism, sorcery, animism, Confucianism, Daoism, Buddhism, and many other factors. Particularly influential in all of this was Daoism. *Dao* means 'path', 'road', or 'way' and referred to the way the universe worked, 'the Order of Nature which brought all things into existence and governs their every action, not so much by force as by a kind of natural curvature in space and time . . .' (Needham, 1956). As well philosophising on this the Daoists became deeply involved in other matters such as alchemy and the search for means of attaining longevity and immortality.

In attempting to understand the universe the ancient Chinese believed that there were two opposing forces operating, *yin* and *yang*, each with its many correlative qualities and functions. For the yin they included, amongst other things, darkness, cold, wetness, softness, quiescence, completion, consolidation, femininity, earth, moon, north, below, squareness, even numbers. For the yang these included brightness, heat, dryness, hardness, activity, incipience, dispersion, masculinity, heaven, sun, south, above, roundness, odd numbers and much else. The yin and yang complement rather than struggle against one another, each being essential for the functioning of the cosmos. In the repetitive cycles of the seasons, days, and other phenomena, the yang and yin ever wax and wane in inverse ratio to one another, without the one ever permanently suppressing the other.

Also of major cosmological importance was the idea of the 'five elements'—wood, fire, earth, metal, and water. The concept of the five elements was not so much one of a series of five sorts of fundamental matter, as of five fundamental processes or qualities which, like the yin and yang, dominated alternately, each being followed by the one it could not dominate. Needham (1956) quotes an author of about 135 BC as saying that wood produces fire, fire produces earth (ashes), earth produces metal (as in ores), metal produces water, and water produces wood (plants need water). He went on to say that wood has its place in the east and has authority over the *qi* (life force or cosmic breath) of spring, fire has its place in the south and has authority over the *qi* of summer, metal has its place in the west and has authority over the *qi* of autumn, water has its place in the north and has authority over the *qi* of winter. Men have no choice but to go by this progression. Earth has its place in the centre and is (as it were) the rich soil of heaven, which brings the five elements and four seasons together. Metal produces water, Needham suggests, either because molten metal was considered aqueous, or perhaps because of the ritual practice of collecting dew on metal mirrors exposed at night.

With these five elements were aligned and associated, in symbolic correlation, everything else in the universe that could be got into a fivefold arrangement, for example the five virtues, the five blessings, the five planets and the five directions (north, south, east, west, and the centre). As well as being associated with the seasons and directions, the elements also became correlated with colours. Thus wood, the east, spring, and green were associated, as were metal, the west, autumn, and white; water, the north, winter, and black; fire, the south, summer, and red; and earth, the centre, and yellow. An interesting manifestation of this arrangement is to be seen at the Altar of Land and Grain in Zhongshan Park, Beijing, on which earths of colours pertaining to the five directions have been appropriately spread (fig. 2.3). Exhibited here too is the concept of symmetry and centrality which played a large part in Chinese thought.

Around this central fivefold order was a larger region comprising all the classifiable things which would only go into some other numerical category, and, as Needham (1956) puts it, much ingenuity was shown in fitting the classifications together. Hence the Chinese number-mysticism or numerology, one of the main purposes of which was to classify all the objects, phenomena, and concepts of the natural and human worlds into sets of items that are the same in number.

The idea of cyclical progressions, epitomised by the behaviour of the five elements and the yin and yang, led to a cyclical view of history. Time was not thought of as a continuum but as a series of 60-year cycles, each year of which was denoted by two characters. The first of these was one of the 10 *gan* (the so-called 'heavenly stems') and the second one of the 12 *zhi* ('earthly branches'). Each of the earthly

Figure 2.3 A view of the Altar of Land and Grain, Zhongshan Park, Beijing, with coloured earths spread to represent the Five Directions

Figure 2.4 *Flowering Plum by Moonlight*, wood-block print from the *Jieziyuan Huazhuan* of 1701

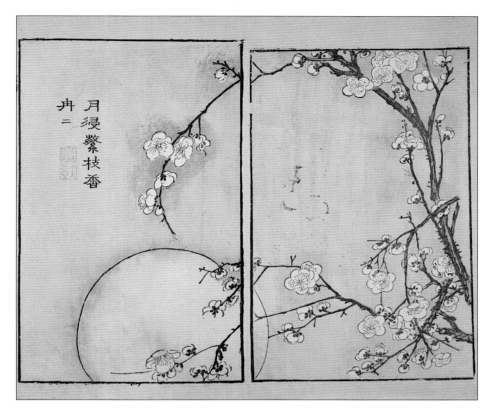

branches was also used to designate one of the twelve 2-hour periods into which the Chinese day was formerly divided. Each of the branches also became associated with an animal. Thus we get the hour of the rat, the year of the ox, and so on. As well as all this a system of prognostication involving the use of these stems and branches was developed.

Superimposed, as it were, on this and on the yin and yang and the five element theories, was the strange system of the *Yi Jing* (Book of Changes), believed to date largely from the Zhou dynasty (c. 1050–221 BC). This book consists of a set of symbols and their explanations. The symbols are made up of sets of lines, some continuous (yang) and others broken (yin). By using all possible permutations and combinations of these, 8 trigrams and 64 hexagrams were formed, which were matched with the months, the seasons, the 24 fortnightly divisions of the year, the compass points, and many other things. As Needham (1956) explains, these symbols were supposed to mirror in some way all the processes of nature, and Chinese medieval scientists were therefore continually tempted to rely on pseudo-explanations of natural phenomena by simply referring the latter to the particular symbol to which they might be supposed to pertain.

In the end all these ideas became integrated with Confucian moral and social values into a single, all-embracing system which retained its prominence until the beginning of this century. That some knowledge of this is helpful in understanding the symbolism which those Chinese inclined to do so could ascribe to the appearance of plants is made clear by a quotation concerning the flowering plum (fig. 2.4) from the second volume of the *Jieziyuan Huazhuan* (Mustard Seed Garden Painting Manual), originally published in 1701. As translated by Sze (1977) from the 1887–88 Shanghai edition, using an old system of transliteration, it reads:

The symbolism (*hsiang*) of the plum tree is determined by its *ch'i*. The blossoms are of the *Yang* principle, that of Heaven. The wood of its trunk and branches are of the *Yin* principle, that of Earth. Its basic number is five, and its various parts and aspects are based on the odd and even numbers. The peduncle, from which the flower issues, is a symbol of the *T'ai Chi* (the Ridgepole of the Universe, the Supreme Ultimate, the Absolute), and hence it is the upright form of the calyx. The part supporting the blossom is a symbol of the *San Ts'ai* (Three Powers of Heaven, Earth, and Man) and consequently is drawn with three sepals. The flower issuing from the calyx is a symbol of the *Wu Hsing* (Five Elements) and is drawn with five petals. The stamens growing in the center of the flower are symbols of the *Ch'i Chêng* (Seven Planets: the five planets with the sun and moon) and so are drawn numbering seven. When the flowers fade, they return to the number of the *T'ai Chi*, and that is why the cycles of growth and decline of the plum tree are nine. All these aspects of the plum tree are based on the *Yang* and therefore are associated with the odd numbers.

The roots from which the plum tree grows are a symbol of the *Erh I* (Two Forms: *Yin* and *Yang*), and this is the reason the trunk is divided into two parts. The main branches symbolize the four seasons and so are composed facing the four directions. The branches symbolize the *Lu Hsiao* (Six Crosswise Lines of the *I Ching* hexagrams) and so have six main 'crossings' (for a complete tree). The tips of the branches symbolize the *Pa Kua* (Eight Trigrams of the *I Ching*) and so have their eight knots or forks. The whole tree with its trunk, branches, and blossoms symbolizes the complete and perfect number (Ten), and therefore ten kinds of plum trees have been designated. All these aspects concerned with the wood parts pertain to the *Yin* and are even numbers. But this is not all.

The front view of the blossom shows the form (*hsing*) of a circle and thus is a symbol (*hsiang*) of Heaven. A flower in back view has angles forming a square and thus is a symbol of the Earth. Branches bending over have an aspect of Heaven, covering the Earth; upright, they appear like the pillars of Earth supporting Heaven.

The stamens are also symbolic. When the blossom is in full flower, it symbolizes the stage called *lao* (ripe) *Yang*, the full development of the flower just before the first step of fading, and the stamens number seven. When the blossom is faded, it symbolizes *lao* (ripe) *Yin*, and the stamens number six. The half-opened flower symbolizes *shao* (lesser) *Yang*, and the stamens number three. When the blossom has partly faded, it symbolizes *shao* (lesser) *Yin*, and the stamens number four.

The bud symbolizes the entity of Heaven and Earth. Its stamens are not yet visible, although their essence (*li*) is already contained within the bud. Therefore one calyx and two sepals are indicated; Heaven and Earth are still an undivided entity, and Man (represented by the third sepal) has not yet appeared. The flower with its stem and calyx symbolizes the beginning of the interaction of the powers of Heaven and Earth; when the *Yin* and *Yang* begin to separate, the cycles of growth and decay and endless mutation are started, symbolizing the materialization and natural development of all things. There are, therefore, eight knots (comings together, connections of branches), nine stages or changes, and ten species of the plum. As may be seen, all these symbols come from Nature itself.

While not going to such lengths, symbolism involving plants is discussed further later in this chapter but it seems appropriate to include here a brief mention of *fengshui* (winds and waters). The basic principle of this procedure is that streams or currents of *qi* run through the earth. These currents are believed to influence the fortunes of individuals and their descendants, depending on how they place their houses and graves in relation to the winds, waters, hills, and valleys of the landscape in which they live.

According to the *Encyclopaedia Sinica* (Couling, 1917), *fengshui* is 'the art of adapting the residences of the living and the dead so as to cooperate and harmonise with the local currents of the cosmic breath'. If these residences are not properly arranged there are likely to be most unfortunate consequences for the living inhabitants or the descendants of those whose bodies lie in the tombs. Conversely, good siting will favour their health, wealth, and happiness. Hence, since changes to the landscape are thought to affect the flow of *qi*, a complex set of rules and principles has been developed to guide building and landscaping activities.

The yin and yang came to be symbolised by configurations of the ground. For instance, high and abrupt eminences were considered yang, and lower, rounded elevations yin. The former ought to be to the left and the latter to the right of any building or tomb, which should preferably be protected by them as if in the crook of an elbow. And in the selection of the site such influences ought also to be balanced, if possible, so as to obtain three-fifths yang and two-fifths yin. As well as this the trigrams and hexagrams, the sexagenary cycle of stem and branches, the five elements, and various other matters are included in the reckoning (Needham, 1956).

Although in many ways *fengshui* is a grossly superstitious system, its influence concerning the heights and forms of buildings, their positioning in relation to the landscape, the avoidance of straight lines, the planting of trees and bamboos as windbreaks, and the value of flowing water adjacent to a house site results in it having a marked aesthetic effect, which Needham (1956) believes accounts for the great beauty of the siting of so many farms, houses, and villages throughout China (fig. 2.5), as no doubt it does for the preservation of pieces of woodland and sheltering belts of trees.

Figure 2.5 (opposite page)

A farm in Sichuan in early spring

It should be noted that, as well as contributing to the sympathetic siting and beauty of buildings, *fengshui* also does much for the comfort and amenity of the inhabitants. Houses are usually built facing due south, towards the sun and sheltered from the supposed evil influences of the north. As a result not only is it warmer in winter but it is possible to estimate the time of day from the angle of the shadows in the courtyard.

Fengshui, needless to say, has had an important influence on the siting and design of gardens, once again the well-being of the owner or occupants being a prime consideration. And the plants used are chosen not only for their appearance, but because they have some feature, shape, colour, or accumulated symbolism suggesting health, wealth, longevity, happiness, or some admirable virtue or other desirable state of affairs.

It seems appropriate to note here, too, that the vital forces have at times been believed to be responsible for the production of beauty and ugliness. For example, this was mentioned by Ouyang Xiu in his *Luoyang Mudan Ji* (Account of the Tree Peonies of Luoyang) of 1034 AD. He said that there is a *qi* of constancy and normality, and when it is manifested in things they have a normal or standard form, being neither particularly beautiful nor particularly ugly. He went on to say that things which express extreme beauty or extreme ugliness are the results of an imbalance in the vital *qi*. As a result he concluded that, since he had observed that the tree peonies within the walls of the city of Luoyang were more beautiful than those of the surrounding districts, that the unbalanced *qi* had collected only within this small area, a state of affairs which he was at a loss to explain (Needham, 1986).

All in all the Chinese have long believed in the essential oneness and harmony of man with the universe, a sentiment which permeates much of their greatest art and poetry. It was believed that 'The harmonious cooperation of all beings arose, not from the orders of a superior authority external to themselves, but from the fact that they were all parts in a hierarchy of wholes forming a cosmic pattern, and what they obeyed were the internal dictates of their own natures' (Needham, 1956).

Perhaps no other people developed so keen a consciousness and awareness of the movement and rhythm of nature as did the Chinese. Use of the lunar calendar meant that the phases of the moon marked the progress of each month. And in the marking of the seasons much attention was paid to the stages of growth, particularly flowering, of various plants, so that, in addition to the guidance provided by the official almanacs, which were issued annually, the times for various agricultural and other operations could be related to these.

Some general beliefs and associations

In earlier times, every one of the principal flowers and trees was said to have a 'fairy' who looked after its welfare, the flower fairies being women and the tree fairies men. The word 'fairy' is an often-used translation of the Chinese word *xian*, which perhaps more accurately should be rendered as 'immortal'. The Chinese concept of the *xian* and the celestial bureaucracy is discussed in some detail by Needham (1974).

The flower fairies feature prominently in *Flowers in the Mirror* (Li, 1965), originally written about 1815, in which Empress Wu orders all the flowers to bloom at the same time as the wintersweet (*Chimonanthus praecox*). This placed the unfortunate flower fairies in an unenviable position because it had been ordained that the flowers should follow a schedule and could not simply be ordered to bloom. In spite of considerable resistance from the flowers themselves the empress eventually got her way, but only after the tree peonies, who had held out after the others had capitulated, had been subjected to a cruel punishment. This story is interesting in that it reflects the long-held belief, related to the Chinese cosmology discussed above and mentioned in the previous chapter, that interfering with the natural order of things is likely to lead to misfortune. This belief is very ancient as it is recorded in the *Yue Ling* (Monthly Ordinances), which dates from between the 7th and 3rd centuries BC (Tun, 1965).

In most parts of the country each month has its flower or flowers, though such floral calendars tend to vary from place to place and time to time. An example, which is taken from a set of Qing porcelain cups by Li (1959), includes flowering plum (*Prunus mume*) for the first month, followed by narcissus, peach, crabapple, tree peony, pomegranate, lotus, osmanthus, chrysanthemum, monthly rose, orchid (*Cymbidium goeringii*), and wintersweet. Likewise Box (1905) details how at a temple in the south-east of Shanghai there were twelve flower 'gods', each of whom presided over one month of the year. The plant for the first month was the plum, followed in succession by the apricot, peach, rose, persimmon, lotus, balsam, osmanthus, chrysanthemum, cotton rose (*Hibiscus mutabilis*), narcissus, and wintersweet. Each 'god' was worshipped in turn by the florists and gardeners, who placed a sprig of the corresponding flower in his or her hand, or planted a specimen of the same in the courtyard in front of the temple.

In the same manner the flowers more or less universally recognised as representing the four seasons are the orchid for spring, lotus for summer, chrysanthemum for autumn, and flowering plum (*Prunus mume*) for winter, though these are very often represented in the company of other seasonal plants and symbols. In addition, special occasions such as births, deaths, and marriages, the passing of examinations, and the New Year all call for appropriate ornamental plants.

A charming evocation of the involvement of plants in the progression of the year is given by George Kates (1967), when describing the house in which he lived in Beijing in the 1930s. He says, for instance, of its inner courtyard, 'From this place open to the heavens, we watched the seasons in their turning majesty, always surrounded by some token of what grew best in each'. Dorothy Graham (1938) has left us a similar picture, pointing out that gardening was simplified by the skill and industry of Chinese horticulturalists who raised the plants and offered them for sale when they were at their best.

A festival known as the Birthday of the Flowers used to be celebrated in certain parts of the country on the 12th day of the 2nd lunar month, and Princess Der Ling (1911) has described the celebration of this event at court during the time of the Dowager Empress Cixi. After the ladies-in-waiting had cut rolls of silk into ribbons, Cixi tied a yellow and a red one to the stem of a tree peony. Then all the court ladies, eunuchs, and servant girls tied red ribbons to every tree and plant in the grounds, a procedure which took up nearly the entire morning. This was followed by a theatrical performance in which all the tree and flower fairies appeared to celebrate the occasion. This festival was supposed to ensure a fruitful season (Williams, 1975). Likewise Box (1902, 1905) gives a description of similar activities during this festival in Shanghai, and notes that, if it rained on the day, it was believed that it would be a bad year for flowers. A similar observation is recorded by Bredon & Mitrophanov (1927), writing about Beijing.

Symbolism

Koehn (1952) points out that in China many thoughts are expressed in a manner not easily understood by those unfamiliar with the Chinese language and the intricacies of Chinese symbolical art. The people of the East have always preferred to reveal their ideas in an indirect manner, and symbolic meanings attached to plants, animals, and inanimate objects have played a great part in this. The visitor to China, or even just to a Chinese restaurant in another country, is confronted by images of bamboo, pine trees, plum blossom, tree peonies, chrysanthemums, dragons, lions, deer, cranes, tortoises, phoenixes, and clouds, to name but a few of the auspicious symbols which surround the Chinese throughout their lives.

Decorative designs with symbolic meanings are found in or on such things as paving, latticework, furniture, doors, windows, household utensils, bronze, jade, porcelain, ornaments, embroideries, and writing paper—in fact almost everywhere it is possible for them to be (fig. 2.2). And the reason why we find these symbols repeated so often is because many Chinese believe, or at least used to, in the importance of repetition, hoping that this will ensure that the blessings of long life, happiness, wealth, and children will come to them in a dangerous and uncertain world. It is impossible, however, for most Westerners to comprehend all the rules, forces, and superstitions that still govern daily life for many Chinese.

In the West symbolism is usually clear-cut. Rosemary, for example, is associated with memory or remembrance, nothing else, whereas in China the symbolism of an object can be tailored to circumstances and rendered appropriate, and very often complex, by a great many factors. For a start, since repetition intensifies meaning, the character for 'happiness', for example, when double means 'greater happiness' and is always used at weddings to wish the couple matrimonial bliss. Not only this but the meaning of characters is often emphasised or given additional nuances by the manner in which they are written. This is why calligraphy, an art poorly understood by those tied to an alphabetical language, is held in such high esteem. Thus the character *shou* (longevity), when elongated expresses the wish for a very long life. Similarly a rounded *fu* (good fortune), suggests complete or perfect good fortune. The Chinese language itself reflects a very different world and way of thinking, as is made clear by Joseph Needham, who is recorded as saying (Richards, 1997) that learning Mandarin got him 'entirely out of the prison of alphabetical words, and into the glittering world of ideographic characters'.

In addition to the sentiments suggested by the manner in which a character is written, its pronunciation often suggests another with similar pronunciation but different meaning.

The Chinese language is conducive to such punning as many characters share the same or very similar pronunciation. For instance the character for a vase is pronounced *ping*, thus a vase, or a picture of one, suggests the differently written but similarly pronounced word for 'peace'. While this may be so for speakers of Mandarin, different possibilities exist in areas where the pronunciation of characters is different. And as with written and spoken words, the shape of an object can suggest much to the Chinese mind. For instance a vase with a long neck (fig. 16.3) suggests the wish for a lengthy peace, while a round form stands for completeness (Koehn, 1952).

As well as all this, significance is attached to numbers, days of the month and so on. Colours, too, have their meanings. Thus red is the colour of joy, essential for all festive occasions, white represents sorrow and also moral purity, yellow used to be reserved for the imperial family, and yellow and orange are associated with prosperity and good fortune. Black, the colour of bruising, is a sign of evil and therefore unpopular (Williams, 1975). Also, as mentioned earlier, the points of the compass are associated with colours. And so it goes on.

In the light of the foregoing it comes as no surprise to find that the plants have become involved in all of this. Some of the components of the symbolic meanings attached to flowers and plants no doubt date from ancient times, while others have emerged later from the superstitious beliefs of the people. However, the elaboration and development of this aspect of culture took place, from the Tang dynasty on, largely at the hands of the 'literati', as the scholars proficient in painting, poetry, and calligraphy came to be called.

Meaning became attached to plants on account of their habitats, their shape, the season of the year in which they flower or are most conspicuous, their involvement in myth, legend, and literature, and, in some instances, their association with particular deities and famous people. Their ability to withstand harsh weather, their medicinal properties, and, as mentioned above, the meaning of characters homophonous with their names have also been given significance (McMullen, 1987). Thus plants used or represented alone, in combination, or together with animals or inanimate objects, can be used to express an endless range of sentiments.

A strange object which became involved in all of this is the *Lingzhi* (fig. 2.6). This name is given to the fruiting body of one or more species of fungi belonging to the family Polyporaceae. These fruiting bodies develop on the roots or trunks of trees and, unlike other mushrooms and toadstools which perish quickly, become woody and enduring. They are mentioned in legends from 3000 years ago as a good omen, appearing, like the phoenix, when a ruler is virtuous and general conditions are favourable. Apparently one appeared in the Han dynasty during the reign of Wu Di, who wrote songs in praise of it, pardoned all prisoners, and gave meat and wine to a hundred households (Koehn, 1952).

The *Lingzhi* was said to grow near springs, in deep valleys, and near the abodes of the immortals. It was believed, for instance, to grow in abundance on the mythical Islands of the Blest, which were supposed to be situated in the Eastern Sea, nearly opposite the coast of Jiangsu, and no doubt the hope of finding it was among the reasons why the emperor Huang Di dispatched an expedition in 219 BC to search for these islands (Needham, 1956). The fungus was considered one of the magical foods that would confer longevity and immortality. It thus became a symbol of these desirable conditions and dried specimens of fungi of this type or carvings representing them have been kept as auspicious objects. It appears frequently in paintings, often in association with pine trees, or in the beak of a crane, itself a symbol of immortality, or in the mouth

Figure 2.6 *Lingzhi and Pine*, wood-block print from the *Jieziyuan Huazhuan* of 1701

of a deer, a symbol of good fortune. Also it resembles in shape the *ruyi*, a curious S-shaped object usually made of jade which is believed to have wish-fulfilling properties, and has come to represent it (Bartholemew, 1985a). Thus, for instance, when in a painting, design, or arrangement the *Lingzhi* is combined with the plant *Rohdea japonica* (see Chapter 20), which is called *Wannianqing* (Ten Thousand Years Green), it is a rebus for *wannian ruyi* (ten thousand years as you wish). A rebus, in case anyone is wondering, is the enigmatic expression of an idea by means of pictures or objects which suggest syllables or words.

Four-character auspicious sayings of this type have been in use since the Han dynasty and pictorial representations of them have existed from the Song dynasty onwards (Bartholemew, 1980). Bartholemew (1980, 1985a, b) gives numerous examples of such rebuses and points out that they were plentiful in the decorative arts of the Ming and Qing, especially during the long reign of Qianlong (1735–96). Because of their decorative qualities and symbolic meanings the garden plants of China play an important role in this tradition. For instance a grouping of *Lingzhi*, narcissus (*Shuixian*), nandina (*Nantianzhu*), and peach (*Tao*), another symbol of longevity, is a rebus for *linxian zhushou* (the blessing of longevity by immortals). Another well known combination (fig. 2.7) consists of magnolia (*Yulan*), crabapple (*Haitang*), and tree peony (*Mudan* or *Fugui*), suggesting the auspicious phrase *yutang fugui* (wealth and rank in the jade hall) (Bartholemew, 1985a, b; Li, 1956a). 'Jade hall' is a term meaning a wealthy establishment and was also applied to the Hanlin Academy in Beijing, whose members were the scholars who had passed the imperial examinations at the highest level.

It should be noted, too, that from early on the beauty of flowers was also considered to represent the beauty of women, as is clear from *Among the Flowers*, a collection of song lyrics of the 9th and 10th centuries AD (Fusek, 1982). Not only were women's names often floral but matching courtesans names with flowers was an accepted compliment.

Figure 2.7 *The Jade Hall Peony* by Xu Xi (active 960–*c.* 975). Above the double tree peonies are the yulan (*Magnolia denudata*), a crabapple, apparently *Malus spectabilis*, and a single white rose, perhaps *Rosa multiflora*. The plant below, to the left of the pheasant, appears to be *Lychnis coronata*.

Figure 2.8 Enveloped in a light fog, pines cling to a pinnacle at the Huang Shan, Anhui

Many flower paintings, too, may be understood as references to women. In short, flowers always bring with them the suggestion of feminine beauty and floral imagery often has unmistakable erotic overtones (Harrist, 1987; Goody, 1993). Certain fruits are similarly evocative.

Trees, too, have come to be seen as metaphors, particularly old evergreen trees, which have long been held in high regard. It is tempting to suggest that this is because much of the country was deforested long ago and any tree which may be useful for timber or firewood is felled as soon as it reaches a useable size. As a result the only old trees one is likely to see are either species which are useless for these purposes, such as the banyan figs in the south, trees growing in temple grounds and other protected situations, or trees retained for *fengshui* reasons. Old trees have at times been worshipped and there is still a genuine respect for them. Bredon & Mitrophanov (1927) record that they were believed to be kindly disposed and to grant petitions, especially for recovery from sickness. Such trees were hung with votive scrolls inscribed by grateful worshippers, had incense burned beneath them, and offerings of food set out before them, as for the images in temples. It was even believed that some of the huge trees on mountains were capable of speech (Bodde, 1991).

As early as the 4th century BC the sight of a gnarled old specimen in the mountains moved the philosopher Zhuang Zi to reflect that it owed its survival to having become useless, producing no flowers or fruit and being unsuitable for timber. He felt that being alive and useless was far preferable to being dead and buried with honour (Barnhart, 1972). The rotting hulk of an old tree still putting forth a few leaves became a powerful symbol and, as Barnhart (1972) has pointed out, it could be said to represent China. The old tree also

came to represent a human image of integrity, dignity, and enduring strength, oblivious of superficial standards. Also such trees suggest both the brevity of human life and its continuity from one age to another. And trees centuries-old standing by shrines long ago came to personify the lives and characters of the men whose memorials they graced. As a result of these developments the old tree has remained in China a powerful symbol which has never lost its appeal.

Perhaps this affection for trees that have seen the events of history and watched over the passing of lives of men is universal but, as Barnhart (1972) points out, nowhere else have they become such an eloquent metaphor, or played so important a role in artistic expression. Gnarled old specimens, particularly pines clinging to windswept mountainsides (fig. 2.8), have traditionally been favourite subjects with the painters, and gardeners and the producers of penjing struggle to create similar images. Chen Haozi, who lived beside the West Lake at Hangzhou and wrote a famous gardening book, *Hua Jing* (Mirror of Flowers), in 1688, wrote 'If a home has not a garden and an old tree, I see not whence the everyday joys of life are to come' (Giles, 1901).

Happily, in gardens and in the grounds of temples, shrines, and palaces in China many enormous and wonderful old trees have survived the vicissitudes of the past (fig. 2.9). Almost all of these have now been listed on heritage registers. To the trunk of each has been attached a small metal label with a number and, in many cases, the name of the species and its age (figs 2.10, 19.122). Particularly renowned specimens are sometimes accompanied by a notice explaining their history and significance (figs 4.3, 4.17). It is notable too that old trees which are dying back, and even dead trees, which in the West would be removed, are retained, very often being used as a support for a vine such as *Wisteria sinensis*. Examples of this can be seen in many places,

including the Forbidden City and the Kong Miao (Confucian Temple) in Beijing (fig. 2.11).

Figure 2.9, 2.10 (opposite page & left) An 800-year-old ginkgo at Jiaoshan, Zhenjiang, and its label

The symbolism associated with most of the individual plants included in this book is mentioned in the entries for the species concerned. However, let us at this stage look at a scroll painted in 1735 by Chen Shu (fig. 2.12), which would have been hung in the main room of a house at New Year, the principal occasion on which plant symbolism still comes into its own. All the objects depicted are auspicious. The wintersweet (*Chimonanthus praecox*) at the top is used as an alternative to *Prunus mume*, representing purity, endurance, and renewal, and indicates that spring is just around the corner. The red-berried heavenly bamboo (*Nandina domestica*) suggests both endurance and good fortune, the narcissus is believed to bring good luck for the ensuing 12 months,

and the apple (*Ping*) behind the pot on the lefthand side suggests peace. The two persimmons represent success in business affairs and, next to them, the lily bulb, *Baihe* (100 United), with its many overlapping scales, brings to mind close friendship and on account of its Chinese name has a multiplying effect on other good wishes. The Chinese olives (*Canarium album*), on either side of the bulb, represent longevity, and the *Lingzhi*, behind the persimmons, also indicates a long life and, as mentioned above, has wish-fulfilling powers. The camellia, too, is an auspicious symbol for the New Year and for spring (Bartholemew, 1985a; Li, 1956a), as no doubt are the pot, another *ping*, and its decoration. Not surprisingly the painting suggests several auspicious sayings and is an elaborate depiction of the wish for health, wealth, and happiness at the beginning of a new year.

It is clear, then, that the grouping of plants and other objects in the garden, in flower arrangements, in poems, in paintings, and as decorative motifs can conjure up a wide range of images, responses, and associations in the Chinese mind. And in the case of paintings the circumstances which led to their execution together with inscriptions by the painter and others can add extra layers of meaning.

In the light of all this it is understandable that, in spite of the amazing richness of the Chinese flora, until recently it has suited the Chinese to cultivate in their gardens, or to use as ornaments for their houses, principally only those plants around which historical, literary, and symbolic associations have accumulated. In this regard it is useful, when trying to form a notion of which plants were most appreciated, not only to search the literature but to find guidance in artistic representations.

Figure 2.11 **A wisteria planted to grow over a dead juniper at the Confucian Temple, Beijing**

Figure 2.12 *New Year's Day* by Chen Shu (1660–1736)

Plants in art

Ornamental plants are to be seen in Chinese landscapes, in depictions of palace and domestic gardens, incidentally in portraits of scholars (fig. 19.74), and in paintings which have plants as their sole or principal subject. For the history of and rationale behind the painting of plants in China excellent introductions are provided by Sirén (1949), Weng (1982), and Barnhart (1983), to name but three, and it is to authors such as these that readers are referred should they wish to delve more deeply into the topic than I have done in what follows.

Painting in China has had a very long history, going back to at least the 3rd century BC, but as the early paintings were done on perishable materials, usually silk, few have survived, though copies were frequently made. The Tang dynasty saw the full flowering of all the arts, with poetry, calligraphy, and painting at the forefront. However, except for symbolic and decorative purposes, animals, birds, and plants did not receive individual attention from artists until the 10th century, when there arose a distinct genre which became known as *huaniao hua* (flower-and-bird painting), in which plants were depicted, very often together with birds and/or insects. This development seems to have occurred at a time when plants came to be admired more for themselves than as components of the general scene. It is interesting to note that deciduous trees do not occupy as prominent a place in this artistic tradition as do the evergreen ones. However, while they are not often used as a principal subject, they are nevertheless a constant element in landscape painting.

Figure 2.13 A wood-block print of bamboo from the *Jieziyuan Huazhuan* of 1701

rightness, strength, and resilience (bowing before the storm but rising again unbroken), humility (its hollow stem symbolising open-mindedness), and integrity (the character for integrity and that for the nodes of the bamboo stem both being pronounced *jie*) (Weng, 1982).

The literati of the Song and Yuan used painting as a medium for personal expression, often using just a brush and black ink to create a stylised impression of a subject rather than a detailed naturalistic depiction. This type of painting is

Although we know that during the Tang there were flower painters with a high level of skill, their works have not survived (Hulton & Smith, 1979). The earliest works in this type which still exist are by artists who were active later in the 10th century, for instance the wonderful example by Xu Xi reproduced here as figure 2.7. These works exhibit a style of accurate realism which was to persist throughout the Song, during which many works of astonishing beauty were created (Barnhart, 1983). The significance of flower-and-bird painting at this time is attested by the inventory of 6387 works in the imperial collection about 1120. Of these 2776 were flower-and-bird paintings, more than twice the number of those of the next largest genre, religious painting (Barnhart, 1983).

It was during the late Song and the following Yuan that the scholar-bureaucrats, through their poetry and painting, attached to plants much of the symbolism that has come down to us, particularly that concerned with gentlemanly ideals. The initial impetus to this way of looking at plants appears to have come from the famous Tang poet Bai Juyi, who in 803 wrote an essay, *Yangzhu Ji* (On Cultivating Bamboo), in which he indicated how the characteristics of the plant could provide the *junzi* (morally superior man) with reminders of proper conduct (Bodde, 1991). As a result of this bamboo came to be seen widely as a metaphor for up-

related to calligraphy. That calligraphy became a major art inseparable from painting may be explained by both the nature of the Chinese written language and the use of the brush and ink for both arts (Weng, 1982).

Bamboo, in particular, lent itself to calligraphic treatment, many of the strokes used to depict it resembling certain strokes used in calligraphy (fig. 2.13). The brush strokes became all important and the result was a fusion of literary and pictorial traditions which became known as *mozhu* (ink bamboo). As Bickford (1996) puts it, the painter reveals 'to those who can read it, his character, feelings, and values through the literary associations of bamboo, the iconographic attributes of the particular pictorial image, and, significantly, through his brushwork'. Although it is hard for the Western mind to grapple with such concepts, this highly developed art joined poetry and calligraphy in the literati repertory as a new means of expression between like-minded men. This almost abstract style of painting, involving ink on paper rather than colour on silk, became a genre of its own, distinct from traditional flower-and-bird painting, and was extended to involve also the pine, plum, orchid, and chrysanthemum as its principal subjects. It is quite beyond the scope of this book to deal further with this remarkable genre, but those who wish to know more about it can do no better than to refer to Maggie Bickford's *Bones of Jade, Soul of Ice* (1985) and *Ink Plum* (1996). The complexity of the symbolism which can be expressed by paintings of plum blossom has already been made clear by the quotation from the *Jieziyuan Huazhuan* given earlier in this chapter.

During the Yuan, a time when the scholar-officials were oppressed, there were many paintings which symbolised moral tenacity under adverse conditions, remaining green or flowering in the depths of winter, as do the bamboo, pine, and plum. Likewise the chrysanthemum, blooming in autumn long after other flowers have withered, and the cymbidium, exuding exquisite fragrance in a remote secluded place, were used to express similar sentiments. Pine, plum, and bamboo were often depicted together as the 'Three Friends of the Cold Season' and the plum, orchid, chrysanthemum, and bamboo were fancifully personified as the 'Four Gentlemen of Flowers'.

The realistic style of bird-and-flower painting persisted during the Yuan dynasty, though under the Mongols the old idea of an imperial academy of court painters was destroyed. It seems to have been during this period that the custom of writing explanatory notes or poems on paintings became widespread. These usually involve the subject in an allusion to some specific idea, circumstance, or occurrence. A good example is the exquisite painting of pear blossoms (fig. 2.1) by Qian Xuan (c. 1235–after 1301), on which the artist's poem evokes the sadness of a court beauty, abandoned and aging, weeping on a verandah behind a locked gate on a rainy night—a metaphor for China in a Mongol prison. One may see the pear blossoms, one may paint them, but all that they have ever meant has vanished (Barnhart, 1983).

Under the native Ming dynasty, painters in the flower-and-bird tradition continued to work for the court but with very little feeling of direct contact with nature, a state of affairs which has largely persisted from then on. Many such works, however, are of great beauty and craftsmanship. During this period many different plants appear in paintings, topics such as 'flowers of the seasons' being popular, and painters producing collections of album leaves of garden flowers. Meanwhile the scholars, particularly in Suzhou in the 16th century, continued to favour working with ink alone, to depict the traditional subjects such as plum blossom, and to combine their paintings with calligraphic inscriptions, affecting to despise the professional artists with their recourse to colour.

The teachers of painting studied the masterpieces of the past and used these as models for their students. In the late Ming and early Qing special books appeared for students containing instructions for the painting of various subjects. Sirén (1949) has drawn attention to the woodcuts after paintings of flowers and fruit by celebrated masters which are reproduced in two well-known handbooks for students of art, the *Shizhuzhai Huapu* (Ten Bamboo Hall Painting Guide), which was published during the first third of the 17th century, and the *Jieziyuan Huazhuan* (Mustard Seed Garden Painting Manual), three volumes of which appeared between 1677 and 1701, a fourth not being published until 1818 (Tschichold, 1952). It is worth noting that these two works contain the earliest known examples of coloured woodblock printing in China (figs 2.4, 2.6, 18.7), remarkable for their masterly use of graduated tones of ink and colour, and which are said to have influenced the development of colour printing in Japan. The *Shizhuzhai Huapu* deals with bamboos, orchids, plum blossoms, and many other flowers and fruits, all represented with the greatest accuracy. In Volume 2 of the *Jieziyuan Huazhuan*, bamboo, plum, orchid, and chrysanthemum, the four plants considered to be of the greatest importance for painters, are dealt with in great detail, much attention being paid to their elaborate symbolism. In Volume 3 more than 70 other plants are depicted, most of which have been identified by Sirén (1949).

In addition to these books the largest collections of plants grouped together in Chinese art are to be found in examples of the 'flowers of the seasons', 'flowers of the twelve months' and 'hundred flowers' genres. These are very often in the form of hand scrolls depicting a large number of plants and are interesting records of plants known to the artists at various times in history. Remember, however, that while most of the plants depicted are garden plants, weeds and wild plants sometimes appear as well. In these paintings the plants are usually arranged according to the seasons, beginning with the first month at the right-hand end and ending with the twelfth at the left. However, the plants are not always arranged exactly in order of blooming, as considerations of composition and the juxtaposition of colours and forms are often taken into account. Nor do compositions of the 'hundred flowers' type usually depict as many as 100 different plants. The expression is a literary allusion, references to the 'hundred flowers' appearing in poems of the Six Dynasties and Tang periods. It was taken up in quite different

Figure 2.14 *Hemerocallis fulva* and *Michelia figo*, a section of the *Song Hundred Flowers*

and culture were devastated during the Taiping uprising and Shanghai became the centre of artistic activities. Then, with the fall of the Qing dynasty in 1911, the country entered a period of political and social change from which it has not yet fully emerged, a period which, until very recently, has favoured neither the arts nor ornamental horticulture.

circumstances in more recent times by Mao Zedong when he said 'Let a hundred flowers bloom and a hundred schools of thought contend'. As with other Chinese usages such as 'ten thousand years', it is merely a way of saying 'many'. An early painting in this genre is that known as the *Song Hundred Flowers*, now in the Palace Museum, Beijing, parts of which are illustrated here (figs 2.14, 16.8, 16.9, 20.6, 20.26, 20.27). This hand scroll, which measures 31.5 cm by an astonishing 1693 cm, is executed in ink on paper and is believed to have been painted during the Song dynasty. Almost as large (41.9 × 1649 cm) is a work in ink and colour on silk attributed to Yun Shouping (1633–90), part of which is reproduced by Barnhart (1983, cat. no.33). A smaller (41.5 × 197.5 cm) example from the 17th century is reproduced in its entirety by Weidner et al. (1988, cat. no. 34). The profusion of flowers in these paintings suggests abundance, an auspicious message typical of such decorative compositions.

Worth noting, too, are the plants depicted on Chinese porcelain and other objects. Bartholemew (1980, 1985a, b), for instance, illustrates or lists numerous examples, including a pair of *famille rose* bowls and a vase of the early 19th century decorated with the 'hundred flowers', and has drawn attention to the messages that numerous botanical motifs on porcelain and elsewhere are intended to convey. Also Krahl (1987) has listed more than 50 plants recognised on Chinese porcelain in the Topkapi Saray.

While the arts continued to flourish during the Qing, particularly during the long reigns of Kangxi (1662–1723) and Qianlong (1735–96), the 19th century was a period of political crisis, during which Qing authority declined and Western interference intensified. In the middle of the century Suzhou, Hangzhou, Yangzhou, and other centres of wealth

Penjing

Although this topic has been discussed in the previous chapter, it seems appropriate to include a few further words about it here, since it is an art form which reflects the appreciation of landscape paintings and old gnarled trees (figs 1.13, 1.14, 19.63). According to Livingstone (1822), dwarfed trees were actually called *gu shu*, ancient trees.

As with everything, there are considerable variations of taste in the world of dwarf trees. Robert Fortune (1844, 1847a) saw them in Guangzhou and Ningbo trained not only to represent old trees in miniature but pagodas, dwelling houses, and various kinds of animals, of which the deer seemed to be the favourite (fig. 19.56). Such creations usually express symbolic meanings based on such things as literary references, the identity of the plant or plants used, and the shape or shapes into which they have been trained. Freeman-Mitford (1900) in Beijing in the 1860s described 'the famous dwarfed trees, trained so as to represent with their branches characters of good omen, such as Happiness, Longevity, etc.'.

Flower arrangement

Chinese flower arrangements, too, cannot avoid having symbolic meanings. But, apart from an article by Ferry (1949) and an excellent book by Li (1956a), little has been written in Western languages about this topic. Li reviews the literature on the subject and points out that Chinese floral art, unlike that of Japan, does not require rigorous adherence to conventions. Nevertheless the importance of line, avoidance of symmetry, and choice of compatible materials are all matters to be taken into account. He gives extensive lists, taken from earlier authors, of flowers and combinations of flowers considered suitable, pointing out that, when two or more flowers are used in an arrangement, it is desirable not only that they should be of the same season but complement each other in colour and form. In relation to this he has given us a translation of some of the advice given in the late 16th century by Yuan Hongdao in his *Ping Shi* (History of Vases). This author calls the most popular and esteemed flowers 'mistresses' and lists for each a number of 'maids', whose charm and beauty complement and enhance that of their superiors. He begins by recommending that *Prunus mume* should be attended by one or more of *Jasminum nudiflorum*, *Daphne odora* and *Camellia japonica*, all of which bloom at the same time, and goes on to give comparable advice for *Malus spectabilis*, tree and herbaceous peonies, pomegranate, lotus, osmanthus, chrysanthemum and wintersweet. Similar information is provided by Du Benjun in his *Pingshi Yuebiao* (Monthly Calendar of Vase Flowers) written in the early 17th century.

In addition to this, as Li points out, a flower arrangement should be designed to fit harmoniously into the room for which it is intended. Just as the growing of potted plants, dwarfed trees, and tray landscapes led in turn to the manufacture of special tables and stands on which to display them, so did the creation of flower arrangements. Likewise, the art of flower arrangement has prompted the production of vases shaped and decorated to allow various different flowers to be effectively displayed (figs 2.15, 16.3).

An excellent example of the Chinese attitude to flower arrangement is given by Shen Fu (1983), who goes into great detail about the niceties of arranging chrysanthemums. He gives advice concerning appropriate numbers of flowers to place in a vase, the number of vases which should be set out on a table, and the heights and arrangement of the stands on which they are placed. Whether the flowers should be crowded or spread out, and whether they should lean towards the viewer or away, all depends on the sense of pictorial composition of someone who knows how to appreciate them. And, in regard to flower arrangement in general, he explains how to make a pin-holder on which to fix the stems and how to prune and bend branches in order to achieve the most tasteful and harmonious effects. He concludes by indicating that maple leaves, bamboo branches, pieces of grass,

and thistles can all be used in decorative arrangements. A single green bamboo twig, if complemented by a few aspen seeds, some leaves of fine grass, and two thistle branches, all of them in proper arrangement, can, he says, have an unworldly beauty.

Artificial flowers and plants

Along with flower arrangements and penjing, artificial flowers and plants were popular in China long before the invention of plastic. Ornamental pots containing small artificial trees with leaves of green jade and fruits or flowers made from turquoise, rock crystal, agate, cornelian, coral, or other semi-precious stones are a common form of ornament in the Chinese house. These are known as *yuhuapen* (pot of jade flowers) and first came into vogue during the Qianlong period (1735–96) (Cheng, 1969). Osbert Sitwell (1949), who visited Beijing in the 1930s, saw them for sale there and said you could watch them being made.

Artificial flowers have a long history in China. For instance, in 1727 Francis Xavier d'Entrecolles wrote to Paris from Beijing describing their manufacture from rice-paper (Bretschneider, 1880), which is made from the pith of *Tetrapanax papyrifer*. Over a century later, Doolittle (1866), who spent 14 years as a missionary in Fuzhou, observed that the Chinese were 'very dexterous in making artificial flowers' and that they excelled in 'making a very large variety'. Examples made of paper or silk are still to be seen in shops, houses, and temples, along with their modern plastic counterparts.

Women used to wear flowers, both real and artificial, in their hair or on head-dresses. Robert Fortune (1847a) observed that the ladies of Fuzhou were particularly fond of flowers for the decoration of their hair, noting that 'The rustic cottage beauty employs the more large and gaudy, such as the red Hibiscus; while the refined damsels prefer the jasmine, tuberose and others of that description: artificial flowers, however, are more in use than natural ones'. In this regard it is amusing to note that, while the Dowager Empress Cixi was fond of adorning herself with living flowers, she obliged the ladies of the court to make do with artificial ones, albeit fashioned from jade and pearls (Der Ling, 1911). Doolittle (1866) also recorded the popularity of flowers, both real and artificial, as hair ornaments, and noted that widows under 40 years of age were not allowed to wear them. In Beijing a market devoted to the sale of artificial flowers,

Figure 2.15 *Peony in a Vase*, Qing dynasty. Amongst various objects depicted here are three corms of the water chestnut (*Eleocharis dulcis*), to the left of the base of the vase.

principally for use as hair ornaments, used to be held on the 4th, 14th and 24th of each month outside the Hademen, a gate which led from the Imperial to the Chinese City (Tun, 1965).

It is interesting to note that the fear that causing plants to bloom out of season might lead to misfortune has at times been extended to the use of artificial flowers. It appears that, in a ceremony called the Beginning of Spring, a presentation of symbolic objects, including artificial flowers, was made to the imperial family in the first month of each year. When this ceremony was elaborated during the Ming, the local people complained that only flowers of the season should be used (Tun, 1965).

Further considerations

In the light of most of what I have so far included in this chapter, it could easily be assumed that intellectual considerations were always uppermost in the minds of Chinese gardeners. However, it seems that in practice they were much like those of other lands. Plants were certainly not appreciated simply for their symbolism or historical associations. Otherwise where would the myriad cultivars of camellia, flowering peach, azalea, rose, peony, orchid, chrysanthemum, and suchlike have come from? Double flowers, new shapes, new colours, variegations, and so on were eagerly sought, and novelties were often jealously guarded or at least very highly priced.

The tree peony craze during the Tang and Song dynasties provides a notable example of this aspect of Chinese ornamental horticulture, as can be gleaned from Ouyang Xiu's *Luoyang Mudan Ji* (Account of the Tree Peonies of Loyang) written in 1034 (Needham, 1986). Apparently if people from elsewhere came looking for a plant of the famous and much prized variety 'Yao Huang' ('Yao Family Yellow') (fig. 12.12), they were likely to be given a graft that had been killed by dipping it in hot water. It is claimed that even until the 1990s some varieties were not permitted to be exported, and that until recently no-one would sell 'Yao Huang' to an American nurseryman wishing to obtain it (Aleksinas, 1995).

Another revealing instance is recorded by Shen Fu (1983), who is believed to have written his *Six Records of a Floating Life* in 1809. He was an orchid enthusiast who vowed never to grow these plants again after someone, who had been refused a piece of his most prized variety, secretly poured boiling water over it and killed it. There is nothing new under the sun, as Philippe Rivoire (1928) remarked after reading a 17th century Chinese account of chrysanthemum culture and finding the procedures to be the same as those followed in France in his time.

In discussing ornamental plants in China, Needham (1986) points out that one should also remember the 'fashionable' aspect of their cultivation. It was the proper thing for successful Confucian officials to take an interest in gardens and plants. And, as Needham puts it, 'an intimate knowledge of plants and their accumulated symbolism provided an opportunity for airing possibly dangerous sociopolitical views within the fabric of apparently innocuous though allusive poetry and literature'.

Matters of class distinction, taste, and snobbery were in evidence also. Needham (1986) says: 'In perfect accord with the ethos and principles of bureaucratic organisation, the scholar-officials developed a mania for classifying flowers and flower varieties into ranks analogous to those of society—emperor, empress, virtuous minister, local magistrate, post master, bandit and so on. This arose in part from the fact that plants and flowers had from early times been emblems of virtues'. As an example Needham draws attention to Qiu Xuan's *Mudan Rongru Zhi* (Record of the Ranks of Tree Peonies) of 1050, in which the author classifies tree-peony varieties by analogy with the ranks of ladies attending upon the emperor. This exercise, as Needham remarks, would have been more profitable had the author directed his energies towards closer botanical description. Numerous other instances can be found in various works in which the garden varieties of various flowers are classified into 'grades'.

There can be little doubt, too, that in the garden some placements or groupings of plants were considered to be more acceptable than others to people of taste. For instance, Craig Clunas (1996) has brought to notice Wen Zhenheng's musings on Suzhou gardens in the early 17th century. Wen lists plants with notes about their suitability and cautions against practices such as the planting of plums in a courtyard, the tastelessness of putting camellias and magnolias together, and the vulgarity of combining peach and willow. Climbing roses, too, were not for people of discernment, at least partly it seems because they required a framework to grow on.

Many of the things Wen Zhenheng fulminated against subsequently became commonplaces of Chinese garden design. One wonders what he would have said of the present practice of constructing large dragons and other creatures from wire and mud and then planting them up with small herbaceous plants, remarkable creations which appear here and there in public places. Examples of this curious practice are illustrated by Zhu (1992). In like spirit are the peacocks, fashioned from cypress foliage and potted marigolds, which I saw in the Mei Yuan, Wuxi in 1995 (fig. 2.16). Whatever opinion one might have concerning the aesthetic qualities of these objects, it has to be remembered that they represent universally recognised symbols of good fortune. It is not clear whether this practice has any antiquity but undoubtedly it has preceded the large 'Puppy', designed in similar vein by the American artist Jeff Koons, versions of which were constructed and planted in Germany in 1992 and Australia in 1995 to critical acclaim.

In considering Chinese attitudes to plants, it is worth mentioning that to an appreciable extent no sharp distinction is made between food, ornamental, and drug plants (Walker, 1944). Fruits and vegetables were not only an essential part of the diet, they also symbolised purity and sobriety. They were the food not only of the common people unable to afford meat, but also of pious Buddhists, hermits, and scholars living in rustic seclusion (Needham, 1984). In the eating of a meal its benefits, real or imagined, with

regard to such matters as health, sexual potency, and longevity are still likely to be regarded as of some consequence, at least by the older generation. An extreme case has been the Daoist belief that certain foods are strengthening and purifying, thus paving the way to immortality. That some provisions, whatever their price, can hardly be described as delicious is thus of minor account.

On the subject of food it is interesting to note that, just as many fruit trees and other food plants are valued highly for their aesthetic qualities, many Chinese ornamental plants can be eaten, at least in times of famine. In spite of the skills of the farmers, agricultural experts, and hydraulic engineers, floods and droughts have caused China to suffer throughout the millennia from periodical food scarcities and full-scale famines. So the botanists produced works indicating which wild or ornamental plants could be eaten, which were dangerous, and which required special treatment to remove poisonous principles. According to Needham (1986) there is no known parallel for this in any other culture.

The earliest and best of these works was written by an imperial prince, Zhu Xiao. It is called the *Jiuhuang Bencao* (The Save-from-famine Herbal) and appeared in 1406. This Ming prince set up private nursery gardens in Kaifeng where he experimented with plants collected from ditches, fields,

Figure 2.16 **Peacocks made of cypress foliage and potted calendulas in the Mei Yuan, Wuxi**

Figure 2.17 Woodblock prints of three forms of *Rosa laevigata* from the *Jiuhuang Bencao* of 1406

and wildernesses. The first edition of his book describes 414 plants, 276 of which had not previously been noted, indicates what parts can be eaten, and is illustrated with woodblock prints (fig. 2.17).

Various similar books appeared in the ensuing centuries and many species also used as ornamental plants were included (Needham, 1986; Needham & Lu, 1968). Bretshneider (1881) lists the plants included in the *Jiuhuang Bencao*, indicating for each whether it is the roots, stems, bark, leaves, young shoots, flowers, fruits, or seeds which can be eaten.

A great many plants are also used in household medicine or in the prescriptions of physicians for preventing, curing, or alleviating human diseases. For example the *Bencao Gangmu*, the famous pharmacopoeia published in 1596 three years after the death of its author, Li Shizhen, includes over 900 plants used in medicine, many of them species otherwise used for food or ornament (Bretschneider, 1881; Read, 1936). It is not surprising, then, that most of the plants included below have been or still are used in traditional medicine. However, since this is a book about ornamental plants, some readers will perhaps be relieved to hear that I have refrained from commenting in detail on their medicinal uses in most instances. Anyone wishing to know more about this might consider consulting one of the more recent Chinese pharmacopoeias, for example the *Zhongguo Bencao Tulu* (Lian & Hu, 1988).

In writing this chapter, in fact in writing the whole book, I have been very conscious that some of what I have had to say will perhaps seem naive or ill-informed to those who are well acquainted with Chinese culture. Nevertheless it is hoped that the brief outline of the involvement of plants with Chinese life and culture which I have given above, however incomplete and imperfect, will serve to provide a background for the better appreciation of the plants which are found in Chinese gardens. It must be pointed out, though, that, just as allusions to classical antiquity, the Bible, Shakespeare, and so on do not always strike a chord with the young of the West these days, not a great deal of China's past culture has been passed on since the establishment of the People's Republic.

CHAPTER 3

THE INTRODUCTION OF CHINESE GARDEN PLANTS TO OTHER COUNTRIES

As might be expected, Chinese plants were taken to adjoining countries early in history (Laufer, 1967). A few found their way overland as far as the Middle East and the Mediterranean and, later, others were taken by seafarers to India, Arabia, and East Africa long before Vasco da Gama rounded the Cape. Also, about the 6th century AD Chinese plants began to be taken to Japan (Li, 1956a), where many of them, for example the tree and herbaceous peonies, became absorbed into Japanese culture and underwent further development. It is not surprising then that various Chinese plants were first encountered by European botanists far from their native home and, as a result, were given misleading specific epithets such as *armeniaca*, *indica*, *japonica*, *persica*, and *syriaca*.

Figure 3.1 *Rosa xanthina*, introduced to the United States from Beijing by F. N. Meyer in 1906

63

Although the discovery of Chinese plants by Europeans and their introduction to the West has been dealt with at length by others (Bretschneider, 1880, 1898; Cox, 1986; Lauener, 1996; Spongberg, 1990, 1993), it seems appropriate for me to say a few words about it here. Apart from such things as the peach, apricot, and daylily, which appear to have been carried along the trade route of the Silk Road to the Middle East and thence to Europe, and one or two other plants brought back by Arab seafarers, few Chinese plants were known in the West until after Europeans had reached China by sea in the 16th century.

Even then, few dried specimens of Chinese plants reached Europe prior to the visit of James Cunningham, who was a physician at the East India Company factories at Xiamen (Amoy) and Zhoushan (Chusan) from 1698 to 1703. Cunningham collected nearly 600 Chinese plant specimens, which were described by others early in the 18th century.

7616

Figure 3.2 The China Aster, *Callistephus chinensis*, from *Curtis's Botanical Magazine* 1898, t.7616

Another important collector was Peter Osbeck, a pupil of Linnaeus, who collected specimens in the neighbourhood of Guangzhou in 1751 and 1752.

A further boost to knowledge about the plants of China came following the visit to the region of the Portuguese Jesuit João de Loureiro. After spending four years in Macao, Loureiro moved in 1742 to what is now Vietnam, remaining there for almost 36 years during which he resided principally in Hue. After a further three years spent in Guangzhou he returned to Portugal, where he devoted himself to completing his *Flora Cochinchinensis*, which was published in Lisbon in 1790. In it he recorded almost 1300 wild and cultivated plants from Vietnam and southern China.

As a result of all these happenings many Chinese plants were known to European botanists by the late 18th century. All the same, very few living plants or seeds were brought back and successfully established before the last decade of that century. Although the Portuguese had reached the Pearl River delta in 1513 and settled in Macao later in the century, apart from a superior variety of sweet orange said to have been sent to Lisbon in 1545 (Cox, 1986) and another in 1635 (Tolkowsky, 1937), they appear not have introduced anything of consequence. One or two plants were raised from seeds sent by James Cunningham around 1700 and *Dianthus chinensis* was probably introduced by the French soon afterwards (Bretschneider, 1880; Cox, 1986). But, in spite of their activity in sending back descriptions of and information about Chinese plants, only rarely did the early Jesuit missionaries send seeds or other living material. Amongst the first we know to have done so was Francis Xavier d'Entrecolles, who sent persimmon seeds from Beijing to Paris in 1736 (Bretschneider, 1880). Whether these germinated and became established is not recorded. However, we do know that a few years later the Tree of Heaven (*Ailanthus altissima*), the Oriental Thuja (*Platycladus orientalis*), the China Aster (*Callistephus chinensis*) (fig. 3.2), and various other plants were raised from seeds sent by Pierre d'Incarville, who was in China from 1740 to 1756, principally in Beijing (Bretschneider, 1880; Cox, 1986). And another missionary in Beijing, J. P. L. Collas (1786a), wrote an article on the Chinese plants which it would be possible and useful to establish in France. The article dealt at length with a comparison of the climate and conditions in various parts of China with those of the different regions of France.

While the Jesuit missionaries were allowed access to

Figure 3.3 **Foreign factories at Canton (Guangzhou), from _China_ by T. Allom and G. N. Wright, Fisher, Son & Co., London, 1843**

various parts of the country, foreign traders were restricted to the subtropical south where, from 1757 on, they were allowed to reside and trade only in one small area of Guangzhou (fig. 3.3) and then only from September to March each year (Whitehead & Edwards, 1974). They were not permitted to bring their wives and families and were obliged to trade with only a few licensed Chinese merchants. From April to August they moved back to Macao where the international merchant community had established itself. However, the European demand for tea, silk, porcelain and other novelties was such that the restrictions were, for the time being, endured.

Under these circumstances it is remarkable that as many Chinese garden plants reached Europe as did in the late 18th century and the first half of the 19th. But Europeans were prepared to go to considerable lengths to obtain the plants with which they had become familiar, both from hearsay and as a result of their depiction in paintings and on wallpaper, porcelain and suchlike, which were easy to transport and arrived long before the plants themselves. Also it was fortunate that the nurseries known as the Fa Tee Gardens on the out-

skirts of Guangzhou were well stocked with plants growing in containers, which could be bought and sent to Europe on the tea clippers. John Barrow (1804), who accompanied Lord Macartney's embassy and visited the area in December 1793, wrote of 'nurseries for propagating the rare, the beautiful, the curious or the useful plants of the country; which are sent to Canton for sale' and noted a great many of the kinds available, including the 'large Peonia, white, red, variegated'. The custom of annually bringing temperate-climate plants such as tree and herbaceous peonies from nurseries further north for sale in Guangzhou ensured that a remarkably representative selection of Chinese garden plants could be purchased there. As a result of this, for example, in the late 18th century Sir Joseph Banks was able to arrange for the importation to the Royal Botanic Garden at Kew of the first tree peony, amongst other things. Further plants were received from William Kerr, whom Banks had sent to Guangzhou in 1803.

Another early European visitor to the Fa Tee Gardens was André Van Braam (1798), who was in China in 1794–95 with the embassy of the Dutch East India Company and who

Figure 3.4 (above) **A nursery in Guangdong**

Figure 3.5 (left) **A cultivar of *Camellia japonica* in a Suzhou garden. Cultivars of *C. japonica* were amongst the most popular of the plants sent to Europe from Guangzhou in the early 19th century.**

made many drawings of Chinese plants. Twenty years on, John Livingstone (1820b), a surgeon who spent time in Macao and Guangzhou with the East India Company, complained that although strangers had earlier had access to the nurseries at all times, their visits had become restricted to three days a month and an admission charge imposed. Later still, Robert Fortune (1844) makes no mention of this, so perhaps the restrictions were removed after the first Opium War. Fortune records that there were about a dozen of these nurseries 'about three miles above the city on the opposite side of the river'. He described how the plants for sale were kept in pots along the sides of narrow paved walks, just as they are in Chinese nurseries today (fig. 3.4), and that there were also areas where stock plants were growing in the ground. He was particularly impressed by the azaleas and remarked upon the large collection of camellias (fig. 3.5), oranges, roses, and other well-known plants which were purchased by the Chinese when in flower. The grouping together of nurseries was apparently common in China, as it was and is with other businesses, and Fortune (1863) later encountered the same state of affairs on the outskirts of Beijing and Tianjin. And in our own time Hazel Le Rougetel (1982) visited a nursery on the outskirts of Guangzhou which had been there for 200 years. The characters inscribed over the gate read *Hua Di* (Flower Land) or *Fa Tee* if one uses the old transliteration of the Cantonese pronunciation, so there seems little doubt that this was a survivor, perhaps the last, from the days when these nurseries gave their treasures to the rest of the world.

Although many plants from this source were sent to Europe, the survival rates of the early consignments were poor, as was commented upon by Livingstone (1820b). He estimated that, owing to the losses in transit, 'every plant now in England, must have been introduced at the enormous expense of upwards of £300'. In the light of this he suggested to the Horticultural Society of London that a properly qualified gardener should be sent out to prepare the plants and that the ship owners and captains should be properly briefed. At the time there was also the suspicion, unfounded it seems, that the Chinese nurserymen were not anxious to allow their treasures to fall into the hands of foreigners and that they killed plants prior to their dispatch. Likewise, foreigners were given to claiming that the failure to germinate of seeds bought in China and sent to Europe and America resulted from their being 'treated' prior to sale. However, Fortune (1844, 1847a, 1850b) defended the good name of the suppliers, stating that the age of the seeds and the depredation of insects was responsible for their poor viability, not boiling or poisoning as people said. Apparently at the Fa Tee Gardens the growers mixed ash with the seeds to protect them from insects and this may have been assumed by some to be done to destroy their viability. Livingstone (1820a) had earlier mentioned the difficulty in southern China of protecting seeds from mildew and insects, and recommended drying them over sulphuric acid.

Perhaps as a result of Livingstone's recommendations, the Horticultural Society sent its own collectors to China, John Potts, who left in 1821 and returned in 1822, and John Parks who made the round trip in 1823–24 (Cox, 1986). These expeditions resulted in the introduction of a number of interesting plants due to improvements in the methods of caring for the plants before and during the voyage. John Potts recorded his appreciation of the assistance given to him by Livingstone and by Thomas Beale, a merchant who lived for 50 years in Macao, where he had accumulated a collection of Chinese plants in his garden.

It appears that John Parks was greatly assisted by John Reeves, the person principally responsible for the improvement in the methods of dispatch and care of plants sent from Guangzhou. Reeves was sent to China by the East India Company in 1812 and remained there until his retirement in 1831. Thanks to his great interest in horticulture many of our best-loved garden plants were introduced during the period he spent in southern China (fig. 3.6). In an obituary notice (*Gardeners' Chronicle*, 1856) one reads that 'Not a company's

Figure 3.6 *Wisteria sinensis* in the Yu Yuan, Shanghai.
W. sinensis was sent to England from Guangzhou in 1816.

ship at that time sailed for Europe without her decks being decorated with the little portable greenhouses which preceded the present Wardian cases; and what was more, Mr. Reeves succeeded in communicating to the captains the enthusiasm which animated himself'. In addition to all of this, for a number of years he employed several Chinese artists who lived in his house and worked under his supervision, producing hundreds of coloured drawings of plants. A collection of these drawings was sent from Guangzhou to the Horticultural Society where it further aroused enthusiasm for the introduction of Chinese plants (fig. 1.5). A second collection was presented by his daughter to the British Museum in 1877.

Britain was not the only country to benefit from Reeves's efforts. For example, records at the Royal Botanic Gardens Sydney show that he sent plants there, as did his friend Thomas Beale (Clough, 1996). After Reeves' retirement

Figure 3.7 *Xanthoceras sorbifolia*, specimens of which were collected in Beijing in 1831–32 by Alexander von Bunge and which was introduced to Paris by Armand David in 1866

others sent further consignments to Australia. No doubt Chinese plants found their way to many parts of the world at this time.

Long before this, apparently starting with Columbus, the Spanish and Portuguese had commenced the exchange of plants between the New World and China. As mentioned in the Introduction and Chapter I, early in the 16th century the Portuguese established a direct route between Asia and South America, and in 1565 the Spanish started sailing between Acapulco and the Philippines (Merrill, 1954). By these routes Asian plants reached America and American plants were transported to Asia.

For early observations on the cultivated plants of Beijing and its environs we are indebted to the Jesuit and Lazarist missionaries who established themselves there from the beginning of the 17th century. Our awareness of these plants was greatly expanded by the visit and observations of the Russian botanist, Alexander von Bunge. Although he spent only 6 months in Beijing, from late 1830 into 1831, from his collection of specimens a considerable number of species were made known to the West for the first time (fig. 3.7).

Subsequently the journeys of Robert Fortune between 1843 and 1861, chiefly in the Yangtze delta region, resulted in many exciting introductions, mainly of cultivated plants. Fortune's travels had been made possible by the concessions ceded by the Chinese after the Opium Wars of 1839–42 and 1856–60. After the first of these wars Hong Kong was formally ceded to the British in 1843, and by 1855 Hong Kong and Shanghai had surpassed Guangzhou as the main centres of China's export-import trade. After the Second Opium War, China was obliged to open up most of the country to Westerners, with the result that the richness of the native flora began to be recognised and the great era of plant hunting in China began. The activities of the professional plant collectors were reinforced by those of missionaries, travellers, and those Europeans employed by the Chinese in organisations such as the Imperial Customs Service, so that a flood of plants not previously cultivated anywhere began to reach the rest of the world. In the face of this, and because a representative selection of Chinese garden plants had by this time been introduced to the West, interest in the plants of Chinese gardens and nurseries declined. Also, once the country had opened to foreign trade, Chinese nurserymen seem to have shown little interest in exporting their plants. This is in contrast to Japan, where the recommencement of trade with the outside world prompted nurserymen to exploit the overseas demand for their plants, and they have continued to do so ever since.

The only evidence I have come across of the Chinese going out of their way to sell plants to Europeans is provided by Fortune (1850e), who records the presence in Shanghai of boats with loads of plants brought from Suzhou and Hangzhou ranged along the river to tempt the eyes of the English. All in all, however, it can be seen as fortunate that so many Chinese garden plants were sent to the West from Guangzhou when they were, and that Robert Fortune was able to collect further north in centres of wealth and culture prior to the upheavals and destruction of the past 150 years.

In the early part of this century what little introduction of cultivated plants from China there has been has resulted largely from the efforts of workers from the United States. Ernest Wilson, who visited China firstly for the nursery firm of Veitch and later for the Arnold Arboretum, and F. N. Meyer, who went on behalf of the United States Department of Agriculture, added greatly to our store of information as well as being responsible for many new introductions (fig. 3.1). Meyer was followed by P. H. Dorsett and W. T. Swingle who carried out similar work. Various others who lived in China for a time, particularly in Beijing, Shanghai, and Guangzhou, also wrote about, catalogued, introduced, and otherwise made known the plants growing in those regions. On the whole, however, except for a very brief period after World War II, the events of the 20th century made such endeavours impossible until after the end of the Cultural Revolution in 1976.

The introduction of a range of cultivars of *Camellia reticulata* from Kunming in 1948 and 1949 caused a sensation. This has been followed in recent years by the importation here and there of others, of tree and herbaceous peonies, and the odd rose, chrysanthemum, and wisteria. Perhaps there have been many more about which I do not know, but their impact must have been very small compared with that of the introduction of the great bulk of Chinese garden plants in the 19th century.

Even so, my own observations suggest that, in spite of everything that has passed, there remain in China cultivated ornamental plants that are not as yet grown elsewhere and which are worthy of wider distribution. Roses, peonies, both tree and herbaceous, and forms of *Malus*, *Paulownia*, *Prunus*, *Syringa*, *Wisteria* and perhaps even *Camellia*, to mention a few, spring to mind. In short a thorough survey of Chinese nurseries and gardens would seem likely still to prove rewarding to the plant hunter.

As might have been expected, the Chinese garden plants which have found favour in the West have largely been those with the more obvious charms. Having left their cultural significance behind, the terrestrial cymbidiums, *Rhodea japonica*, and other plants of modest appearance created little or no impact. On the other hand some plants of this type, such as the liriopes, *Ophiopogon japonicus*, and *Aspidistra elatior*, have proved useful on account of their hardiness and tolerance of difficult conditions. But it was chiefly plants which had already undergone considerable development in China, for example the peonies, azaleas, camellias, roses, and chrysanthemums, which became the most popular, readily giving rise to endless different cultivars in their new homes. Other attractive plants, *Callistephus chinensis* and *Primula sinensis* for instance, threw up many variants once they got into the hands of European breeders.

CHAPTER 4

GINKGO, CYCADS, AND CONIFERS

The ginkgo, the cycads, and the conifers have been included together in this chapter, albeit under separate sub-headings, as these plants are all gymnosperms. The word 'gymnosperm' means 'naked seed', as many readers will know, and is used to describe those plants which do not produce their seeds enclosed within fruits or seed pods, as do the flowering plants. The gymnosperms of today are the modern representatives of a number of distinct groups of seed plants which evolved before the flowering plants. Amongst the Chinese gymnosperms are many species which have become important garden plants. Two of these, the ginkgo and metasequoia, have captured the imagination of people everywhere on account of their being the only living representatives of a group or genus known otherwise only as fossils. Also, as has already been made clear in Chapter 2, several of the species included below have become invested with immense symbolic significance in China.

Figure 4.1 *Pinus tabulaeformis* at the Shuxiang Si, Chengde

It should perhaps be noted that for the purposes of this book I have used the term 'conifer' to embrace members of the families Cephalotaxaceae, Cupressaceae, Pinaceae, Podocarpaceae, Taxaceae, and Taxodiaceae, not all of which produce their seeds in cones. In the Cepalotaxaceae, Podocarpaceae, and Taxaceae, for instance, the seeds are usually borne singly. Nevertheless, in the broad sense, they are generally thought of as being conifers.

GINKGO

The ginkgos are a group of early seed plants which appeared about 200 million years ago, and once formed a conspicuous component of the vegetation of the world. Several genera are known in the fossil record, but only a single species of one genus is present today.

GINKGO Ginkgoaceae

G. biloba 银杏 (Yinxing) *Silver Apricot,* 白果 (Baiguo) *White Fruit,* 鸭掌树 (Yazhangshu) *Duck's Foot Tree*

MAIDENHAIR TREE, GINKGO

Although this fascinating tree is the sole survivor of a once abundant group, with its handsome foliage and clear yellow autumn colour, it is in little danger of becoming extinct. It is now widely planted throughout the temperate world, being generally known as the Maidenhair Tree on account of the distinctive shape of its leaves.

For a long time in the West it was suggested that it had survived only because it had been cultivated and that it was a 'living fossil', specimens appearing almost identical being known from the fossil record. However, it is claimed that trees still occur apparently spontaneously on Tianmu Shan, a little over 100 km west of Hangzhou, but whether these specimens are truly 'wild' remains a matter for speculation. Li (1956b, 1963a) suggests that this area is the centre of origin of the cultivated trees.

According to Lin et al. (1996), the earliest record of the ginkgo is in prose written in the Han dynasty, during which time it became known as a holy tree. However, Li (1956b, 1963a) states that ginkgos were unknown to the ancient Chinese of the Yellow River region. He says that its first indisputable appearance in the literature occurred in the 11th century, as a plant of eastern China south of the Yangtze, and that its 'fruit' was sent north as tribute to the emperor in the

capital, Kaifeng. Subsequently a few trees were planted there, the first record of its cultivation outside its natural range.

No doubt on account of its rarity in the north, the ginkgo became the subject of poems and paintings from the 11th century on. Barnhart (1972) reproduces an album leaf dated 1707, one of a series recording favourite memories of Nanjing, depicting an old tree that had been shattered by lightning more than 1000 years earlier but was still clinging to life. He suggests that we might sense in this tree something of the history of China and its people. As well as being depicted in paintings the ginkgo is occasionally seen as a decorative motif on porcelain (Krahl, 1987).

In the course of history the ginkgo became known as *Yajiao* (also meaning 'Duck's Foot'), on account of the shape of the leaves, and in literary circles as *Yinxing* (Silver Apricot), a reference to the appearance of the 'fruit' (fig. 4.2). Subsequently there were other names in common usage including *Baiguo* (White Fruit), which appears still to be the most widely used colloquial name, while *Yinxing* remains the common literary or botanical name.

Li expresses the view that the presence of old trees in Buddhist and Daoist temples has little or nothing to do with Buddhism or Daoism. He points out that old trees of many species have often been revered and preserved because of a kind of primitive nature worship, the trees being honoured because of their age not their kind. Even so ginkgos do seem to be conspicuous amongst the plantings in temple grounds, where ancient specimens are frequently encountered. These old trees sometimes produce downward growing outgrowths like stalactites from their trunks and lower branches. On reaching the ground these may take root, produce leaves, and give rise to secondary trunks. While I have not noticed these outgrowths on the old ginkgos I have seen in China, Li (1963a) reproduces a photograph of a tree in the west of the country on the trunk of which they are conspicuous, and they are well-known in Japan.

Figure 4.2 (left) Ginkgo 'fruits'

Figure 4.3 (opposite page) The 'King of Trees', a ginkgo said to be 1000 years old at the Tanzhe Si, south-west of Beijing

帝王树

此树又称白果树，别名公孙树，无叶为果树，为裸子
科，落叶乔木，这种树高30多米，树龄为远代科技，树
龄约有一千年左右，清乾隆皇帝曾封这棵树为"帝王"，
为这棵树为帝王树，是当年皇帝封的，树龄是在千多年
以前了。

King of Trees

Lin et al. (1996) list over 130 ginkgos in China with a diameter of over 2 m at breast height and suggest that some of these trees may be 2000 or more years old. What real evidence there is for this suggestion I do not know, but there are certainly many impressive old trees, most of them in the grounds of temples. Amongst notable specimens I have seen are those said to be 800 years old on Jiaoshan (figs 2.9, 2.10), a small island in the Yangtze off Zhenjiang, and examples said to be over 1000 at the Wuta Si on the outskirts of Beijing and a little further away in the Western Hills at the Dajue Si and at the Tanzhe Si. At the last-mentioned temple, for instance, there is the 'King of Trees', given its imperial title by the emperor Qianlong in the 18th century (fig. 4.3). It is on the right as you enter the main courtyard and, whatever the truth concerning its age may be, it is an immense multi-trunked specimen. Its smaller partner to the left, known as the 'Emperor's Wife', is presumably a female and is said to have put forth a new stem each time a child destined to become emperor was born. These trees have often been mentioned by Western visitors including C. E. Hubbard (1923) in his *The Temples of the Western Hills*, and Ann Bridge (1932) in her novel *Peking Picnic* in which, during a visit to the Tanzhe Si, the principal characters went and admired the giant ginkgo and 'his smaller wife'. At the other end of the size scale we find it used in penjing. Fortune (1847a), for example, noted that the Chinese were fond of dwarfing it. Likewise Li (1956a), amongst others, lists it as a plant suitable for this treatment, noting that such trees may be only be 'three or four feet high in one or two hundred years'.

The ginkgo 'fruit' is not of course a true fruit but a seed with a fleshy outer layer bearing a silvery bloom. Seeds were no doubt collected from wild trees and eaten long before the plant was first mentioned in the literature or cultivated. And from the 14th century on the tree is recorded as being widely planted throughout the country. For centuries the seeds have been considered to have medicinal value, being claimed to benefit the lungs, enrich the blood, stimulate the menstrual flow, ease the excretion of urine, and alleviate high blood pressure.

After the removal of the disagreeable fleshy layer, the seeds are still marketed on a large scale and are an important export, being sold in English-speaking countries as 'ginkgo nuts'. As to their qualities, Western poets do not seem to have been moved to wax lyrical on the subject. F. N. Meyer (1911) says they are always slightly roasted before being eaten but

that 'the flavour does not appeal to the palate of the Caucasian race'. This is a statement with which I find it hard to disagree.

It seems likely that the ginkgo was first introduced to Japan, probably from the Yangtze delta region, at about the time it became known in northern China. It is not known wild in Japan but is commonly planted, particularly in the grounds of temples. And in Japan and Korea, as in China, there are enormous specimens claimed to be of great age.

It was not until after Engelbert Kaempfer's observations, during his sojourn in Japan from 1689 to 1692, that the plant was introduced to and became known in Europe, arriving about 1730 (Bean, 1973; Spongberg, 1990, 1993). It was Kaempfer (1712) who gave the plant the name *Ginkgo*, and the origin and spelling of this has been much disputed. However, the spelling is Kaempfer's transliteration of the Japanese pronunciation of the name of the plant as he heard it. It was subsequently adopted by Linnaeus and is now the accepted botanical name. In the 19th century Sir James Smith so disliked this name that he produced one of his own, *Salisburia adiantifolia*, which, invalid though it was, achieved a certain following at the time.

According to Li (1956b, 1963a), it was early noted that there were male and female trees and that they should be planted near one another to ensure fruiting, or that a male branch should be grafted onto the trunk of a female. He also records that the early observers held the curious belief that female trees could produce fruit on their own if planted beside water.

It is a fine ornamental tree which like its compatriots, *Pistacia chinensis* and *Sapium sebiferum*, will reliably produce its excellent autumn colour even in climates where the temperature never falls below freezing. However, because of the disagreeable odour of the fleshy outer layer of the fallen seeds, grafted plants of male clones are best used when the trees are to be grown for ornament. As with most species, ginkgos show considerable variation when raised from seed, and, since the species has proved tolerant of atmospheric pollution, those with an upright habit make excellent street trees. Various selections differing in habit and/or leaf shape have been named in Europe and the United States.

Those wishing to read further about this remarkable plant are referred to the works of Andrews (1996), Holmes (1996), Li (1956b, 1963a), Lin et al. (1996), and Schmidt and Schmoll (1994).

Figure 4.4 (above) *Cycas revoluta* in the Yi Yuan, Suzhou

Figure 4.5 (right) *Cedrus deodara* and *Cycas revoluta* at the entrance to the Song Qingling Museum, Beijing

CYCADS

The cycads are another group of early seed plants. They appear in the fossil record just a little later than the ginkgos and, like them, were once more plentiful than they are now. However, unlike the ginkgos, there are about 11 genera with over 100 species still in existence. The large seeds are produced on leaf-like structures or in cones. As a result of their striking, architectural appearance, which bears some resemblance to that of the tree ferns and palms, many of them have been brought into cultivation. The species described below is the one seen in general cultivation in China, though others may be encountered from time to time. Sampson (1869b), for instance, noted that at least two species of *Cycas* were cultivated in Guangzhou and that they were often confounded. Also, Borrell (1996) records *C. pectinata* (probably a form of *C. circinalis*), *C. rumphii* and *C. siamensis* as ornamentals in the Shanghai area.

CYCAS Cycadaceae

C. revoluta 苏铁 (Sutie) *Reviving Iron*

Curious though the analogy may seem, this plant gets its Chinese name because, like iron, it does not produce flowers. It is also known simply as *Tieshu* ((Iron Tree). *C. revoluta* is usually seen growing in large pots placed beside doorways or at entrances (figs. 4.4, 4.5, 19.104). In the colder parts of the country these are brought indoors during the winter. It is also widely grown in gardens in Japan, where several cultivars have been selected. It is apparently native to the south of that country and to the Ryukus. It is not known when it was introduced to China.

CONIFERS

In spite of their ancient origins, the conifers have remained an abundant and diverse group, dominating the vegetation in cold northern regions and, in many parts of the world, at high altitudes. In the northern temperate zone, where much of the vegetation is composed of deciduous trees and shrubs, the evergreen conifers have long attracted attention. Remaining 'alive' in winter they became endowed with mystique and symbolism. Nowhere is this more obvious than in China, where pines and other evergreen conifers are used metaphorically for incorruptible virtue and for friends who remain constant in adversity (Williams, 1975). An example of this can be seen in an illustration in Tun Li-ch'en's (1965) *Annual Customs and Festivals in Peking*, which shows a gateway on which 'spring couplets' have been pasted. These were pairs of auspicious calligraphic inscriptions, usually on red paper, which were put up in the 12th month. One of the inscriptions in this illustration is *songbai gu renxin* (the pine and the cypress are at the heart of men of old). The other involves the metaphorical significance of *Cymbidium goeringii* and is given under that heading in Chapter 7.

The significance in Chinese culture of trees, particularly old evergreen conifers, has already been discussed in Chapter 2, and the antiquity of their cultivation has been mentioned in Chapter 1. As symbols of dogged perseverance in times of personal hardship the conifers have a history going back at least to Confucius, who recorded the proverb 'Only when the year grows cold do we see that the pine and cypress are the last to fade.' (Waley, 1938). Waley notes that Confucius was only repeating what was probably already an ancient saying and that the character for 'last', *hou*, should probably have been *bu* or *wu*, meaning 'not'. It should be noted, too, that the Chinese name for the cypress, *Bai*, is also applied to junipers and thujas.

Over a thousand years later the Tang poet Du Fu wrote movingly of an old cypress as symbolising the life of a famous statesman of the 3rd century AD, an early example of a long line of works of art paying homage to men and to the past in the form of old trees (Barnhart, 1972). As well as all this, conifers have a lengthy history of use as trees for planting on tombs and graves, to say nothing of their being chosen as the principal trees in gardens and the grounds of temples and palaces. Pre-eminent amongst them are the pines, but junipers, thujas, cypresses, and various other conifers are regarded similarly. As with the ginkgos, the finest old specimens are to be found at temples and shrines where, having escaped the axe, they lend, as Barnhart (1972) puts it, 'an auspicious and majestic air of cool shadow and towering strength'.

The literature of the first millennium BC, which incorporates material from earlier times, indicates that conifers have been cultivated in China for perhaps 4000 years. According to Wang (1988), large terraces for performing sacrifices to the God of Land and Grain were built and ornamented with plantings of the Chinese Red Pine, *Pinus tabulaeformis*, in the Xia dynasty. During the Shang dynasty the Oriental Thuja, *Platycladus orientalis*, was preferred for this purpose and came to be used also for the tombs of emperors. The use of these and other evergreen conifers is said to deter the fabulous creature known as the Wang Xiang, which is supposed to devour the brains of the dead (Williams, 1975). As well as this, branches of evergreen conifers to which old coins and paper pomegranate flowers were tied were placed in vases indoors as an offering to the gods at the Chinese New Year early this century in Beijing (Goody, 1993). Conifer branches were also spread in courtyards as part of the New Year ceremonies (Bredon & Mitrophanov, 1927). Conifers, too, have long been the plants most frequently used for penjing, including a number of species, both native and introduced, which are rarely if ever seen planted in gardens and which are not included here.

Pinus species, *Platycladus orientalis*, and *Juniperus chinensis*, which also has a long history of cultivation, are still the conifers most frequently seen in gardens and the grounds of temples, palaces, and tombs. Although China has a wealth of native conifers, and there may be more still to be discovered, relatively few are planted for ornament. Details concerning the best-known of these are set out below.

CEDRUS Pinaceae

C. deodara 雪松 (Xue Song) *Snow Pine*

DEODAR

While other species of *Cedrus* may now be present in China, the only one to be planted extensively is the deodar. I do not know whether it was introduced to China from India or via Europe but, whatever its origin, it is now conspicuous as a street tree and in other plantings (fig. 4.5). The trees generally do not appear to be of any great age and I have not seen it in traditional gardens.

Figure 4.6 (above)
Cepahalotaxus fortunei with
male cones in the Hangzhou
Botanical Garden

Figure 4.7 (right) At the
Lingyin Si, Hangzhou, from
left to right: *Cryptomeria
japonica, Liquidambar
formosana, Viburnum
macrocephalum,* and
Osmanthus fragrans

CEPHALOTAXUS Cephalotaxaceae

C. fortunei 三尖杉 (Sanjianshan) *Triple-tipped Fir*

FORTUNE'S PLUM YEW

This yew-like tree bearing oval, brown 'fruits' was introduced
to England by Robert Fortune who sent seeds to the Horticultural Society of London from Shanghai in 1844. It seems
to be rare in cultivation in China, though it is native to eastern and central regions of the country (fig. 4.6).

CRYPTOMERIA Taxodiaceae

C. japonica 日本柳杉 (Riben Liushan) *Japanese Willow Fir*

JAPANESE CEDAR

In spite of its name and association with Japan, this species
is also native to China, where it is found across the central
and southern regions from Sichuan to Zhejiang (Rushforth,
1987). It is used in forestry, is frequently encountered in temple grounds (fig. 4.7), and occasionally in gardens. Also, Li

(1956a) says that, with its minute leaves and small cones, it is very effective when dwarfed. Fortune (1847a) encountered it for the first time at the cemeteries of the rich in Shanghai and sent seeds from there to his employers, the Horticultural Society of London, in 1844 (Bean, 1970). The United States Patent Office received a similar shipment in 1858, and the next year young plants were growing in Washinton, D.C. (Spongberg, 1990).

CUNNINGHAMIA Taxodiaceae

C. lanceolata 杉木 (Shanmu) *Fir Tree*

CHINA FIR

C. lanceolata is so widely planted as a timber tree in central and southern China that it seems foolish to omit it from this compilation. The wood is used chiefly for shipbuilding and for houses (Sampson, 1868). It was described in 304 AD by Ji Han, who gave it the name *Shan* (Li, 1979). With needles up to 0.5 cm broad, its appearance evokes the araucarias, to which it is not closely related (fig 4.8). According to Li there are many accounts of huge specimens in the past, and he suggests that in historic times parts of central and southern

China must have been covered with extensive forests of large cunninghamias, perhaps rivalling the sequoia forests of California or the coniferous forests of the mountains in Taiwan. Perhaps because *C. lanceolata* is not on the whole of great beauty, it is rarely planted for ornament, though Borrell (1996) says there are forms with glaucous branchlets and leaves silvery beneath. It was first raised in Britain somewhere around 1700 from seeds sent by James Cunningham (Cox, 1944).

CUPRESSUS Cupressaceae

C. duclouxiana 干香柏 (Ganxiang Bai)

Fragrant [Wood] Cypress

YUNNAN CYPRESS

C. duclouxiana is a conical tree somewhat similar to the Italian Cypress, *C. sempervirens*, but with glaucous foliage. It imposes a distinctly Mediterranean appearance on the countryside in parts of south-western China (fig. 4.9). Armand Delavay collected it in temple grounds in western Yunnan in the 1880s, and it is said to be widely cultivated in temples there (*Curtis's Botanical Magazine*, 1925, t. 9049). Otherwise it is rare in cultivation.

Figure 4.8 (above) *Cunninghamia lanceolata* on the lower slopes of Emei Shan, Sichuan

Figure 4.9 (right) *Cupressus duclouxiana* at the Shilin (Stone Forest), Yunnan

Figure 4.10 (right) *Cupressus funebris* beside a grave on a farm in Jiangsu

Figure 4.11 (far right) Foliage and cones of *Glyptostrobus pensilis*

C. funebris 柏木 (Baimu) *Cypress Tree*

CHINESE WEEPING CYPRESS, MOURNING CYPRESS

This is the principal tree used for planting beside graves (fig. 4.10), hence the specific epithet. Also, according to Koehn (1952), images fashioned from cypress foliage in the form of stags and cranes, symbols of immortality, are often seen in funeral processions. It is said that its use ensures a passage to the 'western happiness world'. It is also planted in gardens here and there. Robert Fortune (1850c, d) thought it the most beautiful tree he saw in the green tea district when he came across it in an old garden. When young it forms a dense conical tree with upright branches, but as it matures the branchlets and smaller branches become pendulous. The adult branchlets are flattened and tend to be borne in one plane, a character more usually associated with the genus *Chamaecyparis*. According to Rushforth (1987), it is native to central China along the Yangtze valley. It was brought back to England by Lord Macartney's embassy at the end of the 18th century (Cox, 1944).

GLYPTOSTROBUS Taxodiaceae

G. pensilis 水松 (Shui Song) *Water Pine*

CHINESE SWAMP CYPRESS

Like *Metasequoia* and *Pseudolarix*, *Glyptostrobus* is a conifer genus endemic to China with only a single species. It was first described in 304 AD by Ji Han who used the name *Shui Song* (Li, 1979). It forms a conical or columnar deciduous tree similar in appearance to the Dawn Redwood, *Metasequioa glyptostroboides*, and to the North American Swamp Cypress, *Taxodium distichum*, which are also found in moist habitats. As with these it produces deciduous leaf-bearing branchlets (fig. 4.11), which are bright green when young and turn reddish brown before falling, and male 'cones' in tassel-like

clusters, usually at the tops of the trees. The wood is soft, light, and of great buoyancy. Li (1979) quotes a record of children along the Pearl River using it as a float attached to their backs when swimming.

According to the *New R.H.S. Dictionary of Gardening* (Huxley et al.,1992), under wet conditions *G. pensilis* does not produce upward-growing pneumatophores or 'knees' from the roots as does *T. distichum*. Likewise Rushforth (1987) says that he has not seen them, even on plants growing in 50 cm of water at Guangzhou. However, Li (1979) says it produces pneumatophores which become quite tall, grow to 30 cm in diameter, and are hollow. Unfortunately I have not had the opportunity of observing for myself where the truth lies in all of this.

Li (1979) is of the opinion that it was the pneumatophores of *G. glyptostroboides* which Ji Han described as a separate plant, *Paomu*, saying that it grows by the side of the *Shui Song* like a parasitic plant. He noted that while still wet it could be carved into pattens (wooden overshoes) as easily as shaving a melon, but that, after it is dried, it became so tough that it was no longer workable. Ji Han went on to say that in 285 AD one hundred pairs were sent from the south to the emperor, who was much interested in their rarity but sneered at the rustic handiwork. According to Li (1979), this wood has also been used as a substitute for cork and as an insulating material in sun helmets.

Wooden pattens have been in use for over 2000 years in China, being made from various light-weight woods. They

Figure 4.12 (left) The trunk of an ancient specimen of *Juniperus chinensis* at the Dajue Si west of Beijing

Figure 4.13 (above) *Juniperus chinensis* 'Kaizuka' in the Huifang Yuan, Zhongshan Park, Beijing

were especially in vogue in the 3rd century BC and, though of decreasing importance, their use persisted through the Tang and Song dynasties. It seems that during the Ming dynasty they gradually passed out of use in northern China though they continued to be used in the south, at least until recent times (Li, 1979).

While it may now be extinct in the wild, *G. pensilis* has been widely planted for timber and to stabilise river and canal banks. It was once thought to have a geomantic effect, bringing good luck and ensuring good crops. In many rural districts branches accompany bridal gifts to symbolise good fortune (Thrower, 1988). Regrettably perhaps, its place has largely been taken in more recent times by *T. distichum*, which is a faster-growing source of timber (Rushforth, 1987). *G. pensilis* is also occasionally dwarfed for use as a table plant in southern China (Li, 1956a).

JUNIPERUS Cupressaceae

J. chinensis 园柏 (Yuanbai) *Round Cypress*

CHINESE JUNIPER

While several junipers are native to China and neighbouring countries, the species most likely to be encountered in traditional plantings is *J. chinensis*. Like the pines, this has long been cultivated and old specimens abound in the courtyards of houses, in gardens, and in the grounds of palaces, temples, and imperial tombs. The bark of some old trees becomes contorted and this adds to the esteem in which they are held (fig. 4.12).

In general appearance the Chinese Juniper resembles the Oriental Thuja (*Platycladus orientalis*), often found growing nearby, but it can easily be distinguished by its cylindrical branchlets and tiny berry-like cones, whereas *Platycladus*, the name of which means 'flat branch', has flattened branchlets together with larger oval cones (fig. 4.19). The name *Yuanbai* refers to the roundness of the cones of the juniper (fig. 4.13).

Although *J. chinensis* has given rise to many named cultivars, its typical form is the one usually encountered in Chinese gardens (figs 5.13, 19.16). Occasionally, however, the cultivar 'Kaizuka' is met with (fig. 4.13). This has more upright branches and is a smaller-growing plant, usually seen as a large shrub. It is known in China as *Longbai* (Dragon Cypress). In nursery gardens at Tianjin Meyer (1916) saw junipers grafted on *Platycladus orientalis*. The Chinese Juniper is readily dwarfed and with its minute leaves is a suitable subject for penjing (Hu, 1982; Li, 1956a).

Old junipers have frequently been depicted in paintings, for example the *Three Juniper Trees* reproduced by Cahill (1978, plates 42–43). This depicts three of seven venerable junipers planted in 500 AD by a Daoist priest at Yu Shan north of Suzhou.

J. squamata 粉柏 (Fenbai) *Powder Cypress*, 翠柏 (Cuibai) *Greenish-white Cypress*

FLAKY JUNIPER

This species, which is found from Afghanistan across to central China, has flaky bark and retains its juvenile leaves. There are several cultivars, ranging from prostrate forms to erect shrubs up to 8 m tall, mostly with bright silver-blue leaves. F. N. Meyer bought plants of an upright form, which the Chinese call 'Fish-tail Juniper', in Beijing in 1908 and sent them to the United States. In the West this form has become the commonest cultivar and has acquired the name 'Meyeri' (Cunningham, 1984). It is not known in the wild but has long been cultivated in China (Rushforth, 1987). The Chinese names given above refer to the glaucous appearance of this species.

KETELEERIA Pinaceae

K. fortunei 油杉 (Youshan) *Oil Fir*

A handsome tree with upright cylindrical cones up to 20 cm in length (fig. 4.14), *K. fortunei* was found by Fortune growing in the grounds of a temple near Fuzhou. He introduced it to England in 1844 (Bean, 1973). It is native to south-east China.

METASEQUOIA Taxodiaceae

M. glyptostroboides 水杉 (Shuishan) *Water Fir*

DAWN REDWOOD

In 1941 the name *Metasequoia* was given to a conifer known only as a fossil by the Japanese botanist Sigeru Miki. Curiously enough in the same year a Chinese botanist visited a village near the border between Hubei and Sichuan and noticed there a deciduous conifer locally called *Shuishan*, similar in general appearance to *Glyptostrobus pensilis*. However, it was not until after specimens were collected in 1944 that it was realised that it was virtually identical to the fossil described by Miki (Li, 1957, 1963a; Spongberg, 1990). Later it was found to be more abundant in the neighbouring region of Hubei province, where as well as growing wild, it is planted along river banks, roads and elsewhere, apparently mainly for aesthetic reasons since, according to Li (1957, 1963a), the wood is not considered to be

of good quality. However, where it is planted in this district the removal of branches almost to the top, presumably for firewood, gives the trees a curious appearance.

It was introduced to cultivation elsewhere when, as a result of the efforts of Dr E. D. Merrill of the Arnold Arboretum, a large quantity of seed was collected in the autumn of 1947 and distributed to botanical gardens throughout the world. Finally it was named and described by H. H. Hu and W. C. Cheng in 1948. It has become popular as an ornamental outside China and is now widely planted in China itself (Borrell, 1996), often as avenues and on the banks of lakes and canals (fig. 4.15). Like the ginkgo, it has been endowed with considerable renown as a 'living fossil'.

Fuller accounts of its discovery, distribution, introduction, and behaviour in cultivation are given by Lauener (1996), Li (1957, 1963a), and Spongberg (1990).

PINUS Pinaceae

While the Chinese have always appreciated the grandeur of their trees, none is held in higher esteem than the pine. No garden or temple ground is complete without it, and it has long been a favourite subject in literature and art, either alone or together with bamboo and *Prunus mume* as one of the 'Three Friends of the Cold Season'.

The Chinese word for pine is *song*, the character for which has been made up from those meaning 'tree' and

Figure 4.14 **Foliage and cones of *Keteleeria fortunei***

'head', as it is regarded as the leading tree. Whether in cultivation or clinging to mountain tops in its natural environment, it has been admired from early times for its size and firm stance, and is regarded as a symbol of long life, hardy and dignified old age, the ability to endure, and the yang principle. Surviving for many centuries in its evergreen beauty, towering over all others, it was seen as a dragon amongst lesser creatures (Barnhart, 1972). Its bark has been been likened to scales and its roots to claws, which enable it to cling to precipices. It is often painted with the sinuosity of a dragon or a soaring phoenix. Morris (1983) even goes so far as to say that the pine was as powerful a symbol for the Chinese as the crucifix for the Western world. It is the king of trees and in ancient times was the official tree for the graves of rulers (Barnhart, 1972; Li, 1963a), though subsequently it was displaced by the Oriental Thuja. A selection of prose, poems, and reproductions of paintings relating to the symbolism of the pine has been published by Lai (1977).

Pine resin was one of the drugs the Daoists, seeking immortality, used to strengthen their bodies when abstaining from cereals in order to rid themselves of corrupting influences. The prestige of the pine tree in this regard was shared by amber and the *Lingzhi* fungus, both of which are intimately connected with the pine. Amber was early recognised as a fossil resin and presumably the quintessence of hardy old age, while the *Lingzhi*, which grew on pine roots, was considered to be immune from decay and sometimes as another transformation of pine resin (Wolters, 1960).

The seeds of pines, known in the West as 'pine nuts', are collected and eaten in China, as they are in many parts of the world. In 304 AD Ji Han wrote of their use in the south, noting that the trees which produced them were similar to other pines but produced seeds of very large size. Li (1979) says

that the only southern pine with large seeds is *P. insularis* (*P. kesiya*), *Dao Song* (Island Pine), and that presumably it must have been this species which yielded the seeds described by Ji Han. In the north the large seeds of *P. koraiensis*, *Hong Song* (Red Pine), are those most frequently used for this purpose (fig. 4.16). No doubt the 'esculent seed of a pine said to come from Tartary', which was served to Clarke Abel (1819) at Tong Xian in 1816, was that of this species. Sweet corn kernels stir-fried with pine nuts form the basis of a dish now characteristic of northern China.

The Chinese pines most frequently seen in gardens are *P. bungeana*, *P. massoniana* and *P. tabulaeformis*. There are of course many other species native to China but it seems that they are rarely cultivated, apart from in botanic gardens or when used in forestry. However, the Japanese Red Pine (*P. densiflora*), Japanese White Pine (*P. parviflora*) and Japanese Black Pine (*P. thunbergii*) are occasionally encountered, very often as penjing.

P. bungeana 白皮松 (Baipi Song) *White-bark Pine*

LACEBARK PINE

According to Li (1968a) this is the only three-needled pine in eastern Asia, but its most distinctive characteristic is its bark, appearing from a distance to be smooth and white. On young trees the bark is mottled with grey, brown, and green, but as they mature the bark appears as a mosaic of patches of white and various shades of grey, resembling in shape the pieces of a jigsaw puzzle (fig. 4.17). It becomes a round-headed tree, very often branching a short distance above ground level to form several trunks, and is one of the most admired of all Chinese plants. The Ming poet, Zhang Shu, some 500 years ago, when viewing a planting in front of a temple through rain and mist, described the trees as white dragons (Li, 1968a).

While examples of this distinguished plant can be found in a great many of the gardens of the Yangtze delta region, the largest and most striking specimens are to be found in the north, the region to which it is native. It is not known whether the beautiful avenue described and illustrated by Hubbard (1923) still exists at the Longmen Si west of Beijing, but Li (1968a) lists several notable old trees, indicating the most renowned to be the one at Mi Xian in Henan, which has been famous since the Ming dynasty. However, more old trees are known to exist in and around Beijing than in any other locality. The specimen best known to Westerners is the

Figure 4.15
(opposite page)
Metasequoia glyptostroboides beside a backwater of the West Lake, Hangzhou
Figure 4.16 (right) Pine nuts (*Pinus koraiensis*) on sale in Chengde

Jiulong Song (Nine Dragon Pine) at Jietai Si south-west of Beijing (fig. 4.18). This famous tree, reputedly over 900 years old, has nine principal trunks and was given this name by the emperor Qianlong in the 18th century. Many visitors have commented upon it since, including G. E. Hubbard (1923), P. H. Dorsett (1931), who visited and photographed it in 1930, and Ann Bridge (1932) who, describing the terrace at the Jietai Si, wrote: 'over it rose an immense white pine, its snowy trunk and branches shining among the great trusses of dark-green needles. The white pine is the most improbable of trees—too good to be true; it is impossible to believe at first that some ingenious Chinese has not sandpapered its smooth trunk and boughs, and then given it several coats of whitewash'. Dorsett managed to collect seed from a tree 'with the whitest bark of all' at the neighbouring Tanzhe Si, but noted that this was not easy. The reason for this was that the priests and attendants were almost continuously watching to pick up and eat any seeds that fell, with the result that the foreign plant hunter found few for himself.

Many specimens can also be seen in Beijing itself, including several in the Forbidden City. More impressive are those in the courtyard of the so-called Round City adjoining Behai Park, one of which is said to have been planted about 800 years ago (Bredon, 1931). More recently whole rows have been planted in Changan Avenue along the walls of the Forbidden City. While these have not yet developed white bark they appear to be growing with a vigour not usually witnessed outside China. It is also planted around temples in Korea (Rushforth, 1987).

P. bungeana was described by Stephen Endlicher in 1847 from a specimen collected near Beijing by Alexander von Bunge in 1831. While the exact locality is not known, Endlicher used the name *Jiulong Mu* (Nine Dragon Tree), which is neither the name of a locality nor the Chinese name for the species but a minor variation of that given by Qianlong to the tree mentioned above. In the light of this Li (1968a) suggests that we may safely presume that the imposing tree at the Jietai Si is the one from which Bunge collected his specimen.

P. bungeana was introduced to Britain somewhere about 1846 from further south by Robert Fortune, who saw only young nursery plants. However, when he visited Beijing in 1861 and saw a magnificent mature specimen in the Western Hills, he was surprised and puzzled as, until then, he 'had not the slightest idea of its extraordinary appearance when full grown' (Fortune, 1863). It is very slow growing and so far has generally failed to live up to expectations when planted outside China. However, this may well be because it has not yet had sufficient time to become large enough to develop its striking characteristics.

P. densiflora 赤松 (Chih Song) *Red Pine*

JAPANESE RED PINE

P. densiflora is native to Japan, Korea, and adjoining maritime Russia (Mirov, 1967; Rushforth, 1987). It is occasionally cultivated in China. It is claimed by some to be native to the north-east of the country, but Borrell (1996) suggests that it is merely naturalised in Jiangsu. He says that there are dwarf varieties with shorter leaves, and it is presumably these that are favoured for penjing.

Figures 4.17, 4.18 (left) The Nine Dragon Pine (*Pinus bungeana*) at the Jietai Si, near Beijing, and its bark

Figure 4.19 The Sleeping Dragon Pine (*Pinus tabulaeformis*) at the Jietai Si, near Beijing

P. massoniana 马尾松
(Mawei Song) *Horse-tail Pine*

In central and south-eastern regions *P. massoniana* occurs spontaneously over a wide area, extending to Taiwan and North Vietnam. It is a two-needled red pine similar to *P. tabulaeformis*, but in its general appearance it rarely forms as handsome a tree as that species. However, it is often encountered in gardens in the south. Borrell (1996) says that there are many variants.

P. parviflora 日本五针松 (Riben Wuzhen Song)
Japanese Five-needle Pine

JAPANESE WHITE PINE

It is not known when or how this was introduced to China. Presumably it arrived from Japan a long time ago. According to Borrell (1996) it is naturalised in Jiangsu. Li (1956a) says it is used for penjing, its short needles rendering it suitable for this purpose, and a fine specimen is illustrated by Hu (1982).

P. tabulaeformis 油松 (You Song) *Oil Pine, Pitch Pine*

CHINESE RED PINE

Widely distributed in northern China, this pine is the one most frequently encountered in gardens, particularly in central and northern parts of the country. It is also an important timber tree. Although Bean (1976) dismisses it as being of no ornamental value in Britain, in China it forms a picturesque tall tree, flat-topped with age (figs 4.1, 19.39). The leaves are usually borne in twos and the fissured bark splits irregularly into plates, orange-brown in their centres, darkening to brownish-grey at their edges. The cones are small, usually less than 7 cm long.

Known to have been cultivated for thousands of years, many fine specimens are still to be seen in gardens and temple grounds, their presence adding greatly to the dignity of these sites. A particular example which has long attracted attention is the *Wolong Song* (Sleeping Dragon Pine) (fig. 4.19) growing out almost horizontally from the same terrace at the Jietai Si south-west of Beijing on which the famous Nine Dragon Pine (*P. bungeana*) is to be found.

P. thunbergii 黑松 (Hei Song) *Black Pine*

JAPANESE BLACK PINE

P. thunbergii is native to the coasts of South Korea and Japan (Rushforth, 1987). It is not known when or from which country it was introduced to China, where it is commonly used for penjing, just as it is in Japan for bonsai. Borrell (1996) records that it appears to be naturalised in Jiangsu.

PLATYCLADUS Cupressaceae

P. orientalis 侧柏 (Cebai) *Sideways Cypress*

ORIENTAL THUJA

Along with *Juniperus chinensis* and *Pinus tabulaeformis*, the Oriental Thuja is one of the three conifers cultivated most frequently and widely in China. At times placed in the genera *Biota* and *Thuja*, *P. orientalis* is now considered distinct and is the only species of its genus. It is found wild from North Korea, across northern China, and down to the drier parts of Yunnan. There is also a small wild population in north-eastern Iran (Huxley et al., 1992). It is easily distinguished from *J. chinensis*, with which it is often planted, on

Figure 4.20 Foliage and cones of *Platycladus orientalis*

Figures 4.21, 4.22 (below & left) A 500-year-old *Podocarpus macrophyllus* at the Lingyin Si, Hangzhou, and the tablet at its base

account of its larger, ovoid cones and flattened branchlets (fig. 4.20). It is to these flattened branchlets that its Chinese name refers.

Cultivated since ancient times, old specimens (fig. 1.2) are abundant in gardens and the grounds of temples and palaces, as well as those of imperial tombs, where they are planted on the grave mounds and in their surroundings. No doubt there will be new plantings at the Western Qing Tombs 110 km south-west of Beijing where, somewhat surprisingly, the ashes of the last emperor were interred in 1995 and where a new mausoleum is to be built close to the tombs of some of his ancestors.

An imposing tree, similar in general appearance to the Chinese Juniper, it is rarely seen outside eastern Asia in its typical form, but rather as one of its numerous cultivars. But in China, as with the juniper, the typical form is planted almost exclusively. According to the *Jiuhuang Bencao* of 1406, it is possible to eat the foliage and cones in times of food shortage (Bretschneider, 1881). It was first raised in Europe from seeds sent by d'Incarville in the middle of the 18th century (Cox, 1944).

PODOCARPUS Podocarpaceae

P. macrophyllus 罗汉松 (Luohan Song) *Lohan Pine*

P. macrophyllus is native to Japan and southern China. It is recognised by its leathery leaves, up to 10 cm long and 1 cm or more wide, even larger on young plants, and, as with many species in the genus, has large seeds borne singly on a swollen receptacle which becomes red and fleshy.

The tree is grown as much for its religious symbolism as its appearance. Apparently the purplish-red seeds of this podocarpus resemble some feature of the outer vestment worn by the lohans. The lohans were the followers of Buddha who, although they had attained the degree of perfection required to enter Nirvana, were entrusted with remaining in the world until the coming of the next Buddha, in order to sustain the teaching and to provide people with objects of worship. They are often depicted on walls of temples or as statues in groups of 16, 18, or even 500.

P. macrophyllus is planted frequently and ancient specimens are to be found in temple gardens. There is, for instance, an old tree (fig. 4.21) at the Lingyin Si, Hangzhou, which according to the tablet at its foot (fig. 4.22) is 500 years old. While the typical form is most frequently seen in temple gardens, in other gardens one also encounters a form known as var. *maki*, which is more compact, slower growing and has smaller leaves. This variant is called *Duanye Luohansong* (Short-leaved Lohan Pine) and is very often seen planted in containers and trained into picturesque shapes.

In paintings from the late 17th century on, *P. macrophyllus* is sometimes depicted with wintersweet and nandina as a variation of the Three Friends of the Cold Season (Bartholemew 1985a, cat. no. 8i; Suzuki 1982, JM1–075; Weidner et al. 1988, cat. no. 39).

P. neriifolius 百日青 (Bairiqing) *Hundred Days Green*

Occasionally cultivated, this forms a tree very similar to the above but usually with longer and narrower leaves. It is widely distributed, being found from the eastern Himalaya to east-central China, south-east Asia and Melanesia. As with so many conifers in China, it is appreciated for its longevity and evergreen habit.

PSEUDOLARIX Pinaceae

P. amabilis 金钱松 (Jinqian Song) *Gold Coin Pine*

GOLDEN LARCH

Together with *Glyptostrobus pensilis* and *Metasequoia glyptostroboides*, this is one of China's three indigenous deciduous conifers. It differs from the true larches (*Larix*) only in minor botanical detail. It grows into a charming, medium-sized tree with bright green spring foliage which turns to a rich gold in autumn. It was originally called *Jin Song* (Golden Pine), but it seems that this name referred to the yellow midrib of the leaf rather than its autumn colour (Li, 1968b). Native to eastern China, it is also seen occasionally in Chinese gardens, where it has been cultivated for at least 1000 years. Schafer (1965) records it as being sent from Zhejiang to Li Deyu's famous garden near Luoyang during the Tang dynasty. It is also frequently seen dwarfed in pots (fig. 4.23), which Li (1956a) says is the result of judicious care from the seedling stage. However, according to Huxley et al. (1992), a dwarf Chinese cultivar known in the West as 'Nana' is used for this purpose. Simply by looking at dwarfed plants in China I was unable to determine whether the dwarf or the wild type had been used.

Figure 4.23 **A dwarfed stand of *Pseudolarix amabilis* at the Flower Nursery, Hangzhou**

P. amabilis was introduced to Britain from Zhejiang province by Robert Fortune in 1853. While he had been acquainted with it in China in gardens or as a pot plant in a dwarfed state, he was not been able to obtain seed until he came across wild trees, the largest of which he estimated to be well over 30 m tall, near the Tiantong Si, south-east of Ningbo (Fortune, 1855a). According to Li (1968b) it must have been abundant until at least the Ming dynasty but by the 18th and 19th centuries it had become rare.

TORREYA Taxaceae

T. grandis 榧树 (Feishu) *Literary Grace Tree,* 香榧 (Xiangfei) *Fragrant Torreya*

CHINESE NUTMEG-YEW

T. grandis is a tree up to 25 m tall, native to eastern China. It was found by Fortune in a garden south-west of Ningbo, into which it had been transplanted from the wild (Nelmes, 1944). The name *Feishu* refers to the association of the tree with literary grace, whereas *Xiangfei* is a name applied on account of the aromatic properties of the seed. The seeds are eaten and several cultivars are known (Borrell, 1996). The seeds of '*Taxus nucifera*' served to Clarke Abel (1819) at Tong Xian near Beijing in 1816 may perhaps been those of this species rather than those of the Japanese *Torreya nucifera*, which had originally been named *Taxus nucifera* by Linnaeus.

CHAPTER 5

BAMBOOS

N o plants are more characteristic of China, more deeply involved throughout the ages with everyday life, or more prominent in Chinese art and technology than the bamboos. Not unexpectedly, this did not fail to impress the early European visitors. Cibot (1777b), for instance, wrote a lengthy article about bamboo, commenting on its cultivation, uses, and representation in art. Subsequently Collas (1786b) described the harvesting, preservation, and preparation of the edible shoots, noting that the price depended upon their size and tenderness. Fortune (1850a), likewise, was impressed by the multiplicity of uses to which bamboo was put—everything from water pipes to hats, furniture, and food.

Many different bamboos are grown in China for their edible shoots and for their culms, as the aerial stems are called. No doubt, too, a considerable number have found their way into Chinese gardens in various parts of the country at one time or another. However, because of the rarity of their flowering and the difficulty of observing the characters of the underground stem (rhizome), identifying bamboos is not for the faint-hearted. To deal

Figure 5.1 A plantation of *Phyllostachys edulis* at Linyan Hill, near Suzhou

89

with the topic thoroughly would require a book of its own, so I shall confine myself here to making some general observations and list only some of the better-known kinds. Those wishing to read about a wider range of species are referred to books such as *Bamboos of China* (Wang & Shen, 1987), *Chinese Bamboos* (Chen & Chia, 1988), *Hong Kong Bamboos* (But, Chia, & Fung, 1985), *Bamboos* (Crawford, 1997), and *Bamboo Rediscovered* (Cusack, 1997).

Much has been written over the centuries about these extraordinary, long-lived grasses, and I am particularly indebted to Needham (1986) for his survey of the Chinese literature on the subject. It is not surprising that the tradition of writing about these plants is longer and more continuous than for any others, stretching back to the *Shi Jing* (Book of Odes), believed to have been compiled between the 11th and 7th centuries BC. Evergreen, pliable and strong, bowing before the storm and rising again, bamboo has been seen to symbolise gentlemanly qualities, lasting friendship, longevity, and hardy old age. As mentioned in Chapter 2, its upright hollow stem was equated with rectitude, resilience, and humility. Its hardiness, vigour, clumping habit, and usefulness were associated with renewal, perseverance, and mutual support. Also, as mentioned in Chapter 2, its nodes are called *jie*, a word homophonous with that for integrity (Bickford, 1996). And a gift of a picture of bamboo, especially at the New Year, suggests *zhubao pingan* (bamboo announces peace and tranquillity) (Koehn, 1952). It is also associated with filial piety as, in one of the 24 classical examples of this quality, a son had a sick mother who longed for a soup made from bamboo shoots in winter. This son wept so copiously on her account in a bamboo plantation that his tears, like the warm rains of spring, softened the hard wintry ground and caused the tender shoots to burst forth (Williams, 1975).

The Tang poets extolled the beauty of bamboo, as did many of those who followed them. It seems probable, too, that bamboo has been depicted more frequently in Chinese art than any other plant. It is one of the few which can be identified with certainty in early painting, emerging as a popular motif in the rising art of flower-and-bird painting in the Tang dynasty. By the beginning of the 11th century the practice of representing it in ink alone had emerged, the ink painters being praised for their ability to capture the essence of the subject. As mentioned in Chapter 2, the structure of bamboo lent itself to representation with brushstrokes of the type used in calligraphy. As a result literary and pictorial traditions became combined in the genre known as *mozhu*, ink bamboo. It seems unlikely that any other plant, except perhaps *Prunus mume*, has become so deeply involved with the culture of any country. Further insights concerning the involvement of bamboo in Chinese culture, particularly in art and literature, are given by Lai (1977).

The bamboos are the subject of the oldest of all the known Chinese botanical monographs, the *Zhu Pu* (Bamboo Manual) of about 460 AD. This is in the form of a poem with four characters to the line and a brief prose commentary after every few lines. It is discussed in detail by Needham (1986), who gives a translation of its perceptive commencement as follows:

Within the vegetable kingdom
There is a thing called bamboo
It is neither hard nor soft
It is neither herb nor tree.

The author, Dai Kaizhi, goes on to describes numerous kinds, where they grow and their uses, together with observations on their flowering and death. Needham observes that when one remembers that this writer was probably the first of all the botanical monographers, one realises the extent of his originality and the standard he set for those who came after him.

A great many books of a practical nature followed and a wealth of technical language developed for describing the various parts of the plant. Remarkably comprehensive was the *Sun Pu* (Treatise on Bamboo Shoots) by the monk Zan Ning, written about 970. In spite of its name it deals with everything to do with bamboos, discussing names, synonyms, cultivation, the harvesting of the edible shoots, and the origins of a total of 98 kinds. He also gives information concerning the medicinal and nutritional uses of bamboo shoots, with elaborate details of their preparation, cooking, and preservation.

Another book to which Needham draws attention is the *Zhupu Xianglu* (Detailed Records for a Treatise on Bamboo) of 1299 by Li Kan. Li Kan began as a painter, so it is not surprising that the first chapter is about how to draw and paint bamboo. But his great achievement was of a botanical nature. Not only did he describe 259 kinds but he observed that there were two distinct types of rhizome, the spreading and the clumping, and gave illustrations of them (figs 5.2, 5.3).

Figures 5.2, 5.3
Li Kan's illustrations
of a spreading
bamboo rhizome
(right) and a
clumping bamboo
rhizome (far right),
published in 1299

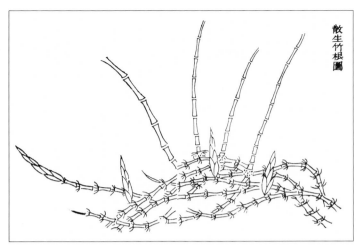

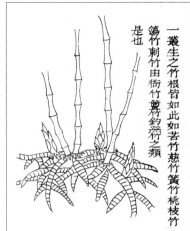

A knowledge of this distinction is valuable for gardeners, as some of the running types (fig. 5.4) can become invasive. Hence, if they cannot be kept confined, it is wise to plant only the clumping types (fig. 5.5). It is, however, not always easy to distinguish between the types, as some of the spreading species stay in clumps when grown in cold climates or under unfavourable conditions.

Figure 5.4 (right)
A running bamboo
in the garden of
Xu Wei's Studio,
Shaoxing
Figure 5.5 (below)
A clump-forming
bamboo in the
garden of Tuisi,
Tongli

It is from the rhizomes that the bamboo shoots arise. These consist of a succession of nodes and internodes within a cover of sheathing leaf-bases (fig. 5.6). Compressed within, as it were, is the ultimate length of the culm. The final height is achieved by the lengthening of each of the internodes and this may occur with astonishing speed. According to Huxley et al. (1992), extensions of up to 1.25 m in 24 hours have been recorded, the culms achieving their final stature in 1–3 months. To keep bamboo groves productive the harvesting of both the young shoots and the mature culms must be carefully managed. Culms intended for timber are usually not harvested until several years have passed (fig. 5.7).

An interesting feature of the bamboos is that most species commence producing their inconspicuous flowers (fig. 5.8) only at intervals of up to and even over 100 years, and this may occur in one year only or continue for many years on the one plant. Sometimes only a few flowers are produced on one culm, while at others a whole culm (fig. 5.9) or even every culm in a clump may be involved. The claim that the plants of any given bamboo species flower at the same time all over the world and then die is certainly not true. However, it is true that very often the culms which produce the flowers and seeds die (Cusack, 1997). Sometimes the whole plant will die, but in most cases new shoots arise (McClure, 1966). It has been suggested that the running types are more likely to regenerate in this manner than the clumping ones. The most likely reason that bamboos of the same species do tend to flower at roughly the same time is that they are all of the same clone, because bamboo propagation is nearly always carried out vegetatively (Crawford, 1997).

Figure 5.6 (above) Bamboo shoots at the Lanting near Shaoxing

Figure 5.7 (opposite page) Mature bamboo culms on sale along the Grand Canal, Yangzhou

Figure 5.8 (above left) Flowers of *Bambusa multiplex*

Figure 5.9 (left) A flowering culm of *Bambusa mutiplex*

the well-known saying *zhumei shuangxi* (bamboo and plum [husband and wife], twofold happiness) (Koehn, 1952). It is also frequently planted in arrangements which suggest spring, with narrow upright pieces of stone or petrified wood placed amongst it to represent permanently the bamboo shoots so characteristic of that season (fig. 5.10). Only occasionally are dwarf species used in the manner of a ground cover as they are more frequently in Japan. However, they are quite often grown as clumps either on their own or in association with rocks. *Indocalamus latifolius* and *I. tessellatus* are small species used in this manner, as are species of *Sasa* (Chen & Chia, 1988).

Amongst the larger bamboo species more commonly seen are forms of *Bambusa multiplex*, *B. beecheyana*, and *B. vulgaris*, particularly in the warmer parts of the country. *Chimonobambusa quadrangularis* may also be encountered and species of *Phyllostachys*, for example *P. aurea* and *P. nigra*, are commonly planted in gardens. Also the giant species *P. bambusoides* and *P. edulis* are occasionally seen in temple grounds and gardens as well as in plantations. All these are native to China.

Occasionally bamboos are used for penjing. Livingstone (1822), for instance, records plants kept only a few inches tall. Apparently a common practice in the dwarfing of bamboos is the premature removal of the sheaths which envelop the young shoots. The resulting exposure to light hastens the maturity of the tissue and thus prevents the full elongation of the internodes. Needless to say, it is the species of small or medium stature which are mostly used. However, certain larger types are sometimes dwarfed, even though such creations can be maintained in a satisfactory form for only a relatively short time (Li, 1956a). More recently details concerning the growing of bamboos as penjing have been given by Wang and Shen (1987), along with illustrations of several species grown in this way. On the whole, however, bamboos do not adapt easily to this type of culture and very often the Heavenly Bamboo (*Nandina domestica*) is used instead (Morris, 1983).

A quite different observation bound to be made about bamboos by any visitor to China is that culms provide ideal surfaces on which to carve names, just as trees, rocks, and so on do at popular sites in the West. The example shown in figure 5.11 records the equivalent of 'Wayne was here in 1979'.

While it is placed in a completely different tribe of the grasses, the Giant Reed, *Arundo donax*, has traditionally been

Figure 5.10 (above) **Stone bamboo shoots in the Imperial Garden of the Forbidden City, Beijing**
Figure 5.11 (left) **A record of a visit to the Guangzhou Zoo**

Bamboo is an essential component of any Chinese garden. The species planted are chosen with regard to size, climate, and taste. Very often bamboo is positioned so that it appears as it would in a painting, associated with rocks or set against the background of a plain wall (figs 5.4, 5.17), which plays the role taken in paintings by the paper or silk. Alternatively it may be placed in some symbolic combination with other plants. For example bamboo (*Zhu*) and plum (*Mei*) together suggest

regarded as a bamboo by the Chinese and I include mention of it here as it may occasionally be seen in gardens. Except for most of the species of *Bambusa*, all the plants mentioned below are hardy to at least -10°C.

ARUNDO Gramineae

A. donax 芦竹 (Lu Zhu) *Reed Bamboo*

GIANT REED

Found typically on the banks of watercourses and growing up to 6 m tall, this large grass can be invasive in climates which suit it. It is, however, very handsome, with its broad, two-ranked leaves borne on stout, arching culms, and variegated forms occur in cultivation. The culms are used for many of the same purposes as those of the true bamboos.

BAMBUSA Gramineae

A largely tropical genus of clump-forming bamboos with species in Asia, Africa, and America.

B. beecheyana 吊丝球竹 (Diaosiqiu Zhu)

Hanging-silk-ball Bamboo

A medium-sized bamboo, with culms up to 12 m tall. It is an important source of commercial bamboo shoots in southern China (Bailey & Bailey, 1976).

B. multiplex 孝顺竹 (Xiaoshun Zhu) *Filial Piety Bamboo*

HEDGE BAMBOO

Widely grown for hedging in south-east Asia, this species (figs 5.8, 5.9, 5.12) produces slender culms 3–15 m tall. Its Chinese name comes from the way the young shoots in the clump come up around their elders as if guarding, protecting, and respecting them. Several cultivars are recognised, including kinds with striped leaves and culms. The best-known cultivar is 'Fengwei Zhu' ('Phoenix-tail Bamboo'). This has shorter culms, more slender than those of the type, and numerous smaller leaves, its appearance suggesting to the Chinese the tail of the phoenix. It is often used for penjing and is known in the West as 'Fernleaf'. Another which is popular for pot culture is 'Rivierorum', which has narrow leaves and arching culms. This is known in China as 'Guanyin Zhu' ('Guanyin Bamboo'). A larger cultivar with yellow culms striped with green is known in the West as

'Alphonse Karr'. Unlike many other species of *Bambusa*, *B. multiplex* is hardy to at least -12°C (Cusack, 1997; Recht & Wetterwald, 1992). The young shoots are edible (Crawford 1997; Cusack, 1997).

B. ventricosa 佛肚竹 (Fodu Zhu) *Buddha's Belly Bamboo*

BUDDHA'S BELLY BAMBOO

Although this species may be quite tall when grown under optimum conditions, the culms are often no more than 0.5–2 m high when it is kept in pots or on poor soil. Under such conditions the leaves are smaller and the internodes are shortened and swollen, hence its Chinese name. On account of this characteristic its appearance in gardens and as a pot plant is most striking. It is hardy to -9°C (Cusack, 1997).

Figure 5.12 *Bambusa multiplex*

B. vulgaris 龙头竹 (Longtou Zhu) *Dragon's Head Bamboo*

COMMON BAMBOO

The culms of this vigorous species are thick and may be up to 25 m tall. It has been cultivated for so long and so widely that its natural distribution is uncertain. It requires a warm climate to grow satisfactorily and has given rise to a number of cultivars. In gardens in southern China it is often seen as the striped form 'Huangjinjianbiyu' ('Green Jade Embedded in Gold'), known in the West as 'Vittata' (fig. 5.13). The spectacular greenish gold, edible shoots, which have black hair patches, are prized by the Vietnamese on account of their bitter flavour (Cusack, 1997). Another cultivar which is popular is 'Da Foduzhu' ('Big Buddha's Belly Bamboo'), known elsewhere as 'Wamin'. This has short and slightly swollen internodes (fig. 5.14) and is planted in the ground or grown in pots. Unlike the somewhat similar *B. ventricosa*, it retains its short internodes, at least on the lower parts of the culms, even when under favourable conditions it may reach as much as 15 m in height (Cusack, 1997).

CHIMONOBAMBUSA Gramineae

An Asian genus of about 10 species with running rhizomes.

C. quadrangularis 方竹 (Fang Zhu) *Square Bamboo*

SQUARE BAMBOO

The culms of this species have prominent nodes, often purplish below, and may be up to 10 m tall. Mature culms are more or less square in cross-section with rounded corners. *C. quadrangularis* has a flowering cycle in excess of 100 years (Mabberley, 1997a). According to Li (1956a) it is sometimes used for penjing. It is hardy to -15°C and the young shoots are edible (Crawford, 1997).

INDOCALAMUS Gramineae

An Asian genus of relatively small, running bamboos.

I. latifolius 阔叶箬竹 (Kuoye Ruozhu)
Broad-leaved Hat Bamboo

The slender culms of *I. latifolius* may reach a height of 2 m and bear large leaves, 10–40 × 1.5–8 cm. Sometimes it is planted in gardens, forming dense mounds, usually amongst rocks. It is hardy to -15°C (Crawford, 1997).

I. tessellatus 箬竹 (Ruo Zhu) *Hat Bamboo*

This species is similar to the above but is usually a little taller (fig. 5.15), up to 2.5 m, and with leaves which may be up to 60 × 10 cm, particularly when it is grown in shade. These leaves are used for hat-making and other purposes in southern China. *I. tessellatus* has a flowering cycle of at least 115 years (Mabberley, 1997a). It can survive temperatures as low as -20°C or -23°C (Crawford, 1997; Recht & Wetterwald, 1992).

Figure 5.13 (opposite page)
***Bambusa vulgaris* 'Vittata'**
Figure 5.14 (left) ***Bambusa***
***vulgaris* 'Wamin' at Guilin**

Figure 5.15 *Indocalamus tessellatus,* flowering peach, and *Juniperus chinensis* in Behai Park, Beijing

and asymmetrical, creating an appearance said by the Chinese to resemble the human face (fig. 5.16), and because of this characteristic it is sometimes dwarfed (Li, 1956a). Several variants are known, including one with green stripes on the stems. The shoots are edible (Crawford, 1997).

PHYLLOSTACHYS Gramineae

Phyllostachys is an Asian genus of medium-sized to large bamboos which are of immense importance as sources of building material and edible shoots. Several species are also grown for ornament. Although they have a running habit and can become very invasive in climates which suit them, under cool conditions they often form relatively compact clumps. The internodes usually have a groove on one side. The species described below are all very cold hardy.

P. aurea 人面竹 (Renmian Zhu) *Human-face Bamboo*

FISHPOLE BAMBOO

In spite of its invasive nature, this species, which can grow up to 10 m tall, is widely cultivated in gardens. The internodes on the lower part of many culms are very short

P. bambusoides 桂竹 (Gui Zhu) *Osmanthus Bamboo*

GIANT TIMBER BAMBOO

This species produces culms up to 30 m tall and 20 cm in diameter, which are glossy dark green, usually without any white powder, and with leaves 9–20 × 2–4.5 cm. Its enormous, rapidly growing spring shoots are a source of wonder, developing into huge leafy culms of great beauty, rather like giant, green ostrich feathers. *P. bambusoides* is the most important timber bamboo of the Orient, several variants are grown in gardens, and the shoots are edible (Crawford, 1997). A cultivar I have seen in Chinese gardens grows to only about 5 m and has yellow culms which are occasionally striped with green. Like *Bambusa vulgaris* 'Vittata' it is known in China as 'Huangjinjianbiyu', but in the West it has been given the name 'Allgold'.

P. edulis 毛竹 (Mao Zhu) *Hairy Bamboo*

MOSO BAMBOO

Almost as large as *P. bambusoides* and equally beautiful is *P. edulis*, which gets its botanical name because it is extensively cultivated for its edible shoots as well as for its large culms. Unlike those of *P. bambusoides*, the stems of *P. edulis* are grey-velvety when young, becoming green or yellowish with age with white powder below the nodes (figs 5.1, 9.12). Also the leaves are smaller, usually 5–12 × 0.5–2 cm. Like *P. bambusoides*, it is sometimes planted in gardens.

P. nigra 紫竹 (Zi Zhu) *Purple Bamboo*

BLACK BAMBOO

The culms of this distinctive species are up to 10 m tall, green at first but turning shiny black as they mature (fig. 5.17). Variants are known in which the culms remain green or are merely blotched with brown or black. On account of its striking appearance it is frequently planted in gardens, even though it is invasive. It is also readily dwarfed (Li, 1956a). Like those of the other *Phyllostachys* species listed above, the young shoots are edible.

Figure 5.16 (far left) Culms of *Phyllostachys aurea*, showing shortened internodes and internodes with grooves on one side

Figure 5.17 (left) Black Bamboo (*Phyllostachys nigra*)

CHAPTER 6

APRICOTS, PEACHES, PLUMS, AND CHERRIES

Many species of *Prunus* are native to China and, although only a few of these have been brought into cultivation there, they are of enormous importance, since they include the stone fruits. Peaches, apricots, plums, and cherries have long been cultivated, and numerous fruiting varieties of each have been developed. Although the beauty of a royal bride is compared to that of flowers of the cherry, peach, and plum in Arthur Waley's (1969, p. 78) translation of a poem in the *Shi Jing* (Book of Odes), believed to have been compiled between the 11th and 7th centuries BC, it is not clear exactly when cultivars were first selected for their ornamental appearance. However, these ultimately became some of the most popular of all garden flowers in China, accumulating a vast amount of mystique and symbolism. In this regard *P. mume* outshines all other *Prunus* species and, it might well be claimed, all other Chinese plants. The peach, too, has become of immense cultural significance. And the red fruits of various species of the cherry

Figure 6.1 **A double pink cultivar of *Prunus mume***

group have been appreciated by poets and artists, being compared to the red lips of famous beauties. Thus these fruits have become symbols for beautiful women (Williams, 1975).

No doubt because of all this, species of *Prunus* are to be found in almost every garden in China. While the cultivars grown principally for their fruit are seen in this situation only rarely these days, this was not always so. During the Ming dynasty, for instance, quite substantial plantings were present in some of the famous gardens of Suzhou (Clunas, 1996), where these trees appear to have been appreciated for both their fruits and flowers. Observations concerning the various species are set out below.

PRUNUS Rosaceae

P. armeniaca 杏 (Xing)

APRICOT

Although the apricot's botanical name implies that it is a native of Armenia, it is believed to have been introduced there from its native home in China. It is mentioned in early Chinese texts and was certainly cultivated for its fruit by about 500 BC (Haw, 1987). Blossoming apricots are mentioned in a 10th century poem quoted by Williams (1975), who also reports that the eyes of Chinese beauties are often compared to the ovoid kernels, just as 'almond eyes' are spoken of elsewhere.

Numerous varieties have arisen in China and in the north are seen planted on terraces on the hillsides. Some of these have been selected for the quality of their flesh and some for their kernels, which are used in the manner of almonds (fig. 6.2). Presumably it was one of these types, which have thin flesh which splits open as it dries, which was sent to Paris by Armand David and described by Carrière (1879) as

Figure 6.2
Apricot
kernels
on sale in
Chengde

Armeniaca davidiana. As far as I know the true almond, *P. dulcis*, is not grown in China. This is probably because the established industry producing apricot kernels, similar in taste and appearance, satisfies the demand.

As well as being grown in orchards, apricot trees are commonly planted in gardens and temple grounds where their flowers are much admired. A weeping form and cultivars with both single and double flowers are known. These are very similar in effect to *P. mume*, though they bloom later and are not held in the same esteem. In fact they are regarded as being much inferior (Bickford, 1996). Nevertheless the apricot was regarded as the floral symbol of the 2nd month (Box, 1905; Bredon & Mitrophanov, 1927). Because of their blossoming at this time, the month of the imperial examinations, they became known as 'successful candidate flowers'. Pictured together with swallows, the Chinese name for which has the same pronunciation as that for banquet, they form a rebus wishing a candidate success in these examinations and the consequent honour of being invited to a banquet by the emperor (Weidner et al.,1988). Apricot flowers often appear in other types of painting, for example a Song album leaf (Hearn, 1996, no. 33) and a 17th century *Hundred Flowers* reproduced by Weidner et al (1988, cat. no. 34).

The first detailed observation by a European concerning the apricot in China appears to be that of Cibot (1780b), who described three classes—those grown for their flowers, those grown for their fruits, and wild apricots. It seems possible from his descriptions that most, if not all, of the flowering kinds he mentioned were cultivars of *P. mume* or *P. triloba*. However, the remainder were certainly apricots, the fruits of the white-flowered, wild plants being harvested only to express oil from the kernels. Amongst the fruiting kinds he noted cultivars with yellow, white, or flesh-coloured fruits as well as the small-fruited kind grown for the kernels, and gave details of their propagation, cultivation, and medicinal uses. He also recorded that apricots were dried, made into pastilles, and used to prepare drinks. Amongst other things it seems that the kernels of wild apricots were believed to cure the bites of mad dogs and that 'milk of almonds' made from the kernels was used by painters to get rid of any suggestion of yellow from white lead. It appears, too, that apricot leaves could be eaten, as they are listed as a famine food in the *Jiuhuang Bencao* (Bretschneider, 1881). However, since various parts of the plant are known to yield hydrogen cyanide, it would seem wise not to eat too many.

Figure 6.4 A double cultivar of *Prunus cerasoides* at the Kunming Zoo

Bretschneider (1898) recorded that '*Prunus armeniaca* with small yellow edible fruits abounds in the mountains of N. China and S. Mongolia'. In 1880 and 1881 he sent stones of the wild apricot to the Arnold Arboretum, Kew, and other botanical gardens, but it is doubtful whether the wild form is commercially available outside China (Lauener, 1996). Wild apricots are still common in the mountains of the north (fig. 6.3, see over).

Meyer (1911) recorded that among the fruiting varieties there were red, orange, yellow, red-and-white, and white-spotted kinds in cultivation, as well as the strain, with several varieties, grown for its kernels, noting that he had never seen a true almond tree in China. He also observed that the cultivars were propagated by grafting onto seedling apricots or the wild peach (*P. davidiana*).

Meyer drew attention to the fact that in China apricots are nearly always eaten either when not quite ripe or made into delicacies. He writes:

> A peculiarity of Chinese taste is that the race as a whole does not care for soft fruits. One may even see high-class Chinese ladies selecting hard apricots from a basket containing a mixture of ripe and green ones and relishing the crunching of the hard fruit between their teeth. At dinner parties, fine-looking but extremely hard pears are served and are keenly relished by the most highly cultured Chinese. For this reason one finds that although raspberries, red currants, gooseberries, and various other small fruits grow wild in the mountains, one never sees them cultivated, and the fruit on these wild bushes is only sparingly picked.

Chinese taste in this regard may have persisted to a considerable extent, as visitors to the country can attest. However soft fruits have gained in popularity—modern strawberry varieties, for example, now being extensively cultivated.

P. cerasifera 櫻桃李 (Yingtaoli) *Cherry Plum*

CHERRY PLUM

A purple-leaved form of the cherry plum is used in public plantings and is occasionally seen in gardens. Whether it is a recent introduction or found its way across from the Middle East long ago I do not know. However, all the plants I have seen appear to have been planted relatively recently.

P. cerasoides 高盆櫻桃 (Gaopen Yingtao) *Tall Pot Cherry*

This species is widespread from the eastern Himalayan region of India, across through upper Burma and northern Thailand into Yunnan. It is somewhat frost-tender and very variable. It is cultivated in Yunnan where there are large-flowered single forms and what appears to be a double variety of this species with deep pink flowers (fig. 6.4,). I have

not seen this anywhere else nor come across any mention of it in the literature, though it is very conspicuous in Kunming in the spring. There is a large planting of it at the zoo. It is worthy of introduction to areas which would be mild enough for it. It may be that the plants included here are forms of the cherry known in China as *P. yunnanensis*, but I have not as yet been able to check this. However, they seem to me to fit comfortably within *P. cerasoides*.

P. davidiana 山桃 (Shantao) *Mountain Peach*
DAVID'S PEACH

This plant, which is common in the hills and mountains of northern China, may perhaps be the wild ancestor of the cultivated peaches. It was named for Abbé Armand David, who was in China from 1862 to 1873, during which period he collected it. Bretschneider (1898) said that it was much planted as an ornamental tree in Beijing. It is still quite common in gardens and public plantings in northern China, where it forms a small tree, often more upright than the true peach, which it very closely resembles. However, it has smoother leaves, and the trunks of many plants bear shiny, reddish-brown bark, something I have not seen on a true peach. The flowers, which appear at the first hint of spring, are almost sessile and range in colour from white to a soft pink. They are followed by small, round, downy fruits (fig. 6.5). In flower it is very charming but, as with so many of its relatives, the display is short-lived.

Meyer (1911, 1916), who saw it growing wild and in gardens at Tianjin and Beijing, reports that it is used as a stock for grafting not only both flowering and fruiting peaches, but plums (*P. salicina*), bush cherries (*P. tomentosa*), flowering plums (*P. mume*), cherries (*P. pseudocerasus*), and even apricots, and that all of them

thrive. He was of the opinion that it had been used for centuries as a stock for stone fruits. He goes on to say that the kernels are used to flavour confectionery and some special dishes, but as they are full of prussic acid only very small quantities are used, and that the small, almost spherical stones are made into rosaries which the Buddhist priests use in their worship. Being deeply and irregularly grooved, they are very ornamental when cleaned and polished.

P. glandulosa 麦李 (Maili) *Wheat Plum*
DWARF FLOWERING ALMOND

Rarely more than 1 m tall, often with upright branches and a tendency to sucker, this small shrub is now well known outside China. There are single and double forms, both white and pink, and when in bloom it is very showy (fig. 6.6). The single forms often produce a crop of small red cherries. Presumably it has long been cultivated in China but does not appear to have reached Europe until the beginning of the

Figure 6.5 *Prunus davidiana* in the garden of the Jingxinzhai, Beijing

Figure 6.3 (opposite page) **Wild apricots blooming on a hillside at Chengde**

Figure 6.6 (right) **A double white cultivar of *Prunus glandulosa* at the Wofo Si, Beijing**

19th century (*Edwards's Botanical Register* 1815, t. 27). A double pink form is depicted in an 18th century *Hundred Flowers* (Capon & Pang, 1981, cat. no. 65) and no doubt this species can be seen in other paintings.

P. mume 梅 (Mei)

JAPANESE APRICOT

Along with bamboo, the plant most intimately associated with art, literature, and everyday life in China is *P. mume*. Its domestication for its fruit dates back to the remotest antiquity, and its flowers were also appreciated at an early period. It is unfortunate, then, that it does not have a suitable name in Western languages. When it appears in paintings or as a decorative motif in pebble paving, on porcelain, or in other ways, it is usually referred to as 'plum blossom' or 'flowering plum'. Likewise the sauce made from its fruits, marketed all over the world for use in Chinese cookery, is known as 'plum sauce'. This is misleading as it is not a plum but more closely allied to the apricot. In fact when grown as an ornamental plant in the West it is usually called Japanese Apricot. Although not originally Japanese, it is widely cultivated in Japan and was first made known to the West by the observations of European visitors to that country. Pressed specimens were collected in China by Dr Clark Abel who accompanied Lord Amherst's embassy in 1815 (Haw, 1987), but living material did not reach Europe until it was introduced from Japan by Siebold about 1844 (Whitehead & Edwards, 1974). Siebold's specific epithet, *mume*, is a transliteration of the Japanese pronunciation of the Chinese name for the plant.

Figure 6.7 Ginger jar decorated with plum blossom (*Prunus mume*)

Flowering in winter, it is generally regarded as the floral symbol of the first month. Unlike the peach it is long-lived, and specimens claimed to be up to 1000 years old are found in gardens and the grounds of temples. Ancient specimens assume a crooked and gnarled appearance and are much admired, the production of flowers and young green shoots by such specimens suggesting longevity and suggesting too that from age there may still come the joys of youth (Koehn, 1952). Because the strongly scented flowers appear on apparently lifeless branches, when snow and ice are still present and no other plants are showing signs of life, it shares the symbolism of the chrysanthemum, which remains in bloom after other flowers are over. It is seen to embody the virtues of purity and ability to endure hardship, evoking, as Needham (1986) puts it, the doughty Confucian hero devoted to ethical principles come what may. Such symbolism is reinforced when the plum is combined with bamboo and pine to represent the 'Three Friends of the Cold Season', the pine and bamboo being evergreen, and the plum putting forth blossoms in winter. These plants frequently appear together in paintings and on textiles and ceramics, symbolising the best qualities of the human character.

In other circumstances the plum, blossoming alone, has been used as a symbol of the recluse, of the separation of friends or lovers, or of a beautiful but lonely or abandoned woman (Bickford, 1996). On a more cheerful note it also calls to mind the coming spring and, as five is an auspicious number in China, representations of the five-petalled flowers are widely used on ceramics and lacquerware and in embroidery, painting, and the designs of lattices, balustrades, windows, and the pebble-paved paths of gardens. The five petals are said, amongst other things, to represent the five blessings—old age, wealth, health, love of virtue, and a natural death (Bartholemew, 1985a). A flowering branch is commonly depicted on ginger jars (fig. 6.7) and vases against a reticulated deep blue background, which is intended to represent the broken ice which has floated down the rivers from the north in early spring (Williams, 1975).

Bickford (1996) says that by the late Song dynasty *P. mume* had come to symbolise China, not only as an emblem of the South but as the embodiment of Chinese cultural values. As a result of this, it was chosen during the 1930s as the national flower, its five petals being taken to represent the five principal races of the country (Chinese, Manchu, Mongol, Uighur, and Tibetan) and also the Five Power Constitution of the Republic. After 1949, however, it appears to have lost this distinction. Nevertheless the flowering plum still commands an extraordinary position amongst the symbolically resonant plants of the country.

It is not surprising, then, that it is one of the most frequently depicted of all subjects in Chinese art and appears in some of the earliest surviving paintings, for instance the early 11th century *Four Magpies* reproduced by She and Yuhas (1973, cat. no.8) and numerous other examples from the Song dynasty. The painting of plum blossom is a specialised branch of art and instructions for it are given in all painting manuals, for example Song Boren's *Meihua Xishen Pu* (The Spirit of Joy: a Treatise on [Painting] the Mei Flower) of 1238, which is mentioned by Needham (1986). This gives directions for the representation of eight stages in the life of the flower. Paintings of this and other plants may be elaborate and in colour or they may be executed extremely economically with black ink, a technique which requires an ability with the brush equal to that possessed by the best calligraphers (figs 2.4, 2.12, 6.8).

The symbolism with which depictions of the flowering plum can become endowed is almost unbelievably complex, as can be ascertained by referring to the passage from Sze's (1977) translation of the *Jieziyuan Huazhuan* (Mustard Seed Garden Painting Manual) quoted in Chapter 2. Anyone wishing to know more about the plant's involvement in art should seek out Maggie Bickford's (1996) fascinating book, *Ink Plum*, and her earlier work (1985), with contributions from others, *Bones of Jade, Soul of Ice: The flowering Plum in Chinese Art*. Numerous reproductions of paintings, along with translations of poems and prose about the plum, may also be found in *Noble Fragrance* by T. C. Lai (1977).

A somewhat different instance of the involvement of *P. mume* with Chinese life is that, at the winter solstice, families would paint in outline a branch with nine flowers and a total of eighty-one petals. This was called the 'nine nines chart of lessening cold'. Each day a petal would be marked so that, by the time the chart was completed, the nine nines of days would be past and spring would be well advanced. Tun (1965) recalls that, as a boy, he carried out this ancient procedure.

During my youth I, too, had an involvement with the flowering plum. A tree of the Japanese cultivar 'Geisha' growing in my family's garden produced a wonderful crop of enticing fruits, bright orange stained with carmine on their

Figure 6.8 *Myriad Blossoms of Jade*—hanging scroll, ink on silk, showing *Prunus mume* in bloom, by Chen Xianzhang, dated 1437

sunny sides. Never before or since have I tasted anything quite like them. The acidity and astringency were such that I wondered whether my mouth would ever recover. I was not surprised, then, to read that *P. mume* was first planted for its fruits, which were used as a source of sour flavour before the invention of vinegar, now to be found on every Chinese table along with the soy sauce (Li, 1959). In spite of the rise of vinegar, large quantities of the fruit are still grown. Sometimes they are eaten fresh, but more often they are preserved and made into a variety of products, notably the plum sauce mentioned earlier, which can be found in most Chinese food stores. Also the flowers are used, or at least have been, for scenting tea (Fortune, 1857), and the kernels and root bark are employed in medicine (Williams, 1975).

Although it had been grown earlier for its fruits, apparently it was not until the 4th or 5th century AD that *P. mume* became known as an ornamental plant. Subsequently, during the Tang dynasty, poets sang its praises and this greatly enhanced its popularity. As with so many plants long cultivated in China, it is not clear to what area the tree was indigenous. Li (1959) says wild trees are still to be found in the mountain regions of central China. He states that records from as early as the 6th century AD indicate that it is native to Shaanxi, from where plants were apparently obtained and domesticated. Needham (1986) quotes other records of trees growing wild in ancient times on a famous pass connecting Guangdong with Hunan, which was accordingly called Meiling Guan (Mei Ridge Pass). However, he also quotes a reference of 863 AD which records an inscription, written by a woman who travelled over the pass with her father, saying that they could not find any *Mei* so they planted 30 beside the road. There were trees still there last century, as Clarke Abel (1819), who was in the area on 20 November 1816, wrote: 'Near the entrance to the pass on the Canton side we saw a species of *Prunus* in full flower, called by the Chinese Mei-hwa-shoo'. Perhaps it still grows there.

Whatever its original habitat may be, *Mei* is a tree associated with the south, where it blooms naturally at the time of the Chinese New Year and where, during the southern Song and the subsequent Yuan, its appreciation reached the proportions of a craze (Bickford, 1996). It was during the Song that the first monographs about it appeared. Also at that time its admirers in Hangzhou and Suzhou sometimes wrote almost pityingly of people in the north who were not acquainted with it.

While in paintings and other representations it is almost invariably the single white form which is depicted (figs 2.4, 6.7, 6.8), praised, and invested with symbolism, coloured variants existed, and both single and double cultivars were enthusiastically cultivated by gardeners. In the *Mei Pu* (Treatise on Mei) of 1186, for instance, Fan Chengda describes his collection of hundreds of trees in more than 12 varieties, most of which he claims to have brought together for the first time on his property near Suzhou. This is the first book in any civilisation devoted entirely to this plant, in fact to any

species of *Prunus*, and several others followed (Needham, 1986). By the late 17th century there were at least 21 distinct varieties and there are still many named varieties of long standing in China. Single, semi-double, and double forms occur, ranging in colour from white to pink and deep carmine. There is a form with purple leaves and even a weeping form, 'Zhaoshui Mei' ('Water Reflecting Mei'), which is known in the West as 'Pendula' and which has lent itself admirably to depiction on hanging scrolls. Needham (1986) records that a very dark kind, 'Mo Mei' ('Ink Mei'), is supposed to have arisen from a graft of the wild form on *Melia azedarach*, an unlikely story if ever there was one, but such claims of successful graft unions between unrelated partners abound in the Chinese horticultural literature, as already noted. The cultivar seen most frequently in China at the present time is a double pale pink (fig. 6.1), apparently identical with that depicted in a drawing in the Reeves collection in the British Museum (Natural History) reproduced by Whitehead and Edwards (1974).

In gardens *P. mume* is planted either singly or as a formally arranged orchard, as in the Guyi Yuan near Shanghai, the Zhuozheng Yuan in Suzhou (fig. 6.9) and, on a much

Figure 6.9 *Prunus mume* orchard in the Zhuozheng Yuan, Suzhou
Figure 6.10 (opposite page) A view through peach blossom on the Su Causeway, Hangzhou

larger scale, in the Mei Yuan at Wuxi. The planting of such ornamental orchards in which many varieties are displayed also became popular in Japan. There was a famous example at Kamata depicted in a woodblock print by Hiroshige, one of his *One Hundred Views of Edo*, and in the garden of Kairakuen at Mito a large number of an original orchard of 10,000 trees still survives.

P. mume is also a favourite plant for penjing and for pot culture, plants in bloom being in demand for the New Year. These often have their branches twisted to suggest deer, storks, or some other good omen, but Li (1959) says 'these appeal to the less learned but wealthy'. Metcalf (1942), writing about the flowers in demand in Guangzhou for the New Year, says that cut branches are not very satisfactory as the flowers soon drop off. My own experience shows this to be true, but it is hard to avoid bringing some indoors in winter to enjoy their delicate beauty and delicious scent.

P. persica 桃 (Tao)

PEACH

Just as the home of the apricot was presumed to have been Armenia, so the peach was thought to be native to Persia, hence its botanical name. As in the case of the apricot, it is now thought that the peach originated in China and found its way along the Silk Road to the Middle East and thence to Europe. It was known to Theophrastus in Greece in the 4th century BC (Haw, 1987). It seems to have reached Italy somewhat later, as in the first century AD Pliny the Elder mentioned that there were several varieties in cultivation there and indicated that they were recent arrivals (Li, 1959). But while the peach may have been grown for over 2000 years in Europe, where it has become the subject of little or no folklore, it may well have been brought into cultivation 3000 or more years ago in China, where it has accumulated almost as much popular symbolic meaning as *P. mume*.

While peach stones have been excavated from archaeological sites in China, some dated 6000 to 7000 years ago, there is nothing to indicate whether these came from wild or cultivated trees. But it seems certain that they were cultivated before 600 BC, as a poem of the period refers to peach trees growing in a garden. Another poem from the same era praises the brilliance of peach blossom, so it is clear that both the flowers and fruit were appreciated in early times. While several cultivars, almost certainly fruiting varieties, were in existence by the Han dynasty, it is hard to be sure when

purely ornamental forms (figs 1.9, 5.13, 6.10, 6.12–6.15) were first developed. However, some are mentioned in poems of the Song dynasty, and a late 16th century work records varieties with 'red, purple, white, double, and bicoloured' flowers (Haw, 1987).

The ancient Chinese believed the peach had originated in the far west of the country in the Kunlun Mountains, traditionally considered to be the home of the deity Xiwangmu, the Queen Mother of the West. She was said to possess an orchard of peach trees, the fruits of which would confer immortality on those who ate them. But since these trees blossomed only once every 3000 years and a further 3000 years had to pass before the fruit ripened (Bartholemew, 1985b) no doubt it was difficult for anyone to test the veracity of this belief. Nevertheless it is clear that the association of the peach with longevity and immortality was established very early. It was reinforced by the story of the Peach Blossom Spring as recounted by Tao Yuanming, who lived from 365 to 427 AD (Barnhart, 1983). According to this tale a fisherman was boating along a stream, unconscious of the distance he had travelled, when he came upon a dense grove of blossoming peach trees lining the bank. No tree of any other kind stood among them and at the end of the grove he found a spring emerging from a cave from which a faint light emanated. Walking through he emerged into a land where time had stopped and farming families were living a cheerful and contented life. He left after a few days but, needless to say, could never find his way back.

Unlike *P. mume*, which was grown in ancient times for its fruit only, the peach was early valued for both its wealth of bloom and its delicious fruits. Li (1959) says that, while the beauty of its blossoms is repeatedly praised in the *Shi Jing*, often in connection with marriage, of which peach blossom is a symbol, their fame and merit is far exceeded by that of the fruits. Sweetmeats shaped and coloured to represent the *Shou Tao* (Peach of Longevity) are, or at least used to be, invariably presented as a gift at any birthday party. The motif of a peach tree growing on a mountain surrounded by waves and bats represents the traditional birthday greeting 'May your happiness be as deep as the eastern sea and may you live to be as old as the southern mountain' (Bartholemew, 1985a). According to Koehn (1952), a peach often takes the place of the character *shou* (long life) and is surrounded by the five bats representing The Five Great Blessings. Also the god of longevity, Shou Xing, is usually depicted as an old man with a characteristically prominent cranium, holding a staff in his left hand and a peach in his right (Bartholemew, 1985b). And Daoist priests are reputed to have believed that if they took doses of the gum exuded by peach trees they would be able to fast forever and become immortal.

It was this pursuit of immortality that led the Daoists away from their original philosophy to dabble in alchemy and to discover many medicinal plants. Meanwhile many tales, legends, and superstitions about peach trees accumulated. Even the wood and the ordinary fruits were considered to possess magical properties. Twigs were employed to ward off evil spirits, peach stones were carved into amulets for a similar purpose, and the gum, in addition to the use mentioned above, was employed in medicine. And the poets throughout the ages celebrated the beauty of the blossom, universally considered to be a symbol of spring and a good omen, as well as being an emblem of marriage (Gulland, 1928). The leaves were listed as a famine food in the *Jiuhuang Bencao* of 1406 (Bretschneider, 1881) but, since they yield hydrogen cyanide, again, eating more than a small quantity may not have proved altogether beneficial.

No doubt as a result of all of this, the peach, like *P. mume*, has long been a favourite subject in Chinese art. It is often taken to be the floral symbol of the third month (Box, 1905; Bredon & Mitrophanov, 1927; Li, 1959), and fruits or fruiting branches appear in many paintings, as for instance in the late Song or early Yuan *Squirrel on a Peach Branch* reproduced by She and Yuhas (1973, pl. 17). Double-flowered forms can be seen in many works of the 'flowers of the seasons' and 'hundred flowers' genres. Likewise both flowers and fruits appear as decorative motifs on porcelain.

Since over the millennia the peach attracted such attention and since so many cultivars were selected and propagated, it is surprising, as Needham (1986) notes, that no medieval Chinese botanist wrote a monograph devoted to it. As he did with the apricots, Cibot (1786a) recognised three main types in China—dwarf kinds grown in pots for their flowers, shrubs grown as ornaments in gardens, and trees grown for their fruits. He discussed their role in literature, art, and tradition, and described methods of propagation and cultivation. He recognised cultivars on the basis of the shape, colour, size, and precocity of their fruits, noting that there were kinds with white, greenish, pale yellow, orange-yellow, or marbled flesh. He also mentioned peaches weighing 'two pounds or even more'.

Well over a century later, Meyer (1911) described Chinese peaches of a variety of shapes and sizes, there being white, greenish, yellow, and red types. He said that some of the red ones looked more like beetroot than anything else. Like Cibot, he described cultivars with very large fruits. For instance, at Feicheng, Shandong, in 1907 he collected 'Stones of the most famous peach of north China. The fruits grow as heavy as one pound apiece and are pale yellowish with a slight blush; meat white, except near the stone; taste excellent, sweet, aromatic, and juicy. Is a clingstone. Has extraordinary keeping and shipping qualities. The branches need propping up on account of the weight of the fruits' (Cunningham, 1984). Fortune (1847a) also encountered a very large peach which was grown a few miles south of Shanghai. He said that it was not unusual to see peaches of this variety 'eleven inches in circumference and twelve ounces in weight. This is, probably, what some writers call the Peking peach, about which such exaggerated stories have been told'.

Other kinds of peach worth mentioning are the curious flat peaches, *Pan Tao*, also recorded by Myer (1911) and regarded as the true peaches of immortality in Chinese mythology (fig. 6.11). These seem to have achieved little

Figure 6.11 **A flat peach, from** *Transactions of the Horticultural Society of London*, 1822

Figure 6.12 Peach blossom on the Su Causeway, Hangzhou

popularity outside China, presumably because their supposed life-prolonging properties have failed to compensate for their awkward shape. Evreinoff (1934), however, records their commercial cultivation in the United States and Russian Turkestan. He described three varieties and recommended them to amateur growers in France for their flavour and for the ornamental value of the flowers and fruits. Another variant is the nectarine, *You Tao* (Oil Peach), which is just a peach with no down on the skin. This also arose in China in ancient times according to Needham (1986). In Meyer's (1911) time, however, nectarines were apparently rare, as he heard of them only a few times.

Although they have a somewhat blowzy air about them, the kinds grown for their flowers are very spectacular. There are singles and doubles, whites, pinks, and reds, as well as striped, dwarf, and weeping varieties. There is also a form with bronze-purple leaves and red flowers. The striped cultivars are apparently chimaeras, as they usually bear some unstriped white and pink flowers.

Flowering peaches are commonly seen in gardens and public plantings and annual peach blossom festivals are held at various sites renowned for these plants, as for example in the gardens which surround the Wofo Si (Temple of the Sleeping Buddha) near Beijing. In the 18th century Cibot (1786a) found them charming planted along paths and on the edges of ponds and canals in the imperial gardens in Beijing. Nowadays they are often seen planted with weeping willows on river and canal banks and along causeways, as at the West Lake in Hangzhou (fig. 6.13) and the Slender West Lake in Yangzhou. No doubt such plantings hark back to the riverside grove leading to the Peach Blossom Spring, though the peaches there grew alone. Perhaps this is why Wen Zhengheng, writing in the early 17th century about matters of taste in the garden, declared peach and willow to be a vulgar combination (Clunas, 1996). Sometimes, as happens with *P. mume*, a collection of cultivars is planted as a small orchard. I have seen examples at the Tanzhe Si (fig. 6.14), south-west of Beijing, and at the Liu Yuan in Suzhou.

Not only is peach blossom a symbol of spring, but flowering branches and potted plants have long been in demand at New Year as symbols of longevity and good fortune. Plants, dwarfed and very often trained into curious shapes (fig. 1.9), are produced for this purpose by nurseries and sold at plant markets or distributed by itinerant plant sellers, as noted by Cibot (1786a) and Fortune (1852). Cibot said that the florists in Beijing in his day offered 18–20 kinds, singles,

Figure 6.13 Peaches and willows on the Bai Causeway, Hangzhou

Figure 6.14 A glimpse of the flowering peach orchard at the Tanzhe Si, near Beijing

doubles, and semi-doubles, with flowers of various sizes and colours. In apparently failing to recognise that these potted plants were probably dwarfed rather than dwarf, Cibot seems to have regarded them as being different kinds from those he saw growing in the ground.

In Guangzhou entire trees are cut off just above ground level like Christmas trees and sold at the time of the Chinese New Year (Metcalf, 1942; Goody, 1993). These trees are cultivated in specialist nurseries by specialist households (Goody, 1993). Not unexpectedly, the red flowered cultivars are the most popular (fig. 6.15). Traditionally these trees are placed in a tall vase tied with a red ribbon. Sometimes nowadays they are decorated with coloured lights as well.

Flowering peaches were taken to Japan and are depicted in Japanese scrolls as early as 1309 AD (Ogisu, 1996), no doubt having been introduced long before this. However, they took a couple of thousand years longer than the fruiting varieties to reach the West, not being known there, at least generally, until well into the 19th century. Since their introduction they have remained universally popular.

P. pseudocerasus 櫻桃 (Yingtao)

CHINESE CHERRY

According to Meyer (1911), *P. pseudocerasus* is cultivated extensively along the Yangtze valley. Haw (1987) also records its cultivation in parts of northern China. Unlike the European cherries, it produces its flowers and fruits in racemose rather than umbellate clusters (fig. 6.16). The small fruits, which ripen early, are about 1.5 cm in diameter. Ingram (1948) says that it has not as yet been found in the wild state and that its ancestry is unknown, but Haw (1987) states that it occurs wild in mountainous areas throughout central and southern China and more sparingly in the north.

P. pseudocerasus has a long history as a fruit tree. Cherry stones have been found in at least one Neolithic site in China, and it seems that in ancient times offerings of cherries were made at ancestral temples. Instructions for planting cherry trees are given in an agricultural treatise of about 530 AD, and there is clear evidence that *P. pseudocerasus* has been grown for its flowers as well as its fruit from at least the Tang dynasty onwards (Haw, 1987).

Figure 6.15 Red flowering peaches, chrysanthemums, and other plants in a display for the Chinese New Year at the Lieshi Lingyuan, Guangzhou

According to Ingram (1948), *P. pseudocerasus* was the first of the oriental cherries to reach Britain, being received from China in 1819 by the nursery of Barr and Brookes in London. It was described and named by Lindley in 1826.

P. salicina 李 (Li)

JAPANESE PLUM

Like the peach and *P. mume*, this Chinese plum is mentioned in some of the earliest surviving texts, and appears to have been cultivated for at least 3000 years (Haw, 1987). There are many cultivars, and Meyer (1911) reported that large red and yellow ones were grown in Shandong. These days several kinds are to be seen for sale in China.

P. serrulata 山樱桃 (Shan Yingtao) *Mountain Cherry*

ORIENTAL CHERRY

With the double-flowered Japanese cherries now so popular around the world, it is interesting that the first plant of this type to reach the West came from China. According to Ingram (1948), it was brought to England from Guangzhou in 1822, where presumably it was collected by Joseph Poole, who was employed by the nursery of Barr and Brookes, to which *P. pseudocerasus* had earlier been sent. It is thus probably the second oriental cherry to reach the West. When it bloomed it was named *P. serrulata* by Lindley in 1830. Ingram suggests that it must have originated as a sport of the indigenous single-flowered wild form of the species, which also occurs in Korea and Japan. According to Haw (1987) it has been cultivated in China at least since the 17th century.

With its relatively small flowers and rather ungainly habit, it is rarely seen nowadays, either in China or elsewhere, gardeners preferring the more spectacular Japanese cherries of the Sato-zakura group, known in China as *Riben Wanying* (Japanese Late-flowering Cherry). Though these are widely planted in China, I have seen only two varieties, one of which is white, tinged with pink, and appears similar to 'Okumiyaku' (fig. 1.7), and the other bright pink and almost certainly 'Sekiyama'. How long these Japanese cherries have been in China I do not know, but the plants I have seen do not appear to be of great age.

P. simonii 杏李 (Xingli) *Apricot Plum*

This small tree from North China was described as a species by Carrière but it is perhaps a hybrid between *P. armeniaca*

Figure 6.16 (above) *Prunus pseudocerasus* at Yangzhou
Figure 6.17 (right) Plumcots

and *P. salicina*. It has white, almost sessile flowers and large, round, dull-red fruits with yellow flesh. It is not known wild (Bailey & Bailey, 1976; Mabberley, 1997) and elsewhere such hybrids are known as plumcots. Meyer (1911) recorded plumcots in orchards in Shandong and said they were called 'Li Xing' (Plum Apricot). He said that the fruits were large, red, and very sweet and aromatic, that there were about three known varieties, and that they were all grafted on seedling plum stock.

If Meyer's description of the qualities of the fruit is correct, it seems that these plumcots deserve to be better known. As it happens, plumcots have appeared briefly in recent summers in my local fruit shop in Sydney. They are red and aromatic, with a flavour intermediate between that of apricots and Japanese plums (fig. 6.17).

P. tomentosa 毛樱桃 (Mao Yingtao) *Hairy Cherry*

DOWNY CHERRY

This downy plant is native to northern China. It forms a much-branched shrub and produces small, round, bright red fruits, which are usually somewhat pubescent and about 1.25 cm in diameter. Both Tun (1965) and Bredon and Mitrophanov (1927) describe the appearance of these cherries on the streets of Beijing early this century in the 4th lunar month. The fruits are collected from both wild and cultivated plants according to Meyer (1911), who says that the plant is propagated by layering, division, or grafting on *P. davidiana*. The fruits are also mentioned by Bretschneider (1898) and Williams (1975). Ingram (1948), who quotes Siebold as saying that in parts of China it is cultivated for its fruit, speculates that this is why it was carried to Japan some 50 years earlier than the time of his writing. However, according to Mikinori Ogisu (1996), *P. tomentosa* is depicted on a Japanese scroll of 1309 AD, so presumably it was introduced to Japan earlier than this.

P. triloba 榆叶梅 (Yuye Mei) *Elm-leaved Mei*

FLOWERING ALMOND

While *P. mume* will grow in protected situations in Beijing, the hardier *P. triloba* is more frequently seen in the north of the country. Fortune (1863) observed that it was one of the few cold-hardy plants offered by the nurserymen of Tianjin and Beijing, and Bretschneider (1898) said that a form with single, pink flowers was much cultivated in Beijing. It seems to be used in the north as a substitute for *P. mume*, and is even given the same name. Dorothy Graham (1938) certainly

Figures 6.18, 6.19 (left & below) Double cultivars of *Prunus triloba* at the Imperial Summer Villa, Chengde

had this impression, regarding the elm-leaved plum as the harbinger of spring, the first of the 'hundred flowers'. She also noted that it never bore fruit, which seems to be true of the double forms usually seen in gardens nowadays. Sirén (1949) noted that in Part 3 of the *Jieziyuan Huazhuan* of 1701 it is called *Xue Mei* (Snow Mei), a name which perhaps refers to its ability to thrive in cold climates or to its early blooming.

P. triloba blooms profusely early in the season and there are both single and double forms (figs 1.9, 6.18–19, 6.20). The leaves, which develop as the flowers fade, are often more-or-less 3-lobed, have serrate margins, and bear very conspicuous parallel veins, which give them a pleated appearance, particularly when young. A double flowered variety was introduced to Britain by Fortune in 1855, and it was upon this that Lindley founded the name in 1857 (Lauener, 1996). Cultivars of this species are conspicuous in Beijing in early spring and even more so in Chengde, about 250 km to the north-east, both in the streets of the town and in the gardens of the Imperial Summer Villa.

P. × yedoensis 东京樱花 (Dongjing Yinghua)
Tokyo Flowering Cherry

TOKYO CHERRY

The Tokyo Cherry was recorded in cultivation in the lower Yangtze valley region by Steward (1958), having presumably been imported from Japan at some stage. It is seen occasionally in other parts of China but is uncommon.

Figure 6.20 **A single cultivar of *Prunus triloba* at Chengde**

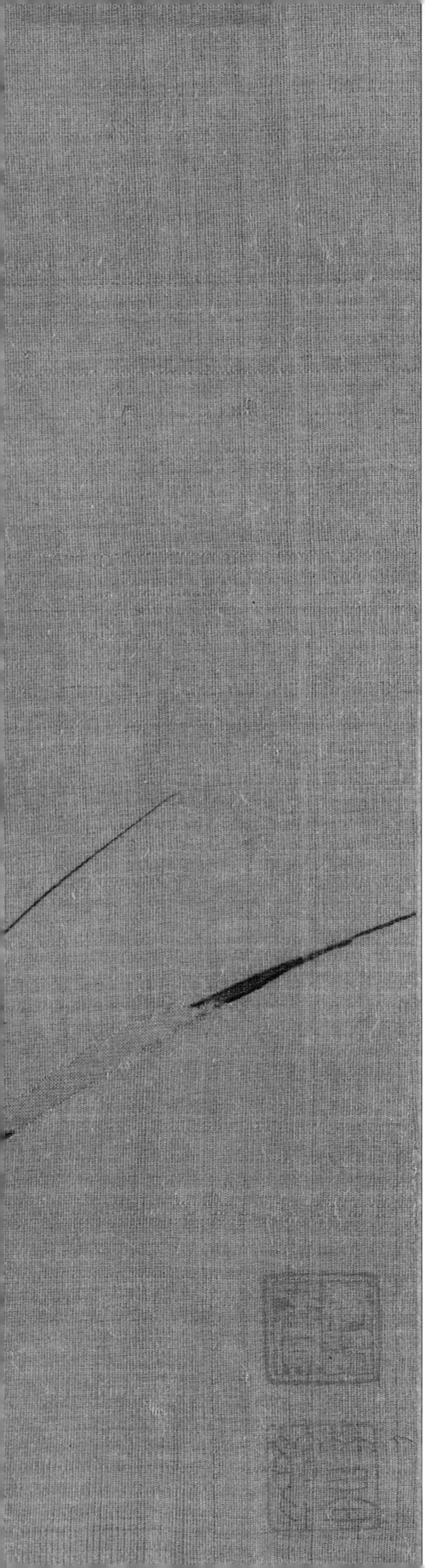

CHAPTER 7

ORCHIDS

While China has well over 1000 orchid species (Hu, 1971; Chen & Tang, 1982), relatively few of them have become popular with gardeners there, a state of affairs paralleled by many groups amongst the country's native plants. The modest terrestrial species of *Cymbidium* were the first to attract attention, and in China these are still the most esteemed of all the orchids. However, occidental visitors, to whom the word 'orchid' conjures up visions of spectacular blooms, may well find it hard to respond to the charms of the specimens which Chinese connoisseurs admire, since most of these produce only small, yellowish-green or brownish-green flowers.

The Chinese word for orchid is *lan*. However, it is not clear exactly what it meant in the ancient literature, except that it applied to some fragrant plant or plants thought to be able to ward off evil. It is believed that Confucius (551–479 BC) may have been influential in attaching the name solely to orchids. It is evident that he knew of their habitat in the wild from his saying: 'The chih-lan that grows in the deep gorges does not withhold its

Figure 7.1 *Cymbidium ensifolium*, an album leaf by Ma Lin (active 1216–55), ink and colour on silk

fragrance because of lack of appreciation. The superb person strives for self-discipline, maintenance of principles and establishment of virtue. He does not alter his integrity because of poverty and distress' (Hu, 1971). He also wrote: 'The association with a superior person is like entering a hall of chih-lan. In the course of time one becomes accustomed to the superior ways of life and gets used to the fragrance'. This statement is taken to indicate that in his time people cultivated the plants. The long and short of all this, according to Hu (1971), is that the cultivation of cymbidiums in the northern centres of culture, together with the far-reaching influence of Confucian teaching, combined to fix the name *lan* to *Cymbidium*. While in due course it came to refer to orchids in general, it is primarily the cymbidiums that the word *lan* brings to mind for the Chinese.

During the period 317–420 AD the Wang family of Zhejiang, who had the controlling power, built the Lanting (Orchid Pavilion) on a site near Shaoxing where the King of Yue had formerly cultivated orchids (fig. 7.2). It was to serve as a gathering place for scholars and is famous for a gathering held there in 353 AD by Wang Xizhi, a renowned calligrapher. This pavilion and its garden have continued to be a place of pilgrimage for scholars and calligraphers ever since, and a few cymbidiums can still be seen there.

During the Song dynasty many people migrated southward with the retreating government, settling in the warmer provinces of the south, where orchids were more abundant and where their cultivation was easier. At this time orchid growing became the hobby of many retired scholars and officials, as did the painting of them and the composition of poems about them. For almost 800 years Zhejiang and Fujian have remained the primary centres of orchid cultivation and have supplied markets all over the country (Hu, 1971).

A short piece about cymbidiums was written by Huang Tingjian about 1090 (Needham, 1986), but it was not until 1233 that Zhao Shigeng, in Fujian, published the first known monograph on orchids, the *Jinzhang Lan Pu* (Treatise on the Orchids of Jinzhang). He described 22 kinds and gave advice concerning choice of varieties, aspect, shading, watering, fertilisers, potting mixtures, propagation, and control of pests and diseases. From his descriptions it appears that, as well as cymbidiums, he grew representatives of other genera such as *Phaius* and *Calanthe*. This book was followed in 1247 by the *Lan Pu* (Treatise on Orchids) of Wang Guixue, also of Fujian. He described 37 different kinds, again apparently including

genera other than *Cymbidium*, and gave instructions for cultivation, propagation, and so on. In the introduction he gives the reasons for his love of cymbidiums, saying that whereas, amongst the most admired plants, the bamboo has nodes (*jie*, a character homophonous with another meaning 'integrity') but no flowers, the flowering plum has flowers but no leaves at flowering time, and the pine has leaves when 'in flower' but no fragrance, the orchid has all of these at flowering time and is a symbol of the perfect personality. Both these books have been translated into English by Watling (1928) and make interesting reading. After this apparently there was a long lull in the literature until the late Ming and Qing, when further detailed works about orchids appeared (Li, 1959), and perhaps these reveal more. From the end of the 18th century on, when Chinese nurseries were frequented by Westerners, further monographs appeared and these paid more regard to the orchid floras of the provinces (Hu, 1971; Needham, 1986).

While the *Cymbidium* species cultivated by the Chinese are modest in appearance, their scent is considered to be the most perfect amongst all flowers, the 'ancestor of all fragrances', permeating the air without one being fully conscious of it. It is compared to the breath of beautiful women and the fame of great men (Koehn, 1952; Williams, 1975). Also it is recorded that, when two people decided to become friends for life, they wrote their family trees in books called Gold Orchid Books which they exchanged. A picture of an orchid in a pot suggests friendship with perfect men, and a painting of an orchid and a spray of osmanthus conjures up the wish that your fame equal the fragrance of these flowers (Koehn, 1952). There can be no doubt that, when it comes to accumulated mystique, symbolism, and general involvement with Chinese culture, the cymbidiums are up there with the pine, bamboo, flowering plum, peach, and chrysanthemum.

Over the years many horticultural varieties have been selected, the choicest commanding high prices. These vary in the colour and form of the flowers and the characteristics of the foliage. There are even forms which are admired principally for their variegated leaves. Williams (1975) noted that when orchid-gatherers came in from the hills, a fancier would often buy a whole boat-load in the hope of obtaining a novelty.

In the warmer climate of the south many more kinds were brought into cultivation. Morrison (1822), for instance,

Figure 7.2 The Goose Pond at the Lanting (Orchid Pavilion) near Shaoxing

lists several and the months in which they bloomed in Guangzhou, but it is impossible to identify most of them with certainty from their Chinese names. Westland (1894) has left us a clearer picture, recording that he had never seen terrestrial orchids grown so well as he did in Guangzhou. He was particularly impressed by the unusually tall *Arundina chinensis* and *Phaius tankervilliae*, which had been noted there a century earlier by John Barrow (1804), and in remarking on the epiphytes he heaped particular praise on *Renanthera coccinea.* It seems, however, that amongst the Chinese in Guangzhou the cymbidiums remained the favourites, in spite of the flamboyance of some of the other orchids available. Watling (1928) noted that in the gardens of old Chinese families in that city there were cymbidiums propagated from plants grown in the same gardens by their ancestors 300 years earlier.

Orchid plants are almost invariably grown in pots, very often specially designed for the purpose. These pots are kept in shade houses or in some other sheltered area and brought into a courtyard, garden building, or house when the plants are in bloom (fig. 7.3). Large collections are maintained in public gardens, for example the various botanical gardens, and others which I have seen are at the so-called Flower Nursery, a little to the west of the lake in Hangzhou, and at the Orchid Garden (fig. 7.4), not far from the railway station in Guangzhou, where one can admire the orchids and partake of orchid tea. As well as this, orchid exhibitions are mounted in various parts of the country, for example in the Huifang Yuan in Zhongshan Park, Beijing, where collections are displayed in both spring and autumn.

The use of orchids in medicine has a long history in China. The topic has been covered in varying degrees of detail by Hu (1971), Chen and Tang (1982), and Needham (1986). An early record of such use is that of Ji Han in 304 AD, who recorded that what was probably *Luisia teretifolia* was an antidote to a powerful poison made from insects (Li, 1979). It appears that the three main types used in Chinese official medicine are *Baiji* (*Bletilla striata*), *Tianma* (*Gastrodia elata*) and *Shihu* (*Dendrobium* and *Flickingeria* species). However, a great many others are used in folk medicine. Hu (1971), for instance, records 21 species in 15 genera, including some of the cymbidiums.

Tianma (Heavenly Hemp) is only one of several names given to the chlorophyll-less *G. elata*, to the tubers of which many remarkable qualities have been attributed. Since it is not cultivated I shall say no more about it here, but some of

Figure 7.3 **Cymbidiums in a courtyard at the Baoguang Si near Chengdu, along with *Chimonanthus praecox*,** *Podocarpus macrophyllus*, *Sophora japonica* **and** *Primula sinensis*

**Figure 7.4 (right) The Orchid
Garden, Guangzhou**
Figure 7.5 (below right)
Arachnis flos-aeris

the dendrobiums, known collectively as *Shihu*, are grown from time to time. According to Hu (1971) the name *Shihu*, the meaning of which is not clear, is a case of *baizi*, a character wrongly written instead of another of the same or similar sound. It seems that the name originally given to *Dendrobium* species found growing on rocks was *Shihuo* (Rock Living). It was reasoned that such plants must have unusual strength to be able to establish themselves on such a substratum and draw nourishment from it. It was further reasoned, therefore, that a medicine made from them would strengthen the weak. Hu (1971) lists 15 species of *Dendrobium* amongst the common types of *Shihu* on the market, while Chen and Tang (1982) record 17. Most of these are sold as the dried stems of plants collected from the wild, a state of affairs which threatens some species with extinction. However, prescriptions occasionally call for fresh material and so some shops keep potted plants of the specified kinds, for example the well-known *D. nobile*.

In addition to those already mentioned, many other orchids are grown in China. But the traditionally-admired species of *Cymbidium* are still the most frequently encountered and have the greatest appeal for the Chinese orchid connoisseur. And at the present time, when monographs about garden plants are rare in Chinese bookshops, one is most likely to find those about cymbidiums, illustrated with numerous colour photographs. Hence I shall confine myself here to the cymbidiums and one or two other well-known kinds.

ARACHNIS Orchidaceae

A. flos-aeris 指甲兰花 (Zhijia Lanhua) *Fingernail Orchid,* 蜘蛛兰 (Zhizhu Lan) *Spider Orchid*

A. flos-aeris is a robust terrestrial orchid from south-east Asia, which is occasionally seen in cultivation in southern China. It was recorded in Guangzhou by Barrow (1804), while he was accompanying Lord Macartney's embassy to China in 1793–94, and was noted by other early visitors. The leaves are two-ranked and the branching stems may reach up to 2.5 m or more in height. It produces large panicles of fragrant, yellow-green flowers horizontally spotted or striped with maroon (fig. 7.5).

ARUNDINA Orchidaceae

A. chinensis 竹叶兰 (Zhuye Lan) *Bamboo-leaf Orchid*

Like the above, *A. chinensis* is a terrestrial orchid with upright stems bearing leaves in two ranks. It is apparently native to southern China and is cultivated there. It was noted in Hong Kong by Fortune (1849a) and greatly admired later in the century in Guangzhou by Westland (1894). The short-lived flowers, which have lavender-pink petals and a lip blotched with carmine, resemble small cattleyas. In general appearance it is similar to the better known *A. graminifolia* but is smaller in all its parts.

BLETILLA Orchidaceae

B. striata 白芨 (Baiji)

As used here the character *bai* means 'white' but, according to Hu (1971), the *ji* forming part of the officially recognised name of this species is a further case of *baizi*, the mistaken use of a character instead of another of the same sound. He says the character pronounced *ji* which was originally used was a different one to that used nowadays and meant 'chicken', the plant being called White Chicken, an allusion to the fleshy, white rhizomes. These are employed in medicine, being credited with strengthening weak lungs, ameliorating the effects of tuberculosis, and inducing a gain of

weight. According to Li (1959), they are also incorporated in a paste used in manufacturing porcelain.

B. striata is a deciduous species native to China, Korea, and Japan and is one of the most easily grown of all terrestrial orchids. However, in spite of its long history of medicinal use, according to Haw (1985c) it does not seem to have been much cultivated, as an ornamental or otherwise, until comparatively recent times. Supporting this opinion is the fact that, as far as I know, it appears only in paintings of relatively late date. Li (1956a, pl. 12), for instance, reproduces a photograph of a silk tapestry based on a painting by the emperor Qianlong, showing plants in bloom in a penjing arrangement. Li (1959) says that it is often planted for its flowers in Beijing, where it is known as *Zi Lan* (Purple Orchid) or simply as *Lanhua* (Orchid Flower). In the 18th century it was recorded in the south by Loureiro (1790), and later Fortune (1844) saw it being grown in the Fa Tee Gardens. The large, pleated leaves are decorative and the upright stem bears several carmine flowers (fig. 7.6). A pale pink and a white variant also occur, as do forms with variegated leaves.

CYMBIDIUM Orchidaceae

As mentioned above, in spite of their low-key appearance, the terrestrial cymbidiums were admired for their scented flowers and graceful foliage, and became involved in symbolism and metaphor. Like *Prunus mume* and bamboo, they became a favourite subject with the painters, often represented in black ink (fig. 7.7). The graceful foliage lends itself well to the long single brush-strokes of which the Chinese painters are so fond (Koehn, 1952).

The species most admired have always been *C. ensifolium* and *C. goeringii*, both of which flower in the spring. These are the orchids so frequently depicted in paintings and mentioned in literature. Translations of poems and prose about them are given by Lai (1977), who also reproduces a selection of paintings of these species.

Figure 7.6 *Bletilla striata*

Figure 7.7 (left) A wood-block print of *Cymbidium goeringii*, from the *Jieziyuan Huazhuan* of 1701

Figure 7.8 (below left) A cultivar of *Cymbidium ensifolium* at the Yungu Hotel, Huang Shan, Anhui

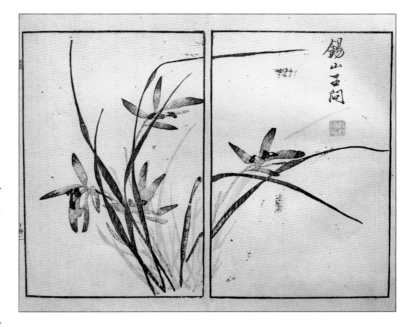

Also commonly grown in China are *C. floribundum*, which flowers a little later, and *C. kanran* and *C. sinense*, which do not come into bloom until the autumn. However, selection of unusual forms has taken place with the result that some clones in cultivation are now far removed from their wild ancestors. In addition to this, while hybridisation has not, at least until recently, been used in China to produce variants, naturally occurring hybrids have been brought into cultivation. These two factors have made the classification of some of the cultivated specimens very difficult (Du Puy & Cribb, 1988). As a result of all this there are cultivars with white or unusually coloured or shaped flowers and others with variegated leaves.

The large-flowered cymbidium cultivars so popular in the West have not, it seems, so far captured the imagination of the Chinese. However, one might expect that the modern hybrids with small flowers in dense racemes would appeal to Chinese taste, in spite of their lack of scent.

C. ensifolium 建兰 (Jian Lan) *Fujian Orchid*

C. ensifolium occurs widely in the Asian region. It is a terrestrial species with small pseudobulbs, which are often more or less subterranean, and produces racemes of 3–9 flowers in spring on the apical third of the scape. These are usually scented, up to 5 cm in diameter and greenish yellow, often with purplish stripes, and with a spotted lip (fig. 7.8). The petals are somewhat incurved but do not closely cover the column. There are many cultivars, the whites and those which lack red pigment being highly valued.

It seems likely that it was *C. ensifolium* which Huang Tingjian referred to as *hui* about 1090 AD. Needham (1986) quotes him as saying (concerning *lan* and *hui*) 'Planted in sandy soil they come up luxuriantly. If watered with tea they have a beautiful scent. In these characteristics they are the same. When a single stem produces a single flower, and the perfume is penetrating, the orchid is of the *lan* (type), when

a single stem produces 5 or 6 flowers and the perfume is not strong, it is of the *hui* (type)'. From this we can deduce that these names were used to distinguish the single-flowered *C. goeringii* from what was almost certainly *C. ensifolium*, a species also brought into cultivation early and which produces scapes bearing 3–9 flowers. Needham suggests that the *hui* was *C. floribundum*, but this seems improbable, as it is scentless and produces scapes with from 15–40 or more flowers. It also seems unlikely that the *hui* concerned was *C. faberi*, known today as *Hui Lan*. If *C. faberi* is cultivated at all in China, this must certainly be a rare occurrence.

Along with *C. goeringii*, *C. ensifolium* is frequently depicted in art from the Song dynasty on. Examples reproduced here are the charming album leaf by Ma Lin (fig. 7.1) and a section of the *Song Hundred Flowers* in the collection of the Palace Museum, Beijing (fig. 16.9).

C. floribundum 多花兰 (Duohua Lan)
Many-flowered Orchid

C. floribundum is native to southern China, including Taiwan. Unlike the other species listed here, it usually occurs on rocks, but has also been observed as an epiphyte on *Liquidambar formosana* (Dupuy & Cribb, 1988). It has very small pseudobulbs and blooms in late spring, producing arching racemes of up to 40 or more scentless flowers. These are usually rusty brown or green, with the sepals spreading and the petals incurved. However, there are many named variants in cultivation in Japan and China, where it has been grown for several centuries (Dupuy & Cribb, 1988).

C. goeringii 春兰 (Chun Lan) *Spring Orchid*

This species is widespread in the southern parts of the mainland and in Taiwan. It is also found in the Ryukus, Japan, Korea and, as a rarity, in north-west India (Dupuy & Cribb, 1988). This is the most famous and admired of all the Chinese orchids, much celebrated in poetry and painting (fig. 7.7). Its symbolic significance is summed up by the inscription *chunlan junzixing* (the spring orchid embodies the essence of the superior man), which forms half of the 'spring couplet' mentioned in the introduction to the section on conifers in Chapter 4.

   C. goeringii is a terrestrial species with small pseudobulbs, which are more or less subterranean, and is not at all showy, the strongly scented flowers usually being borne singly on scapes which are much shorter than the arching, grass-like leaves. Occasionally scapes with up to four flowers are seen. The flowers, 4–5 cm in diameter, are usually green or brownish green, with spreading sepals, forward-pointing petals and a lip blotched with red (fig. 7.9). It has been in cultivation in China for at least 1000 years and many variants have arisen. Those with variegated leaves or white or unusually shaped flowers are particularly prized. It is also cultivated in Japan.

C. kanran 寒兰 (Han Lan) *Cold Orchid*

C. kanran is another terrestrial species found in southern China, including Taiwan, and in the Ryukus, southern Japan, and South Korea. It has narrowly ovoid pseudobulbs, often conspicuous, and slender leaves. The tall scapes usually bear 5–12 flowers, widely spaced and held above the leaves

(fig. 7.10). The flowers are strongly scented, the sepals and petals usually being green, sometimes striped with red, with the petals incurved, almost forming a hood over the column. The sepals are narrow, tapering and widely flaring, giving the flowers a spidery appearance which distinguishes it from all other species. It has been grown for many centuries and is known in a wide range of colours in cultivation (Dupuy & Cribb, 1988). In China it commences flowering in autumn. Presumably its Chinese name refers to its blooming in autumn and winter.

C. sinense 墨兰 (Mo Lan) *Ink Orchid*

C. sinense gets its Chinese name on account of the dark colour of the flowers of most forms. The sepals are usually purplish brown with the petals slightly paler and incurved, often forming a hood over the column. The scapes are tall and erect, usually with 8–25 strongly scented flowers held above the leaves (fig. 7.11). It is a robust, terrestrial plant with small pseudobulbs and comparatively bold, broad foliage, and is found from north-east India and Burma, across northern Thailand and southern China to Taiwan and the Ryukus (Dupuy & Cribb, 1988). It is variable over its range, and in cultivation there are many forms. It commences flowering in autumn and, along with *C. kanran*, often appears in exhibits of orchids at this time.

PHAIUS Orchidaceae

P. tankervilliae 鹤顶兰 (Heding Lan) *Crane's Head Orchid*

P. tankervilliae is a terrestrial orchid with a wide distribution, being found in India, Sri Lanka, China, south-east Asia, and Australia. It has presumably been cultivated in China for several centuries, as it was introduced from there to England by Dr John Fothergill about 1778 (*Curtis's Botanical Magazine*, 1817, t. 1924). Loureiro (1790) also recorded it as being cultivated in China. It has ovoid pseudobulbs from which arise elliptic to lance-shaped leaves up to 1 m in length. The tall scapes bear racemes of large, downward-facing flowers, the petals and sepals are usually white outside and brown within, and the lip is blotched with carmine. The Chinese name alludes to the similar colouring of the crane's head.

   Fortune (1849a) remarked on this orchid in Hong Kong and Westland (1894), referring to it as *P. grandifolius*, later

Figure 7.9 A cultivar of *Cymbidium goeringii*

Figure 7.10 A cultivar of *Cymbidium kanran* at the Huifang Yuan, Beijing

Figure 7.11 A cultivar of *Cymbidium sinense* at the Huifang Yuan, Beijing

wrote: 'I have never seen terrestrial orchids grown so well as they do them in Canton, particularly *Phajus grandifolius* and *Arundina Chinensis*. *Phajus grandifolius* is generally grown in clumps in twelve-inch pots, and they produce more than a dozen spikes of flowers from three to four feet high'. Watling (1928) was similarly impressed, recording that it grew well in this climate, and that pots containing seven flowering spikes, each with 16 or more flowers, were common.

RENANTHERA Orchidaceae

R. coccinea 绯红火焰兰 (Feihonghuoyan Lan)
Crimson Flame Orchid

Along with *Arundina chinensis*, *Phaius tankervilliae*, and other subtropical orchids, *R. coccinea* (fig. 7.12) appealed to foreigners far more than the modest cymbidiums favoured by the Chinese. John Lindley (1828), for instance, let his pen run away with: 'The work of the missionary Loureiro, published in 1790, a drawing in the possession of the Horticultural Society of London, and the reports of travellers who have visited China, have been, up to the present time, the only evidence to Europeans of the existence of this truly magnificent plant, the beauty of whose blossoms surpasses everything known in the vegetable world'. Likewise Westland (1894), who was familiar with it in Guangzhou, said: '*Renanthera coccinea* is perhaps the most striking of Chinese epiphytes, where if grown on semi-deciduous trees or pollard-stumps, or ordinary blocks exposed to sun it will flower

Figure 7.12 *Renanthera coccinea*, from *Curtis's Botanical Magazine* 1830, t. 2997

luxuriantly. Its brilliant reddish-brown panicles measure from two to three feet in diameter and dangle around the tree from top to bottom in a blaze of cardinal bloom'.

These descriptions require little elaboration from me, other than to point out that it is native to south-east Asia, and that in warm climates it flowers most of the year.

MAGNOLIAS AND THEIR ALLIES

Magnolia denudata is one of the best known of all Chinese flowering trees and, along with the purple-flowered M. liliflora, has been cultivated in gardens there for over 1000 years. However, apart from these, and apart from one or two species of the closely related genus Michelia and occasional plants of Liriodendron chinense, very few members of the magnolia family have found their way into traditional Chinese gardens. This seems surprising when you consider that China is the country with the largest number of species.

In recent years a collection of some 80 species of Magnoliaceae has been established at the Kunming Botanic Garden and some previously uncultivated species have been planted as street trees in Kunming (Sun et al., 1996). Also a collection of about 130 species in 11 genera has been planted in the South China Botanical Garden, Guangzhou (South China Institute of Botany, 1997). Thus it seems probable that in the years to come more species than those listed below will appear in Chinese gardens and other plantings.

Figure 8.1 *Magnolia denudata*, the Yulan

129

LIRIODENDRON Magnoliaceae

Like *Magnolia*, *Liriodendron* is one of the interesting genera which have species in both eastern Asia and eastern North America. There are only two known species, *L. tulipifera*, the American Tulip Tree, and *L. chinense*, the Chinese Tulip Tree. Both produce lobed, truncate leaves and terminally-borne, green flowers.

L. chinense 鹅掌楸 (Ezhangqiu) *Goose Foot Catalpa*, 马褂木 (Maguamu) *Mandarin-jacket Tree*

CHINESE TULIP TREE

The history of the discovery of the Chinese equivalent of the American Tulip Tree has been summarised by Li and Wright (1954). It was first collected at Lu Shan in northern Jiangxi in 1873, and in 1886 Hemsley described it as var. *chinense* of *L. tulipifera*. In 1903 Sargent raised it to specific rank. It seems likely that this handsome tree has only recently found its way into traditional Chinese gardens. It is certainly not common but I saw a young tree in bloom in the Yi Pu, Suzhou (fig. 8.2). Ernest Wilson introduced it to cultivation in the West from Hubei in 1901 (Lauener, 1996).

Unless there is some Chinese breed of goose with unusual feet, the name *Ezhangqiu*, while evocative, does not conjure up as accurate a picture of the shape of the leaves of the

Figure 8.2
Liriodendron chinense in the Yi Pu, Suzhou

Chinese Tulip Tree as does *Maguamu*. The simplest way to describe them is probably just to say that they are slightly exaggerated versions of those of its American counterpart. The flowers, too, are very similar.

MAGNOLIA Magnoliaceae

As mentioned in the introduction to this chapter, although there are many species of *Magnolia* native to China, only *M. denudata* and *M. liliflora* are widely cultivated as garden plants. Perhaps because of their relatively inaccessible habitat or the conservatism of Chinese gardeners, the wonderful tree magnolias of western China do not seem to have found their way into traditional Chinese gardens. Until recently the only other indigenous species to have been cultivated appear to be *M. officinalis*, the bark of which is used in medicine, *M. campbellii*, *M. delavayi*, and *M. coco*. On the other hand the large-leaved evergreen *M. grandiflora* from North America is occasionally seen in late Qing gardens and has become popular as a street tree and in parks.

M. campbellii 滇藏木兰 (Dian Zang Mulan) *Yunnan and Tibet Magnolia*

M. campbellii is a variable species native to the Himalayan region. The Chinese name given above presumably refers to subsp. *mollicomata* which occurs in Tibet and western Yunnan. I have included it here because Sun et al. (1998) record a cultivated tree in Nan-Hua County, Yunnan, which they say is 600 years old.

M. coco 夜合花 (Yehehua) *Night-closing Flower*, 夜来香 (Yelaixiang) *Night-coming Fragrance*

A slow growing shrub, rarely more than 2 m tall, it has evergreen leaves to 18 cm long and small white flowers. Reputedly native to Java and southern China, it is a popular garden flower in the warmer regions on account of its fragrance. The flowers open for a short period during the day and often close early in the evening, hence the name *Yehehua*. In spite of this characteristic, it apparently is most strongly scented at night, as its alternative name indicates. Besides its ornamental value, the flowers are used for flavouring tea (Urban Services Department, 1971). It is one of the plants which was recorded from China by Loureiro (1790).

Figure 8.3 (right) *Magnolia denudata* in the wild at Huang Shan, Anhui

Figure 8.4 (above) Old trees of *Magnolia denudata* beside the Tang pagoda at the Baoguang Si, near Chengdu

M. delavayi 山玉兰 (Shan Yulan)
Mountain [= wild] Magnolia, 优昙花 (Youtanhua) *Youtan Flower*

M. delavayi is an evergreen species from Yunnan and southern Sichuan, in general appearance resembling a large-leaved *M. grandiflora*. The flowers, however, are unimpressive. Although up to 20 cm in diameter they are narrow-petalled, greenish-white or yellowish-white, shortlived, tend to open at night, and are only slightly fragrant. I have included it here as Sun et al. (1998) record trees cultivated in temples in Yunnan with ages up to 700–800 years. It cannot stand severe frost and presumably has only been grown in the south-west of the country close to its native habitat. The Magnolia Society has recently awarded Dr Sun, the Deputy Director of the Kunming Botanical Garden, a research grant to study the ecology and pollination of this species (Callaway, 1998).

The derivation of the name *Youtanhua* is not clear. However, *Tanhua* is the name given to the night-blooming cactus *Epiphyllum oxypetalum* and *You* means 'excellent', so the name possibly alludes to this plant, which is also called *Yuexiameiren* (Beauty Under Moonlight).

M. denudata 玉兰 (Yulan) *Jade Orchid*

YULAN

M. denudata (fig. 8.1)is native to east-central China and still occurs wild there (fig. 8.3). It is a favourite garden plant and has been cultivated at least since the Tang dynasty (Li, 1959). Old trees are still to be seen in some temple grounds and gardens (fig. 8.4), while younger specimens are found wherever one goes. Barrow (1804) noted plants for sale in the nurseries near Guangzhou in 1794. It was once popular as a pot plant and there is an interesting photograph of potted plants in bloom late last century in the Guild Garden at Chongqing in Mrs Archibald Little's *Intimate China* (1901).

Cibot (1778b, 2) gave an excellent description of the plant, even noting the threads by which the seeds hang from the fruits as they are shed. He stated that its cultivation had begun in the Tang dynasty and that it was brought from the south to Beijing during the Ming. He recorded that it was propagated by seeds and suckers, and that the florists raised it in boxes and brought it into flower in greenhouses for the New Year. He also described how in the gardens of Beijing and its environs it survived the winter only with the trunks and branches wrapped in thick bandages of woven straw enclosed in strong mats. This would not seem to have been necessary, as these days it appears to survive there without such protection. However, the winters may have been colder in Cibot's day.

With its creamy-white flowers borne on the bare branches at the end of winter, the yulan is regarded as an emblem of purity, feminine sweetness, and beauty. It is sometimes called *Yingchunhua* (Welcoming-spring Flower), on account of its early blooming (Williams, 1975. Plants bearing flowers with some purple colour towards the bases of the petals are seen in some Chinese gardens (fig. 8.5). Whether all these are the true species or hybrids with *M. liliflora* is hard to tell.

M. denudata has been a favourite subject with painters throughout the ages (Bartholemew, 1985a, b; Kao, 1988; She & Yuhas, 1973), appearing in some of the earliest surviving flower-and-bird paintings (fig. 2.7), and is frequently used as a decorative motif on porcelain and in embroidery. On account of its symbolism, as has been indicated in Chapter 2, it is frequently represented together with crabapple and tree peony as a rebus suggesting the auspicious phrase 'wealth and rank in the jade hall'.

Besides being valued for its beauty, the bark and buds are said to have medicinal properties and the flowers are used in making scent (Wang & Ma, 1995). Also Cibot (1778b, 2) noted that the pulp surrounding the seeds was dried, powdered, and taken in the manner of snuff for medicinal purposes. That the flowers were edible was noted by Zhou Lujing in his *Rucao Bian* (Monograph on Uncultivated Vegetables) of 1582 (Needham, 1986), and Cibot (1778b, 2) later recorded that in the south the flower buds pickled with vinegar and ginger were used as an accompaniment for rice. Li (1959) says that the petals are sometimes dipped in flour and fried in oil until crisp and that they have a slightly sweet and fragrant taste. The edibility of magnolia petals is unfortunately also discovered from time to time by birds and animals. In Australia, for instance, possums sometimes eat the flower buds in winter and various birds have been known to ruin the spring display.

M. grandiflora 荷花玉兰 (Hehua Yulan)
Lotus-flower Magnolia

BULL BAY, SOUTHERN MAGNOLIA

Robert Fortune (1850e), on visiting Mr Beale's garden in Shanghai, was of the opinion that it promised to be to that city what the well-known garden of his father had been in Macao. Fortune (1852) noted that *M. grandiflora* had been introduced there and that its fine green leaves and noble flowers were much admired by the Chinese. Perhaps it is due to this admiration that it now appears as a street tree and in parks, gardens, and temple grounds all over the Yangtze delta region (fig. 8.6). Whether Mr Beale's tree was the first to be planted in China I do

Figure 8.5 Magnolias in the garden of the Taihua Si in the Western Hills, Kunming

Figure 8.6 The entrance to the Tiantong Si near Ningbo, with *Magnolia grandiflora* on the right. Robert Fortune stayed in this temple and collected here and in the neighbourhood in 1850 and 1853.

not know, as Fortune (1847a) had also recorded it in the East India Company's garden in Guangzhou. Although there are now many mature specimens these do not appear to be of any great age, so it seems likely that it was not introduced until the 19th century.

M. liliflora 紫玉兰 (Zi Yulan) *Purple Magnolia)*, 木兰 Mulan *(Woody Orchid)*

This species, although cultivated for a long time in Chinese gardens, seems to have attracted less admiration than *M. denudata*, for which it is frequently used as a grafting stock (Li, 1959). Fortune (1849b) was surprised to find it forced into bloom and for sale in the flower shops of Shanghai in January, no doubt as an ornament for the New Year. It was introduced to Europe from Japan, where it is widely grown, but is believed to be a Chinese plant, probably native to the central part of the country. According to Johnstone (1955), it is doubtful whether it still exists there as a truly wild plant.

It appears in paintings from as early as the 13th century, for example in the *Song Hundred Flowers* in the collection of the Palace Museum, Beijing (fig. 16.9). Other paintings of *M. lilifora* from the Ming and Qing dynasties have been reproduced by Barnhart (1983, cat. no. 21), Weidner et al. (1988, cat. no. 11) and Wang and Barnhart (1990).

In cultivation it forms a spreading, much-branched shrub rather than a tree, and tends to send up suckers from the base. A proportion of the upright, narrow-petalled flowers begin to open before the leaves appear and others continue to do so afterwards, with the result that it has a long flowering period. The form most commonly seen in the West has very dark flowers and is sometimes sold with the misleading name of *M. 3 soulangiana* 'Nigra'. As it occurs in China the flowers are not usually quite so richly coloured (fig. 8.7). It was, of course, the introduction to Europe in the late 18th century of *M. denudata* and *M. liliflora* which led to the raising there of hybrids between them, the forms known as *M. 3 soulangiana*, which are so popular today.

133

M. officinalis 厚朴 (Houpu) Thick [bark], Plain [wood]

Although rarely seen in gardens and not particularly ornamental, this magnolia is cultivated for its bark which commands high prices as it is used for making a famous tonic preparation. This is why it was given the specific epithet *officinalis*, which means 'sold in shops', generally those of apothecaries. Large quantities of the bark are sent each year from central and western China to other parts of the country (Li, 1959). It is used principally for treating fevers and was one of the 'driving away heat' ingredients of the water ices mentioned by Bredon and Mitrophanov (1927) as being sold early this century during the 5th month in Beijing. The flower buds are also used in medicine (Forrest, 1995).

M. officinalis is a deciduous tree which bears very large leaves, up to 40 **3** 20 cm, and produces its flowers in summer. These are quite large too but, appearing after the leaves and being of an uninteresting off-white colour, they create little effect. As well as the type originally described, a form with notched leaves has been distinguished as var. *biloba*, but Callaway (1994) suggests that such a distinction is unnecessary as individuals with this type of leaf occur in various populations along with unlobed forms. Since the bark of both kinds is used as a herbal remedy, most of the wild population has been destroyed by over-harvesting (Forrest, 1995).

MICHELIA Magnoliaceae

The species which up to the present have been placed in the genus *Michelia* are tropical or subtropical and have been cultivated in southern China for longer than the magnolias of temperate origin (Li, 1959). Their scent is their great attraction and they are sometimes grown in greenhouses in the

north. However, it should be noted that, while the plants included here differ from the magnolias described above in producing their flowers laterally in the axils of the leaves, rather than terminally, modern botanical opinion holds that there is no valid reason for retaining *Michelia* as a separate genus.

M. alba 白兰 (Bailan) White Orchid

This large-leaved, evergreen tree, which grows to about 9 m in height, is widely planted in south-east Asia. It is similar to *M. champaca* but as the flowers are usually smaller and almost white (fig. 8.8), they are inconspicuous. However, they are strongly scented and for this reason the plant is cultivated in tropical and subtropical areas. Both Metcalf (1942) and Thrower (1988) record that women wore the flowers as a hair ornament.

M. champaca 黄兰 (Huanglan) Yellow Orchid

CHAMPACA

Widely cultivated in India and south-east Asia, this large-leaved evergreen tree is grown in China at least as far north as Hangzhou. In favourable climates it will form a tree up to 25 m tall. The small, narrow-petalled flowers (fig. 8.9) may be white, yellow, or almost orange, and have traditionally been used as temple offerings and hair decorations, and in garlands and corsages. Herklots (1936) says that the flowers were sold in Hong Kong by hawkers who called out 'Heung pak laan' (fragrant white orchid). Morris (1983) says that a tea perfumed with white orchid flowers is a specialty of Suzhou. Whether these references relate to *M. champaca*, *M. alba*, or both is unclear.

Figure 8.7 (opposite page) *Magnolia liliflora* in the garden bordering the South Lake, Hangzhou

Figure 8.8 (far left) *Michelia alba*

Figure 8.9 (left) *Michelia champaca* in a restaurant garden on the shore of the West Lake, Hangzhou

Figure 8.10 *Michelia figo* in the Zhuozheng Yuan, Suzhou

M. figo 含笑 (Hanxiao) *Smiling Face*

PORT-WINE MAGNOLIA, BANANA SHRUB

This evergreen, small-leaved shrub is native to southern China. Although hardier than *M. champaca*, it cannot be grown out of doors in the northern part of the country. It is widely cultivated in the warmer parts where it grows to about 4.5 m (fig. 19.11). The small yellowish flowers are variously tinged, streaked, and margined with purple. In the colour and marking of the flowers the selection usually grown in China differs somewhat from that introduced to Australia and perhaps other countries. The banana-like scent, which is the chief appeal of the plant, is very faint in the morning but intensifies as the day wears on. Fortune (1844, 1847a) noted it in the Fa Tee Gardens, where it perfumed the air in spring, and Herklots (1936) wrote that it was much cultivated in southern China and that the flowers were sold in Hong Kong for use as hair ornaments. It is recorded that the plant is propagated by air-layering and that the flower buds are sometimes used to perfume hair oil and tea (Urban Services Department, 1971).

The Chinese name of the plant suggests the loving smile of a modest girl (Williams, 1975), supposedly because the flowers do not open fully, suggesting the mouth when smiling. This is certainly the case under most circumstances but, as shown in figure 8.10, the flowers are capable of opening widely when they are exposed to sunlight.

Schafer (1967) says that *M. figo* does not seem to have been noticed by the Tang poets, but this is not surprising as it is a southern plant. However, it seems probable that it has been in cultivation for over 800 years, as according to Hargett (1988–89) it was amongst the plants grown in the Emperor Huizong's garden at Kaifeng before its destruction in 1127 AD. Also it is figured in the *Song Hundred Flowers* in the collection of the Palace Museum, Beijing (fig. 2.14).

Figure 8.11 *Michelia maudiae* in Taiziwan Park, Hangzhou

M. maudiae 深山含笑 (Shenshan Hanxiao)

Deep-in-the-mountains Michelia

A handsome small tree, its large, scented, pure white flowers, rival the appearance of those of the similar but better known *M. doltsopa* (fig. 8.11). It is native to Guangdong, Fujian, and southern Zhejiang (Herklots, 1936). Although only recently brought into cultivation in China, it is planted in Hong Kong (Thrower,1988), and I have seen it in flower in Taiziwan Park, Hangzhou, and in a courtyard of the Tiantong Si near Ningbo.

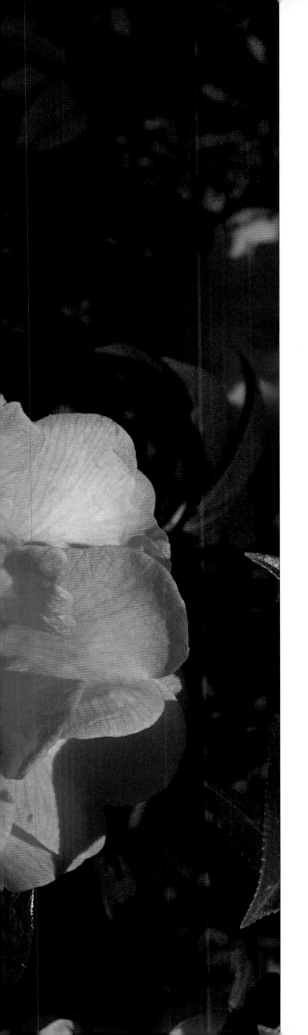

CHAPTER 9

CAMELLIAS

Having earlier declared my intention of dealing with some of the best-known Chinese plants more or less in the order in which they bloom, I should probably have placed this chapter earlier in the book. However, since so many plant groups have representatives which bloom in winter and early spring, I found myself unable to decide on a sequence which accurately represented the situation. The problem is that in much of China temperatures rise rapidly as winter ends. The spring is thus often a here-today-and-gone-tomorrow affair, with the flowering periods of many plants overlapping in a manner which does not occur in places where the temperatures rise more gradually.

Although many species of *Camellia* occur in China, comparatively few were brought into cultivation there until recently. The most important camellia of all, *C. sinensis*, the tea plant, has long been grown for its leaves. Other species have been exploited for their oil-bearing seeds and, of these, *C. oleifera* is the most important. As far as ornamentals are concerned, it is unusual to encounter in Chinese gardens anything other than cultivars of *C. japonica* and *C. reticulata*.

Figure 9.1 *Camellia reticulata* 'Liuye Yinhong'

Apart from the collections in modern botanical gardens, only a handful of the indigenous species are definitely known to be or to have been used as ornamental plants. To these must be added forms of *C. sasanqua*, which are said to be grown in Chinese gardens, presumably having been imported from Japan. However, according to T.J. Savige, an international authority, it is likely that there were or are in China ornamental cultivars of other species but that, owing to lack of horticultural interest and the destruction of records during the Cultural Revolution, it is difficult to obtain definite information. Savige considers it likely that forms of *C. chekiangoleosa*, a close relative of *C. japonica*, were used, and what appear to have been garden varieties of *C. oleifera* were introduced to Britain from China last century. As well as this it seems that various wild species are planted from time to time in gardens and courtyards in the areas where they occur. There are records for instance of *C. taliensis* and *C. pitardii* being used in this manner (Chang & Bartholemew, 1984; Pang et al., 1995). However, I have not included details about these species here.

In China the name for the ornamental camellias is *Chahua* (Tea Flower), or *Shancha* (Mountain Tea). Needham (1986) points out that this indicates that for centuries the relationship of these camellias to the tea plant was clearly recognised. There is little mention of them in the early Chinese horticultural literature, detailed studies not appearing until the Ming, although the existence of many cultivars at this time indicates that they must have been in cultivation much earlier. Also, Pang et al. (1995) mention that, according to historical records, camellias were cultivated at Huiji in the Wutong Garden of Fu Chai (who reigned 495–476 BC) in the state of Wu, a kingdom in the Yangtze delta region. They go on to say that camellia cultivation became very popular in southern China in the Tang dynasty, when trees were planted in the gardens of officials and in the courtyards of Buddhist temples. One might hazard a guess that these were probably forms of *C. japonica*. However, they also report that two camellia trees are shown in a Tang scroll painting concerning the history of the Zhongxing Kingdom in Nanzhao, a region which is now part of Yunnan, and these may have been forms of *C. reticulata*. Anyway, whatever they may have been, it seems that the appreciation and cultivation of camellias began as early as that of most of the other favourite garden plants. And flowering, as most of them do, in winter and early spring, they have been adopted as auspicious symbols for the Chinese New Year (fig. 2.12).

The earliest horticultural monograph on camellias, now lost, was probably that of Zhao Bi, who is known to have passed the imperial examinations in 1472. Then in about 1495 there came the *Yongchang Erfang Ji* (Notes on Two Ornamentals at Yongchang) by Zhang Zhichun, which has survived. This deals with 20 kinds of rhododendrons and 36 camellias (Needham, 1986). As far as I know, this has not been translated into any Western language, but it would be interesting to find out exactly what the author said, as Yongchang (now Baoshan) is in far western Yunnan, between the Mekong and the Salween, not far from Tengchong, the neighbourhood of which is believed to be the original habitat of *C. reticulata*. The next book about camellias is Feng Shike's *Dianzhong Chahua Ji* (Notes on the Camellias of Yunnan), thought to have been produced in the late 16th century, in which are enumerated 72 varieties, mostly of *C. reticulata*. The author was impressed by the size of the trees, by the waxy corollas varying widely in size and colour, by their value as cut flowers, and by the length of the blooming period (Needham, 1986).

The camellias of Fujian in the south-east came up for examination by an official who chose the pseudonym Pu Jingzi and finished his *Shancha Pu* (Treatise on Camellias) in 1719. He described 43 kinds, including some obtained from Japan, where their culture had also become fashionable. It seems likely that most if not all of these were cultivars of *C. japonica*. Over two centuries later came the *Diannan Chahua Xiaozhi* (Survey of the Camellias of Yunnan), published in 1930 by Fang Shumei, a traditional Chinese scholar who lived in Kunming (Bartholemew, 1982). Most of the material in this book is compiled from earlier literary sources together with some of his own observations. His list of named varieties includes those he knew to be growing in Yunnan as well as others from older references, some dating back to the Ming dynasty. Although it is not now possible to identify all of these, particularly since he did not distinguish between *C. japonica* and *C. reticulata*, at least 15 of the names have been identified as *C. reticulata* cultivars still grown in Yunnan. Several others are clearly forms of *C. japonica*, including 'Baiyupei', long known in the West as 'Alba Plena'. The book remains an important source of information about the period before Chinese camellias began to be studied with modern horticultural and botanical methods.

Not surprisingly, many foreign visitors remarked on the

Chinese camellias. Fortune (1847a) was greatly impressed by those he saw in Fujian, saying that the district around Fuzhou seemed to be 'the great Camellia garden of China, and in no other part of the country did I ever see these plants in such perfect health, or so beautifully cultivated'. But then he visited only a few parts of China and saw nothing of the inland provinces, where in Chongqing, for instance, Mrs Archibald Little (1901) saw 'Camellias of infinite variety' late last century. Curiously, none of the foreign visitors to Yunnan, even the famous plant-collectors, seems to have noticed its wealth of varieties, particularly the magnificent cultivars of *C. reticulata* which are grown so widely there.

According to Li (1959), the propagation of camellias in China depends to a great extent on grafting on seedling stocks. It is said that the tea plant is also sometimes used as a stock, but only for yellow-flowered forms. What is meant by this is unclear, as yellow-flowered species have only recently been brought into cultivation and I do not propose to deal with them here. Probably the 'yellow' forms cultivated by the Chinese were merely 'yellowish', in the manner of the tree and herbaceous peonies which have been described as 'yellow' in the Chinese literature for close on 1000 years. While visiting a nursery near Shanghai Fortune (1852), for instance, bought a 'yellow' camellia which had flowers of anemone form, the 'outer petals white, the inner primrose yellow'. This was a cultivar of *C. oleifera* now known as 'Jaune'. 'Brushfields Yellow', an Australian cultivar of *C. japonica*, is also of this type, and there are other modern cultivars which are white with a faint yellowish tinge.

Although camellias have probably been cultivated in China for over 1000 years, they did not begin to reach the West until the 18th century. The first to arrive was apparently the single red form of *C. japonica*, said to have been in cultivation in England before 1739 (Booth, 1830). However, according to Booth, the first double varieties were not received until 1792. Many more camellias arrived during the next 40 years, mostly forms of *Camellia japonica* and principally from Guangzhou. No attempt was made to use their Chinese names and they were given European ones. Also, having come from the tropics, they were grown in conservatories or greenhouses. They caused a great sensation and they reached the height of their popularity in the 1830s and 40s. Seedlings were raised and many new cultivars were introduced. Later in the century, and on into the 20th, introductions of *C. japonica* and *C. sasanqua* from Japan added greatly to the range of cultivars available in the West.

In the early part of this century interest in camellias was given a boost when George Forrest introduced *C. saluenensis* and wild forms of *C. reticulata* from Yunnan. The raising of interspecific hybrids in the 1930s, particularly those involving *C. saluenensis* and cultivars of *C. japonica* and *C. reticulata*, led to the appearance of many new and charming camellias, the qualities of which were not generally recognised until after World War II. At the same time there came the revelation that in Yunnan there were numerous spectacular garden varieties of *C. reticulata*, many of which were introduced subsequently. Then came a great renaissance in camellia growing and breeding all over the world. Further *Camellia* species became available to breeders and a flood of new hybrids began to appear. Camellia societies sprang up, conferences and shows proliferated, and thousands joined the ranks of enthusiasts. Then from about 1980 on it became possible for plant collectors to visit China once more and the Chinese themselves became interested in their wild species. Consequently, in recent years many species from the west and south of the country have been introduced to cultivation for the first time, and a renewed burst of hybridisation is under way. The number of cultivar names, including synonyms and errors, has now reached an astonishing total of more than 32,000 (Savige, 1993) and looks set to continue to increase considerably.

CAMELLIA Theaceae

C. edithae 东南山茶 (Dongnan Shancha)
South-eastern Camellia

C. edithae is related to *C. reticulata* and occurs in Jiangxi, Fujian, and Guangdong. I am indebted to T. J. Savige for drawing my attention to a cultivar of this species called 'Heimudan', the existence of which came to light at the international symposium on *C. chrysantha* held in China in 1994, when it was mentioned as the parent of a hybrid called 'Jiaoyang', the other parent being *C. japonica* 'Danzhi'. 'Heimudan' ('Black Peony') has dark red, semi-double or open peony-form flowers 8 cm in diameter. There is a colour photograph of it in Pang et al. (1995, p.89). It has been introduced to Australia by Bob Cherry of Kulnura, NSW, who has collected an astonishing array of plants from the milder parts of China during more than 20 expeditions.

C. grisjii 长瓣短柱茶 (Changban Duanzhu Cha)
Long-petalled Short-styled Camellia

This species, which has small, white, scented flowers, occurs in Fujian, Guangxi, Hubei, Hunan, and Jiangxi. Its cultivar 'Zhenzhu Cha' ('Pearl Camellia'), which has small, somewhat papery leaves and formal double white flowers about 4 cm in diameter (fig. 9.2), is occasionally seen in Chinese gardens. For instance there is a large plant near the entrance to the

Flower Nursery in Hangzhou. 'Zhenzhu Cha' is amongst the camellias introduced to Australia by Bob Cherry. Feng et al. (1986) record that several cultivars of *C. grisjii* are growing in the Kunming Botanical Garden.

C. × hiemalis—*see below under* C. sasanqua

C. japonica 山茶 (Shancha) *Mountain Tea*

Linnaeus gave this camellia the specific epithet *japonica*, as it was first described from Japan by Engelbert Kaempfer (1712), though it is also native to China. It is not clear whether camellias mentioned by the early writers in China were forms of *C. japonica*, but it seems likely that most of them were. It is certainly the most commonly cultivated species there and is still the only one found in the gardens of most parts of the country, growing both in the ground and in pots. In the north, however, one only sees plants in containers, which are moved indoors in winter.

Robert Fortune (1847a) saw *C. japonica* growing spontaneously on the island of Putuo Shan in 1844. He met with many specimens up to 9 m in height but noted that they were 'only the well-known single red'. Single reds, presumably forms of the wild species, are the kinds most commonly seen in gardens in China to-day (fig. 9.3). No two appear exactly alike and presumably they are raised from seed. No doubt such plants are the least expensive, whereas singles of other colours and cultivars with semi-double, anemone-form, or double

Figure 9.2 (left) *Camellia grisjii* 'Zhenzhu Cha' in the Flower Nursery, Hangzhou
Figure 9.3 (below) *Camellia japonica*, wild type, in Hangzhou

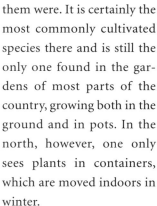

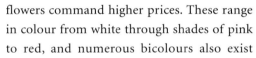

Figure 9.4 (right) A cultivar of *Camellia japonica* in Hangzhou

Figure 9.5 (above) *Camellia japonica*, a penjing specimen in the Yu Yuan, Shanghai

flowers command higher prices. These range in colour from white through shades of pink to red, and numerous bicolours also exist (figs 3.5, 9.4). Many of these appear to be the same as or very similar to those originally exported to Europe and elsewhere from Guangzhou. Which cultivars arose in China and which were brought there from Japan is not clear. If Pu Jingzi imported camellias from Japan before 1719, as mentioned above, others perhaps arrived earlier still. Likewise, Chinese cultivars may well have gone to Japan.

As already mentioned, it is not clear when *C. japonica* was first cultivated in China. However, since it appears to be native to the densely populated areas of the east of the country and has conspicuous red flowers, it seems unlikely to have been overlooked by early gardeners. The red flowers have been admired standing out against the snow, and its blossoming under these conditions was commented upon by the Song poet Lu You (Wang & Ma, 1995). It appears in paintings from at least the beginning of the 11th century, and later as a decorative motif on porcelain (e.g. Bartholemew, 1985a, cat. no. 54; Li, 1959, plates 16–18). While the early illustrations are usually of the single red type (e.g. Na et al.,1969, no.15), in the hanging scroll *Four Magpies* (She & Yuhas, 1973, no. 8) of the 11th century, a single white is shown. Thus it is clear that this variant was known at the time and may well have been cultivated. In due course *C. japonica* gave rise to a very large number of cultivars, which became popular throughout the country, even in the north where they were brought indoors in winter. In Yunnan, however, they have had to compete with the cultivars of *C. reticulata* and are seen less frequently there, being grown principally as stocks onto which their more spectacular cousins are grafted (Yu, 1950). Cultivars of *C. japonica* are also used for penjing (fig. 9.5), though when grown in this manner it is the shape of the plant which is of consequence rather than its flowers.

As noted earlier, camellias have become auspicious symbols for the New Year and for spring. In Guangzhou the flowers of *C. japonica* were favoured as an offering to the temple gods when making special requests for the New Year, although for some reason red and white forms were not often seen there at this season (Metcalf, 1942). Small plants in bloom are still offered for sale in southern China at this time of year. It seems doubtful, however, that camellias are now used in the manner described by Clarke Abel (1819), who wrote that they were 'cultivated in such profusion in Canton, that their petals are sometimes used at feasts to strew upon the table, so as perfectly to cover every part unoccupied by dishes'. Apparently, too, women never used the flowers to ornament their hair because the buds open long after their initial appearance on the plant, a characteristic suggesting a long wait for the birth of a son (Metcalf, 1942). According to

Figure 9.6 (left)
Camellia maliflora, from *Edwards's Botanical Register* 1821, t. 547

Figure 9.7 (far left)
A semi-double form of *Camellia oleifera*, from *Edwards's Botanical Register* 1827, t. 1091

Wang and Ma (1995), the wood is good for carving and the flowers are used in medicine.

In view of its long history of cultivation, it is not surprising that there are in China trees of *C. japonica* which are claimed to be of great age. Pang et al. (1995) list several, including a plant of 'Jiangxue' on Lao Shan near Qingdao, Shandong, said to be about 600 years old.

C. × *maliflora* 櫻花短柱茶 (Yinghua Duanzhu Cha)
Cherry-flowered Short-styled Camellia

This bushy shrub has small, double, pale pink flowers about 6 cm across (fig. 9.6). It is very much after the style of the modern miniature camellias and appears to have arisen in cultivation in China. Its parentage is unknown although it has been suggested that, on account of its abnormal ovaries and stamens and its sterility, it must be a hybrid (Chang & Bartholemew, 1984). It was introduced to Britain from Guangzhou in 1818 (Sealy, 1937) and was first described as *C. sasanqua* 'Palmer's Double' by Sims (1819). According to Savige (1993) it can no longer be found in China. Nevertheless I have included it just in case someone should come across it again. I have not seen it in any Chinese garden.

C. oleifera 油茶 (Youcha) *Oil Camellia*

This fragrant, white-flowered species is very similar to *C. sasanqua* and has often been confused with it. It was once widely grown south of the Yangtze River for its seeds from which an oil was pressed (Chang & Bartholemew, 1984). The first full account of it in Western literature is that by Sir George Staunton, who was in China in 1793–94 with Lord Macartney's embassy (Sealy, 1954). He identified it as *C. sasanqua* and said that the petals were sometimes used for scenting tea (Booth, 1830). It was later described as a distinct species by Clarke Abel (1819), who saw it under cultivation in China in 1816. As mentioned in the introduction to this chapter, it seems that cultivars of this species were also used as ornamental plants.

The flowers of *C. oleifera* are white and those of the usual type are said have 5–7 petals (Savige, 1993), though an introduction made in 1820 for the Horticultural Society of London is illustrated with 7–8 (*Edwards's Botanical Register* 1825, t. 942). Other forms brought into Britain early last century from Guangzhou as *C. sasanqua* have more and presumably were garden plants. The plant which became known as 'Lady Banks's Camellia' was sent by William Kerr in 1811

(Sealy, 1954) and is illustrated with 12–13 petals (*Edwards's Botanical Register* 1825, t. 12), while another, imported for the Horticultural Society in 1823 and described as '*Camellia sasanqua flore pleno*', is shown with as many as 18 (fig. 9.7). Booth (1830) was of the opinion that this was the same as the small, white, double camellia represented amongst the drawings in the collection of the Horticultural Society. Although I have not seen it, I assume the drawing to which he referred was the same one mentioned by Synge (1953), which was amongst those sent from Guangzhou by John Reeves early last century and is now in the Lindley Library of the Royal Horticultural Society.

The cultivar 'Jaune', mentioned in the introduction to this chapter, was obtained by Fortune in 1848 and sent to Standish and Noble of Bagshot, England, who put it out for sale (Savige, 1993). It is illustrated in colour by Ferguson (1997). Whether any cultivars of *C. oleifera* are still grown as ornamentals in China I do not know, but if they are it seems likely that this practice is rare.

C. reticulata 南山茶
(Nan Shancha) *Southern Camellia*

The flowers of cultivars of *C. reticulata* have about them something of the sumptuousness and mystique of the tree peonies, and our debt to the generations of Yunnanese gardeners who long ago developed these wonderful plants is considerable (figs 9.1, 9.8, 9.9). For me they still have a magic that has not been equalled by the much-praised hybrids derived from them.

The early books about the camellias of Yunnan have already been mentioned, and the history of their cultivation has been summarised by Yu (1950), Feng et al. (1986), Needham (1986) and

Pang et al. (1995). From the information available it seems probable that *C. reticulata* has been cultivated in Yunnan from at least the Sui or Tang dynasties, and both Yu (1950) and Feng et al. (1986) quote references to this species said to date from the 11th century on. Also it seems likely that the camellias of the south, which were described in 1175 by Fan Chengda as having flowers twice as large as those of the central provinces (Needham, 1986), belonged to this species. Even so, mention of the Yunnan camellias in the literature remained rare until the Ming, when closer contact was established between Yunnan and central China. From that time on the Yunnan camellias were praised in poems for their longevity, their evergreen habit, and the beauty of their flowers (Feng et al., 1986). It is clear that during the Ming there were already many cultivars with single and double flowers in shades of red and pink. It has been pointed out, too, by Pang et al. (1995) that in Yunnan there is a concentration of old camellia trees, principally cultivars of *C. reticulata*, along the ancient trade route from Burma, through Kunming and

Figure 9.8 *Camellia reticulata* 'Damanao' ('Large Cornelian')

north towards Chengdu. They list and reproduce photographs of huge specimens of various cultivars, some claimed to be as much as 650 years old. Feng et al. (1986) also list ancient specimens still growing in Yunnan. Wonderful examples can be seen, for instance, in the gardens of temples in and around Kunming and Dali.

Amongst the ancient trees the cultivar 'Shizitou' ('Lionhead'), a double red, is frequently encountered (Feng et al.,1986; Pang et al., 1995), but perhaps the most picturesque is the example of 'Wanduocha' ('Ten Thousand Flower Camellia') at the Yufeng lamasery near Lijiang. This tree is reputed to be 500 years old and to bear at least 4000 flowers each year between February and April. It is said that during the Cultural Revolution a monk risked his life by secretly keeping it watered (Buckley et al., 1994).

So popular is *C. reticulata* in Yunnan that it has been chosen as the floral emblem of the province, and it is the most important plant in the local horticultural trade (Yu, 1950). With its handsome green foliage and red flowers, it is regarded as a symbol of good fortune and wealth. In the old days plants of this camellia were included as parts of marriage dowries, and at New Year a vase of the red flowers was one of the offerings every family presented to its ancestors.

Because they come from so far inland, the cultivars of *C. reticulata* have been slow to reach the outside world. According to Andoh (1972), between 1673 and 1680 a semi-double variety was introduced to Japan, and a woodblock print of 1695 shows a form which he says is almost certainly the same as the cultivar 'Captain Rawes'. He also records that in the library of the Imperial Household there are drawings, of unknown date, of others. However, he suggests that the varieties growing in Japan at the time of his writing were probably introduced during the preceding 100 years, perhaps via Taiwan, which China had ceded to Japan in 1895.

The existence of these sumptuous ornamental plants became known to the West only because Captain Rawes brought an unnamed semi-double variety to England from Guangzhou in 1820. The same cultivar was introduced again in 1824 by John Parks and it first flowered in 1826. It was referred to by John Lindley (1827) as 'Captain Rawes's Camellia' and described as a new species, *C. reticulata*. The conspicuous veining of its leaves is what prompted Lindley to choose the specific epithet *reticulata* for this plant. It later became known as *C. reticulata* 'Captain Rawes', a cultivar which has not been identified with any of those at present

found in China. It must, however, have been grown there for well over 100 years before its introduction to England if, as suggested by Andoh (1972), it was taken to Japan in the late 17th century. It has been recently been reintroduced to China and given the name 'Guixia' ('Returning Rosy Clouds'). However, since no original Chinese name has been found for it, it should, according to the *International Code of Nomenclature for Cultivated Plants* (Trehane, 1995), continue to be called 'Captain Rawes'.

The reverse of this state of affairs applies to the second cultivar of *C. reticulata* to arrive in the West. This was a large red formal double, believed to have been introduced by Robert Fortune about 1850, which was named *Camellia reticulata flore pleno* (Lindley, 1857). A century later it emerged that it is still grown in Yunnan under its traditional name 'Songzilin' ('Pine Cone Scales'), which is what it should be called (Chang & Bartholemew, 1984).

While the wild form was introduced to cultivation from seeds collected between 1913 and 1925 by George Forrest in the extreme west of Yunnan, it seems extraordinary, as mentioned early in this chapter, that none of the famous plant collectors who visited Yunnan made any mention of the wealth of cultivars there. As a result, 'Captain Rawes' and 'Songzilin' remained the only ones known in the West until after World War II, even though Hu (1938) had earlier revealed that 'Over seventy varieties, all of great beauty, are cultivated in Yunnanfu (Kunming), the capital of the province'. He suggested that 'It would be worth while in the interests of horticulture to get all these varieties introduced into Europe'. The events of the years which followed, however, made this impossible. It is interesting to note, too, that Synge (1953) says that among the drawings of camellias sent from China by John Reeves to the Horticultural Society in the early part of last century there are several which may possibly represent varieties of *C. reticulata*, so perhaps more than two cultivars were known in south-eastern China at the time. If it has not been done already, it may well be worth examining these drawings in the light of present knowledge concerning the numerous cultivars of this species.

In China itself, as mentioned earlier, attention was drawn to the Yunnanese camellias in 1930 by the publication of Fang Shumei's *Diannan Chahua Xiaozhi* (Survey of the Camellias of Yunnan). Amongst the 72 kinds listed at least 15 of the names have been identified as cultivars of *C. reticulata* still grown in Yunnan (Bartholemew, 1982).

The first detailed horticultural investigation of these plants was made by Yu (1950). He visited gardens and temples in an attempt to find the varieties mentioned in the old literature, but found the task difficult as the old literary descriptions were inadequate for diagnostic purposes. However, he succeeded in identifying 18 of them and his descriptions and photographs greatly stimulated enthusiasm for these remarkable plants. He also described their cultivation and propagation, saying that they were very difficult to grow from cuttings and that Chinese gardeners inarched them onto seedlings of *C. reticulata*, *C. pitardii*, and *C. saluenensis*, or onto cutting-grown plants of *C. japonica*. Nowadays cleft grafting is used (Feng et al., 1986).

Quite remarkably, considering the turbulent times, 18 cultivars were imported from the Kunming Institute of Botany and established in California in 1948 and 1949. A smaller number were sent to Australia, and in due course most of these cultivars became widely distributed. The history of these introductions has been documented by Peer (1951). In more recent times other cultivars have been sent from Yunnan to various parts of the world. Also, in the past 50 years the number of cultivars has greatly increased, largely due to the efforts of enthusiasts at the Kunming Institute of Botany (fig. 9.9). In 1980, for instance, 105 were recorded (Yu & Bartholemew, 1980). Two years later, Bartholemew (1982) said that, of 106 cultivars of Yunnan origin, 39 were old cultivars and the remaining 67 were newly developed or had been collected from the forests near Tengchong, where double and semi-double forms have been found growing wild. Subsequently Feng et al. (1986) recorded 120 cultivars, and no doubt by now many more have been bred or selected.

Camellia rosiflora　玫瑰连蕊茶 (Meigui Lianrui Cha)
Rose-flowered Camellia with Joined Stigmas

This species has small, soft pink flowers only about 3.5 cm in diameter and is native to Hubei, Jiangsu, Sichuan, and Zhejiang (Chang & Bartholemew, 1984). Although there had been a plant of *C. rosiflora* of unknown origin at Kew for some time, when the grafted portion of a camellia brought from Guangzhou in 1822 by John Potts died, the stock grew up and was recognised as the same species (Hooker, 1858). Presumably it was cultivated in Guangzhou, at least as a stock, though it does not seem to be grown in Chinese gardens now. It has been suggested that it may have been a parent of *C. × maliflora*.

C. sasanqua　茶梅 (Chamei) *Mei Camellia*

C. sasanqua gets its Chinese name on account of its blooming at the same time as *Mei* (*Prunus mume*). As mentioned above, forms of this Japanese species are reported to be cultivated in China (Chang & Bartholemew, 1984; Feng et al., 1986; Li, 1959; Savige, 1993). Unlike most camellias, *C. sasanqua* is scented. The flowers are said by Williams (1975) to have been used to flavour tea, but he may well have been quoting Sir George Staunton, who had identified as *C. sasanqua* the plant which was later described as *C. oleifera*.

Figure 9.9　A recently introduced cultivar of *Camellia reticulata* at the Kunming Institute of Botany

Figure 9.10 (left) *Camellia × hiemalis* (?) in a planting beside the West Lake, Hangzhou

Figure 9.11 (above) *Camellia sinensis*, the tea plant

While no doubt cultivars of *C. sasanqua* were introduced along with other Japanese camellias, they do not seem to be common in Chinese gardens nowadays. I have not seen them in any traditional Chinese garden, but Feng et al. (1986) say there are several cultivars at the Kunming Botanical Garden.

The only camellia of this type which I have come across in China was in a planting on the shore of the West Lake at Hangzhou in 1996 (fig. 9.10). From its flowers it appeared to me to be very similar to if not identical with the cultivar 'Kanjiro', for which Savige (1993) records the Chinese synonym 'Kancilang'. To the untutored eye this appears be a form of *C. sasanqua*, but it is considered to be a cultivar of *C. × hiemalis*, believed to be of hybrid origin, probably derived from *C. sasanqua* and *C. japonica*. It is not known in the wild and was described and named in Japan in 1940 from a penjing plant said to have been taken there from Shanghai in 1925. However there is also evidence that it originated in Japan (Chang & Bartholemew, 1984). Thus it is not clear whether it was taken from Japan to China rather than the other way round. Whatever the truth of the matter may be, it is certainly not common in China nowadays, though it has been collected by others from Hangzhou.

The phytochemical evidence of Parks et al. (1981) indi-cates that many of the plants at present regarded as cultivars of *C. sasanqua* are derived from hybrids between *C. sasanqua* and *C. japonica* that have introgressed back to *C. sasanqua*. In addition, as compared with wild plants, they tend to bloom later in the autumn or early in the winter, exhibit some fusion of petal and filament bases, and have pink and red shades of floral pigmentation and relatively coarser branch and foliage texture. Chang and Bartholemew (1984) indicate that, unless reasons can be found to do otherwise, the name that should be applied to these hybrids is *C. × vernalis*, in which presumably *C. × hiemalis* should be included. But, because most gardeners see them all simply as 'sasanqua camellias', I have left things alone for the time being. After all there are many cultivars at present placed in *C. japonica* and *C. reticulata* which appear to be similarly genetically contaminated.

C. sinensis 茶树 (Chashu) *Tea Tree*

TEA

It is to the Chinese that we owe the custom of tea drinking and tea is undoubtedly the best known of all their plants (fig. 9.11). Hence, although it is not usually grown in gardens, its small white flowers being inconspicuous, I felt I must include it. No doubt it is because of its modest appearance that it is rarely represented in art, although it is depicted in the *Song Hundred Flowers* in the collection of the Palace Museum, Beijing.

There is much argument concerning the real home of the tea plant. It seems probable that it originally occurred in an area extending from Assam to Yunnan, and that the Chinese were the first to use and domesticate it, perhaps in Sichuan. It is believed to have been used as a medicine in the Shang dynasty and that its use as a beverage commenced in the Zhou (Chang & Batholemew, 1984). However, although it may have been known earlier under different names, the word *cha* was not in use before the Han, and growing tea for use as a beverage apparently did not commence until the 3rd century AD. It seems probable that it was not until the 6th or 7th centuries that tea drinking became widespread and common amongst people of all classes (Li, 1959; Needham, 1984), and in 760 a resident of Hubei, Lu Yu, wrote a treatise on tea (Haw, 1985a). Later we find the leaves listed as a famine food in the *Jiuhuang Bencao* (Bretschneider, 1881).

The popularity of tea has never been challenged in China. Every second person carries with them a small vacuum flask, or even a screw-top jar, from which they take a sip every so often. And in those parts of the country where it can be grown, tea plantations remain a picturesque and characteristic feature of the countryside (fig. 9.12).

The fine teas of the south-eastern coastal provinces became China's principal export to the West. When tea was first introduced to the West it came from Fujian where its name was pronounced 'tay', and this sound has given rise to its names in several European languages. And it was on the tea clippers involved in this trade that most of the popular Chinese garden plants were brought to Europe and other parts of the world.

A form of *C. sinensis* with pink flowers and reddish maroon foliage was found growing in Japan, where it has been given the name 'Benibana Cha' (Savige, 1993). I have been told by Mr Savige that this 'Red Tea' is an old Chinese garden plant, though it is doubtful whether it is still found in China today.

Figure 9.12
A small tea plantation outside the Huanglong Si, Hangzhou, viewed through a stand of *Phyllostachys edulis*

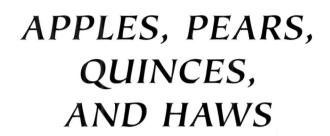

CHAPTER 10

APPLES, PEARS, QUINCES, AND HAWS

Since I have put the stone fruits in Chapter 5, it seems logical, even though they are placed in several genera, to put the pome fruits together here. These also belong to the rose family but to that section of the family which produces this type of fruit. Not only do many important fruit trees belong to this group but also some of the most beautiful Chinese garden plants. Most notable are the crabapples (*Malus* species) and the flowering quinces (*Chaenomeles* species), both of which have been cultivated in China for centuries. It was early recognised that these were closely related and they were placed together in a group of plants called *Haitang*. Li (1959) says that this name suggests that the Chinese may have believed that these plants came from abroad, as *tang* is an ancient name for a group of plants related to the pears and *hai*, meaning 'sea', is generally used for things imported from overseas. However, there can be little doubt that the plants placed in this group originated in China.

Figure 10.1 A red cultivar of *Chaenomeles speciosa*

That the *Haitang* were related to *Mei* (*Prunus mume*) was recognised as early as the 3rd century AD and about 790 they were described as 'the holy immortals among the flowers' by Jia Dan in his *Baihua Pu* (Treatise on the Hundred Flowers) (Needham, 1986). By the 11th century they had become favourite subjects for poets and painters (Li, 1959). The first monograph devoted to them appeared in the 11th century and another followed in the 13th. After this there do not appear to have been any further monographs, though *Haitang* continued to be described and discussed in the horticulural literature. The Chinese Quince (*Pseudocydonia sinensis*), the Chinese Hawthorn (*Crataegus pinnatifida*), and the Chinese Pears (*Pyrus* species and hybrids) also have a long history in China.

Ernest Wilson (1913) recorded that in western Hubei the leaves of several species of wild pear and apple were used as a source of tea by the poorer classes and that the infusion, called *hong cha* (red tea), was of a rich brown colour, very palatable and thirst-quenching.

CHAENOMELES Rosaceae

Blooming in winter and early spring, the two *Chaenomeles* species described below have become favourite garden plants wherever they can be grown. The characteristic arrangement of their short-stalked flowers on the bare twigs is immediately evocative of oriental art. In China they are grown both in the ground and as penjing. It seems that both species were first introduced to Europe and elsewhere from Japan, even though *C. speciosa* is native to China. Subsequently many hybrids have arisen.

As well as being widely grown in gardens in China, the flowering quinces frequently appear in paintings, at least from the 16th century on, though it is difficult to assign the depictions to a particular species (e.g. Capon & Pang, 1981, cat. no. 65; Barnhart, 1983, cat. no. 26; Weidner et al., 1988, cat. no. 39; Bartholemew 1985a, cat. no. 8d). Bartholemew (1985b) points out that a flowering quince sometimes replaces the crabapple in the grouping with tree peonies and white magnolia to suggest the rebus *yutang fugui* (wealth and rank in the jade hall) mentioned in Chapter 2. Bartholemew (1985a, b) also says that in Chinese-American households in San Francisco the flowering quince has become the symbolic flower for the Chinese New Year, as the peaches, so favoured in southern China, are not yet in bloom at that time.

C. japonica 日本木瓜 (Riben Mugua) *Japanese Quince*

JAPANESE FLOWERING QUINCE, JAPONICA

C. japonica is a thorny, low-growing, deciduous shrub with flowers usually in some shade of red (fig. 10.2). White and semi-double forms are also known. A native of Japan, it is now grown widely in China, where my own observations suggest that it is planted more commonly than *C. speciosa*. Meyer (1911) says a dwarf form grown in Hangzhou and Suzhou is, or was, known as *Luo Haitang* (Trailing Haitang). He also says that around Beijing *C. japonica* is known as *Bei Mugua* (Northern Quince) and that there is another variety called *Muli* (Woody Pear), which has larger fruits. He states that these aromatic fruits are used to perfume rooms, but mostly by the poorer classes, as they are very much cheaper than those of the Chinese Quince (*Pseudocydonia sinensis*), which is used for the same purpose. However, Tun (1965), writing of the 10th month in Beijing at the beginning of this century, said of the Japanese Quince that 'Its fragrance, in comparison with the quinces from the south, is even more penetrating'. So, if by 'the quinces of the south' he meant the fruits of *P. sinensis*, those of *C. japonica* would seem to have been the better buy.

Figure 10.2 *Chaenomeles japonica* in the Wang Yuan, Hangzhou

Figure 10.3 (far right) *Chaenomeles speciosa* in fruit

Figure 10.4 (above) *Crataegus pinnatifida* in bloom at Gongwangfu, Beijing

Figure 10.5 (right) Fruits of *Crataegus pinnatifida* on sale at Chengde

C. speciosa 皱皮木瓜 (Zhoupi Mugua)

Wrinkled-skinned Quince

CHINESE FLOWERING QUINCE, JAPONICA

This Chinese species has been in cultivation for centuries and, just as *C. japonica* was introduced to China, so *C. speciosa* reached Japan and became popular there. Accordingly, it also became known in the West as 'Japonica'. In late winter and early spring it produces showy flowers which may be single or double and white, pink, or red (fig. 10.1). It is sometimes called *Tiegeng Haitang* (Stuck-on-the-stem Crabapple), a name which is self-explanatory when one sees the almost sessile flowers and fruits (fig. 10.3). The fruits, like those of *C. japonica*, are fragrant and have no doubt been brought indoors on this account. According to Li (1959), in China it was mainly planted for its fruits which were candied. The fruits and roots are also said to possess medicinal properties (Wang & Ma, 1995).

CRATAEGUS Rosaceae

C. pinnatifida, which is grown commercially for its fruits, is the only species of *Crataegus* I have seen in traditional Chinese gardens, though apparently other species are also grown commercially. Wilson (1913), for instance, records passing through orchards of *C. hupehensis* in north-western Hubei, noting that it is one of several kinds cultivated in China.

C. pinnatifida 山楂 (Shanzha) *Mountain Hawthorn*

With its bright green, lobed leaves, corymbs of white flowers, and clusters of large scarlet fruits, *C. pinnatifida* makes a handsome ornamental plant (fig. 10.4). While it is planted only rarely in gardens and used only occasionally for penjing, in commercial orchards it is a conspicuous feature of the northern Chinese countryside, where it is grown for its fruits (fig. 10.5). Though these fruits are mentioned as a famine

Figure 10.6 *Malus baccata* at the Imperial Summer Villa, Chengde

food in the *Jiuhuang Bencao* (Bretschneider, 1881) and are sour and unexciting when eaten raw, they have become popular when cooked with sugar, made into jellies and drinks, or candied. Sellers of the toffee-coated or candied fruits threaded onto slivers of bamboo are still a common sight in and around Beijing. Varieties have been selected for the size of their fruits and these are propagated by grafting on seedling stocks (Meyer, 1911).

CYDONIA—see *Pseudocydonia*

MALUS Rosaceae

As with many garden plants, several species of this genus have been described from cultivated specimens. These plants had probably been in cultivation for hundreds of years before being named by Western botanists. They include double-flowered forms and forms with large or distinctively coloured flowers. In some cases the original wild types are not known at all, in others the wild type has subsequently been described as a variety of the cultivated form, and some appear to be hybrids. As with so many groups of Chinese garden plants, it is clear that much further taxonomic work remains to be done before the position is clarified. As a result, when one is confronted with these plants in Chinese gardens the task of assigning individual specimens to a species is not always easy, particularly when it comes to the

white flowered types, which vary considerably in the size and shape of their flowers. Hence it is my hope that readers will take this into account when reading what follows. Also it seems likely that several species and hybrids are grown to produce the crabapples sold on the streets, usually in candied form. In the old days in Beijing it was believed that when one ate them on a winter night they would disperse the effects of coal and charcoal fumes (Tun, 1965).

The flowering crabapples frequently appear in paintings, very often in combination with tree peonies and *Magnolia denudata*, suggesting the rebus *yutang fugui*, mentioned under *Chaenomeles* above. Also, amongst other things, apple blossom is regarded as an emblem of feminine beauty (Williams, 1975). The flowers depicted are usually five-petalled and are very variable in form, so that it is usually impossible to assign them to any species with certainty. They are also sometimes employed as a decorative motif on porcelain.

M. baccata 山荆子 (Shanjingzi)
Mountain Chaste Tree Crabapple

SIBERIAN CRABAPPLE

Native to north-eastern Asia, this species forms a shrub or small tree which bears clusters of long-stalked, white flowers (fig. 10.6), which are followed by fruits only about 1 cm in diameter. *M. baccata* has apparently hybridised with other species and, as mentioned above, there is thus considerable variation amongst the white-flowered crabapples seen in China. In the true species the calyx teeth are not retained on the fruit. In most other crabapples the calyx teeth remain. *M. baccata* is occasionally seen in Chinese gardens and is used as a stock on which other kinds are grafted (Li, 1959; Meyer, 1911).

M. halliana 垂丝海棠 (Chuisi Haitang)
Hanging-silk Crabapple

Perhaps the most handsome of all the Chinese crabapples, the Hanging-silk Crabapple gets its name because of its pendulous flowers borne on long, slender stalks. As seen in Chinese gardens it has semi-double, pink flowers opening from red buds (fig. 10. 7), and usually forms a much-branched shrub rather than a small tree. It is not known when it was first cultivated in China, but it can be recognised in paintings dating from at least as early as the beginning of the 17th century (e.g. Capon & Pang, 1981, cat. no. 65; Kao, 1988, cat. nos 57, 86; Weidner et al., 1988, cat. no. 34). Liu (1982) says that it is particularly appreciated in Suzhou gardens for its intertwined stems. It is also used for penjing (Hu, 1982).

Li (1959) records that it is native to western China and is especially abundant in Sichuan, where it has been a favourite subject with the poets. However, since the name *M. halliana* was given to a Japanese cultivated plant with semi-double to double flowers, its origin and affinities are unclear. It was introduced to New England from Japan by G. R. Hall in 1861 and has been given the cultivar name 'Parkmanii' (Fiala, 1994). What is claimed to be a wild type was collected in Kyushu by Ernest Wilson and sent in 1919 to the Arnold Arboretum. It is known as var. *spontanea*, although Fiala (1994) says it is probably a hybrid. Perhaps it may even emerge that *M. halliana* is a Japanese plant long cultivated in China.

The form introduced by Hall from Japan has often proved difficult to grow in Western gardens, but in Chinese gardens the form usually seen is vigorous, floriferous, and apparently undemanding. So also are forms introduced from China in 1978 by Donald Edge of the National Arboretum, Washington, DC (Fiala, 1994), and it is to be hoped that a vigorous cultivar of this charming crabapple will soon become generally available.

M. hupehensis 湖北海棠 (Hubei Haitang) *Hubei Crabapple*

This species is occasionally grown in China, principally for its leaves which are used as a tea substitute. It is also used in some places as a grafting stock (Li, 1959). It forms a larger tree than the other species included here, becoming 12 m or more tall under favourable conditions. It is perhaps the most handsome of all species of *Malus* but, although rivalling the

Figure 10.7 *Malus halliana* in Beihai Park, Beijing

best of the cherries when covered with its long-stalked white flowers (fig. 10.8), it is rarely grown in Chinese gardens. It was introduced to Britain by Wilson who sent seed from Hubei in 1900 (Lauener, 1996). As grown in the West it comes true from seed, even when grown near other species which flower at the same time, as the seeds are produced without the need for pollination.

M. × micromalus 西府海棠 (Xifu Haitang)
Western Palace Crabapple

Here I must admit to having become confused. Various writers, including Li (1959) and Needham (1986), have identified the Western Palace Crabapple as *M. floribunda*. Li says

that among the cultivated forms of this species there are semi-double and double varieties and that it is the most widely planted species in China. This is not the case now and I doubt that *M. floribunda* was ever widely grown, if at all. In China today the name Western Palace Crabapple is applied to a plant or plants with large white flowers borne on relatively short pedicels and identified, at least in some places, as *M. × micromalus* (fig. 10.9). This species is usually said to have large pink flowers which do not fade (den Boer, 1959; Fiala, 1994; Huxley et al., 1992). I have not seen anything like this in China but, as my travels have been far from exhaustive, this probably does not mean much. Since it is suggested that it is a hybrid between *M. baccata* and *M. spectabilis*, I presume that it is not impossible that there are both white- and pink-flowered seedlings about. Some of the single white crabapples seen in Chinese gardens might even be forms of *M. × robusta* (*M. baccata* × *M. prunifolia*) or perhaps even just variants of *M. spectabilis*, of which Li (1959) says there is a white variety.

M. prunifolia 楸子 (Qiuzi)

M. prunifolia has white flowers which open from pink buds. A north-east Asian species, it is grown principally for its fruits, which are treated in much the same manner as those of *Crataegus pinnatifida*. Meyer (1911) suggests that the small, sour, mostly red apples seen in northern China may

Figure 10.8 (left) *Malus hupehensis* in the Mei Yuan, Wuxi

Figure 10.9 (below) A white crabapple at the Beitang (North Church), Beijing

Figure 10.10 (right) *Malus spectabilis* at the Tanzhe Si, south-west of Beijing
Figure 10.11 (below right) *Malus spectabilis* at the Wofo Si, Beijing

be selections of *M. prunifolia* or have originated from its hybridisation with *M. baccata*. In fact Fiala (1994) suggests that *M. prunifolia* itself is probably a hybrid. A form described as *M. prunifolia* var. *rinki* has large, single, pink flowers and, according to Steward (1958) is cultivated in China as an ornamental. Professor Zhang Zhiming, who has provided the Chinese names used in this book, says it has small purple fruits and is called *Binzi*.

M. spectabilis 海棠花

(Haitanghua) *Flowering Crabapple*

This is the typical *Haitang*, cultivated for over 1000 years. It is not known in the wild and is probably a hybrid (Fiala, 1994). It is a robust plant and forms a small, upright tree growing to about 8 m. The flowers of the form usually encountered are dark pink in bud, fading as they open, up to 5 cm in diameter, single or semi-double, and borne on relatively short pedicels (figs 10.10, 10.11). The typical form is to be found in gardens in many parts of the country, particularly in and around Beijing. It is also some-times used for penjing (Hu, 1982), and the fruits and leaves are men-tioned as famine foods in the *Jiuhuang Bencao* (Bretschneider, 1881).

In the Shilin (Stone Forest), Yunnan, there are numerous trees of a double crabapple which appears to belong to this species but have more

Figure 10.10 (right) *Malus spectabilis* at the Tanzhe Si, south-west of Beijing

Figure 10.11 (below right) *Malus spectabilis* at the Wofo Si, Beijing

Figure 10.12 A richly coloured form of *Malus spectabilis* at the Shilin (Stone Forest), Yunnan

richly coloured flowers (fig. 10.12). What this really is I do not know, but if it has not been introduced to cultivation elsewhere it would certainly be worth doing so. As mentioned above, Li (1959) says there is also a white form. I have not encountered this, but some of the single white crabapples described above under *M.* × *micromalus* look as if they might equally well be placed in *M. spectabilis*.

M. spectabilis is believed to be the first flowering crabapple to have been cultivated and is probably the one mentioned by poets of the Tang dynasty. Also what is considered to be this species is often depicted in paintings and on porcelain, very often along with *Magnolia denudata* and tree peonies. In the painting reproduced as figure 2.7 the flowers are single, as they are in almost all depictions, although this may well be no more than a case of artistic preference. As happened with *Prunus mume*, it is likely that the elite regarded five-petalled flowers as being more tasteful and auspicious than the double ones.

The Manchu prince, Pu Ru, who lived at Gongwangfu in Beijing earlier this century, used to give a party there each year when the crabapples were in bloom. Judging from old

photographs these trees were the typical semi-double *M. spectabilis*. When Osbert Sitwell (1974) went to one of these parties in April 1934, he enjoyed looking down on the blossom from the tops of flights of stone steps. Dorothy Graham (1938) recorded that these trees were planted on a terrace at the northern end of the lake, and that they had been pruned so that they were multiple-trunked. When I first visited this garden in 1994 no sign of them remained, but on returning in 1996 I found that they had been replanted.

Malus cultivars 苹果 (Pingguo)

APPLE

The common fruiting apples are clearly of complex hybrid origin, so I have steered clear of any of the botanical names at present applied to them. Apples are grown widely in those parts of China where the climate suits them and, from time to time, the Chinese cultivars have been assigned to *M. pumila* (Steward, 1958) and *M. prunifolia* (Meyer, 1911). However, there seems as yet to be no reason to regard them as being different in origin to those cultivated in other parts of the world.

While apples are not usually grown in gardens, visitors to China will see the orchards and encounter the fruits in restaurants and shops. André Van Braam (1798), who visited the country in 1794–95 with an embassy from the Dutch East India Company, wrote: 'At *Pe-king* I never saw more than one sort of apple, which is of a very indifferent quality, mealy, of an insipid flavour and more calculated to please the eye than the taste'. Likewise Fortune (1863), while in Tianjin, said that the apples there were 'very beautiful to look upon but the taste is sweet, without flavour, and almost insipid'. The same goes for the Chinese apples one finds for sale these days. They are enormous and look absolutely wonderful (fig. 10.13, see over), but their flesh is soft and tasteless. It is not surprising then to find imported varieties such as 'Delicious', sitting beside them on the fruit stalls, albeit at a much higher price. However, there must be, or have been, indigenous kinds which appeal to Western taste, as Fortune (1857) saw orchards of an apple near Ningbo which had fruits 'about as big as a golden pippin, but excellent in flavour; indeed the only kind worth eating in this part of China'.

On account of the similar pronunciations of the characters *ping* (apple) and *ping* (peace), the gift of a few apples suggests the idea of 'peace be with you' (Williams, 1975). Likewise, an apple in a painting, as in figure 2.12, can carry with it the same sentiments. The apples in this painting and in the 17th century woodblock print reproduced by Li (1956a, pl. 16), appear to be much the same as those grown in northern China today.

PSEUDOCYDONIA   Rosaceae

This genus was established to accommodate the Chinese Quince, which at times has been placed in *Chaenomeles*, *Cydonia* and *Malus*. Plants placed in these genera have been recognised as closely related for at least 900 years. For example, Shen Li, writing in the 11th century, stated in his *Haitang Ji* (Treatise on Haitang) that when the quinces and apples come into flower it is clear that they belong to the same group (Needham, 1986). The Chinese Quince has been left in the genus *Cydonia* by Phipps et al. (1990) in their checklist of the subfamily Maloideae, and there seems to be little reason for disagreement with this decision. However, since it is still usually to be found under *Pseudocydonia* in recent horticultural works I have followed suit.

P. sinensis 木瓜 (Mugua) *Tree Melon*

CHINESE QUINCE

The Chinese Quince grows into a substantial tree, 10 m or more in height. It is deciduous or semi-evergreen, and the trunks of old specimens, the bark of which peels off in large flakes, are most handsome (fig. 10.14). The pale-pink flowers (fig. 10.15) are borne singly and are followed by strongly aromatic, yellow, egg-shaped fruits up to 20 cm or more long. No doubt it was the size of the fruits which led Geil (1911) to comment on 'tree melons' when writing about Chengdu. Meyer (1911) also noted that the fruit can be very large, sometimes even 25 cm long and weighing 4.5 kg. He recorded that it was never eaten by the people of northern

Figure 10.14 (right) *Pseudocydonia sinensis* in the Yi Yuan, Suzhou

Figure 10.15 (far right) *Pseudocydonia sinensis* in the Guyi Yuan, Jiading

China but used exclusively for perfuming rooms, whereas the foreign residents of the time cooked them. Apparently the fruits were also used for scenting clothing, as this is recorded by Tun (1965), who said that the fragrance lasted a month. Whether anybody still uses them for these purposes I do not know but, in relation to the perfuming of rooms, Shen Fu (1983), writing in Suzhou early last century, says that they should not be allowed to sweat. Apparently it was considered acceptable to eat them in times of famine, as they were listed in the *Jiuhuang Bencao* (Bretschneider, 1881). Like the crabapples, *P. sinensis* has been cultivated for a very long time and beautiful old trees can be found in many gardens. Also, it is occasionally depicted in paintings (Wang & Barnhart, 1990).

It is interesting that Cibot (1778b, 14), who usually showed considerable perception regarding some of the wilder claims of the Chinese horticulturalists, apparently believed them when they said they grafted quinces on oranges. He said that this gives them an oblong fruit the size of a small melon, of which the colour, the flesh, the seeds, the fragrance, the taste, and the juice combine the qualities of the two species. He called them 'Les Oranges-Coings' ('Orange-Quinces').

PYRUS Rosaceae

While several species of *Pyrus* are native to China, some of which have become popular ornamentals in other countries, it is doubtful whether, apart from the fruiting pears, any of these are at present grown in traditional Chinese gardens. However, *P. betulifolia*, *P. ussuriensis*, and *P. calleryana* are

Figure 10.13 Pears and apple bought in Beijing

used as rootstocks for the fruiting types (Cunningham, 1984; Huxley et al., 1992; Meyer, 1911, 1916).

P. betulifolia 杜梨 (Du Li) *Puckered Pear*

P. betulifolia is said to have got its Chinese name because the small fruits become puckered and edible after frost. The flowers, leaves and fruits were listed as famine foods in the *Jiuhuang Bencao* (Bretschneider, 1881). Bretschneider (1898) said that it was much cultivated in Beijing gardens. Perhaps it can still be found there, though I have not seen it. As mentioned above, it is one of the species used as a rootstock for the fruiting pears.

P. pyrifolia 沙梨 (Sha Li) *Sand Pear*

SAND PEAR

P. pyrifolia produces small hard fruits and gets its name on account of the gritty texture of its flesh. Apparently there are cultivated forms with larger fruits of more acceptable texture. It is believed to have been involved in the development of the Chinese pears described below. Whether examples of this species are amongst the pear trees grown in Chinese gardens I do not know, but as it is very handsome in bloom this may well be so.

Pyrus cultivars 梨 (Li)

CHINESE PEARS

While the European pears are usually considered to be derived from *P. communis*, the Chinese pears are somewhat different and are said to have been developed from hybrids between *P. pyrifolia* and *P. ussuriensis* (Huxley et al., 1992). *P. pyrifolia* occurs in central and western China, where the fruits are sometimes eaten, whereas *P. ussuriensis*, which is native to north-east China and eastern Siberia, has small unpalatable fruits.

Tun (1965), writing of Beijing at the beginning of this century, recorded the appearance there in the 9th month of 'Duckling Yellows', pears shaped like quinces and like yellow ducklings in colour. Most of the Chinese pears produce yellow fruits which are oval rather than the typical pear shape. These vary considerably in size, some being huge (fig. 10.13), but whether any of the kinds I have seen is the same as that described by Tun I do not know. Although these pears look marvellous, for those used to European pears most of them are a disappointment, like the Chinese apples. The flesh is

Figure 10.16 **Pear trees in bloom in the mountains north-east of Beijing**

crisp and has little flavour, as Lay (1846) noted of pears from Shandong which came on the market in Fuzhou in January 1845. Sitwell (1949) refers to them as being crisp and appreciated for scent rather than flavour. However, while the Chinese apparently prefer these hard pears, there are reported to be a few varieties with softer and more flavoursome flesh (Meyer, 1911). Perhaps the enormous yellow pear which Van Braam (1798) ate during his visit in 1794–95 was one of these, as he wrote: 'Before it is pared it seems hard, but when eaten, it is juicy melting, and of an agreeable taste'. And Fortune (1863) said that one of the pears encountered in Beijing was 'high-flavoured and melting' and 'a most welcome addition to the tables of the foreign residents'. This was the first pear of this type which he had come across in China and he was surprised that it was unknown in Tianjin, no great distance away.

The Chinese pears grow into handsome trees with glossy leaves. With their large white flowers and good autumn colour, many of the cultivars are first-rate ornamental plants, exceeding even their European counterparts in this respect, and are a delightful feature of the countryside (fig. 10.16).

Occasional trees are found in gardens as well as in the orchards, blossoming pear trees having been admired for hundreds of years. They have been the subject of paintings from at least the early Yuan dynasty on, and a splendid example is reproduced as Figure 2.1 in Chapter 2.

The pear tree is long lived, and as it has been known to bear fruit when as much as 300 years old, it has become one of the symbols of longevity. The Duke of Shao is recorded as having dispensed impartial justice under a pear tree in 1053 BC, so the pear also symbolises wise and benevolent administration. Since the characters for 'pear' and 'to leave' are both pronounced the same way, it is customary to avoid eating pears at weddings or before farewelling anybody. It seems, too, that the laxative effect of eating a large amount of pear flesh has been noted by the Chinese. And as a result of the founding of a college of music in a pear orchard by one of the Tang emperors, members of the theatrical profession are known as 'brethren of the pear orchard' (Williams, 1975). Williams also says that the flowers and bark are used in medicine and that the wood was formerly much used for making printing blocks and wood-cuts.

CHAPTER 11

LILACS

Although they cannot be said to have achieved in China anything like the popular acclaim accorded the other plants to which I have devoted separate chapters, lilacs (*Syringa* species) are such a conspicuous feature of Chinese gardens, particularly in the north of the country, that I felt it was justifiable to single them out. Visitors to Beijing in early to mid April will see and smell them at every turn, as will those who find themselves in Chengde two or three weeks later. No doubt it is the same in other parts of northern China. Even in the Yangtze delta region it is not unusual to find a lilac or two in gardens.

Lilacs are known collectively as *Dingxiang*. Their popularity, it would seem, flies in the face of the old superstition that a lilac should not be planted in a private residence because the character for nail is also pronounced *ding*, which can be construed as implying family strife (Williams, 1975). Perhaps this is why lilacs, compared with most other popular garden plants, are rarely mentioned in the literature and seem not to have attracted the attention of artists. The only paintings of plants I could find which appeared to be depictions of lilacs were

Figure 11.1 A lilac in the garden of the Imperial Summer Villa, Chengde

163

of kinds with very short panicles of flowers, possibly forms of *S. meyeri* or *S. pubescens*. An example from the 17th century is reproduced by Wang and Barnhart (1990), and another from the 18th by Capon and Pang (1981, cat. no. 65).

While a dozen or more species are native to China, only a few of these have been cultivated in traditional Chinese gardens. In my experience it is rare to see any lilacs other than *S. oblata* except in some of the botanical gardens, where collections have been assembled in recent times. There is a good range of species in the Beijing Botanical Garden, for instance, but I shall confine myself here to *S. oblata* and one or two others, even though it is not certain that all of these found their way into traditional plantings.

SYRINGA Oleaceae

S. laciniata 裂叶丁香 (Lieye Dingxiang) *Cut-leaf Lilac*

CUT-LEAF LILAC

S. laciniata was found growing wild in south-eastern Gansu by Meyer in 1915. I have seen plants in Chinese gardens which were not in bloom but had both entire and laciniate leaves and appeared to be this species.

As known in the West, *S. laciniata* has flowers of a rather washed-out mauve. However, in the gardens of the Imperial Summer Villa at Chengde I saw plants of this type with flowers of a much darker colour and which did not appear to be producing any laciniate leaves. These appeared to be close or identical to what is regarded elsewhere as *S. × persica*, a lilac said to be a hybrid of *S. laciniata* but the true affinities of which are somewhat obscure. The plants at Chengde have formed dense shrubs with narrow, entire leaves and many more-or-less upright branches arising near ground level.

Figure 11.2 *Syringa meyeri* **in Behai Park, Beijing**

Small axillary panicles of richly coloured flowers are produced prolifically along the upper parts of the previous season's growths, producing a display rivalling that of the best forms of *S. oblata* (fig. 11.1).

Fiala (1988) is of the opinion that *S. laciniata* became crossed with another species, probably *S. oblata*, and that the hybrid found its way to Persia long ago. After being cultivated there it was eventually given the name *S. × persica*. Whether the plant at Chengde is the same as this, another similar hybrid, or just a form of *S. laciniata* I do not know.

S. meyeri 蓝丁香 (Lan Dingxiang) *Blue Lilac*

The name *S. meyeri* was given to a cultivated plant bought along with the Meyer Lemon on 31 March 1908, at Fengtai on the outskirts of Beijing by Frank N. Meyer, who sent it back to the United States Department of Agriculture. Meyer noted that it was 'A small-leaved lilac bearing many panicles of purple flowers … Used much in forcing; quite rare and expensive' (Cunningham, 1984). He also recorded that one of the forms he sent was white (Fiala, 1988). In cultivation outside China it rarely grows to more than 1.5 m tall, forming a much branched, twiggy plant, which has small leaves and is somewhat pubescent. The flowers are profusely borne in axillary panicles which are usually no more than 8 cm long. The edges of the petals are incurved. Fiala (1988) records Joseph Hers, who collected it in China in 1920, as saying that it is strongly scented and rarely found on its own roots, mostly being grafted on privet. Presumably it was once a popular pot plant in the north of the country, although that is certainly not the case today. It has recently been found wild in Liaoning (Green & Chang, 1995).

The only plant I have seen in China which might be attributed to this species is in Behai Park in Beijing (fig. 11.2). Li (1959) says that *S. meyeri* blooms in late May and early June and again in August and September, which suggests an affinity with *S. pubescens* subsp. *microphylla*.

S. oblata 华北紫丁香 (Huabei Zi Dingxiang)
North China Purple Lilac

This is very common in gardens, temple grounds, and street plantings, particularly in and around Beijing. Bretschneider (1898) reported that, while it was much cultivated in that city, it had not been observed wild in the nearby mountains,

Figure 11.3 (above) *Syringa oblata* in the garden surrounding the Wofosi, Beijing, in autumn

Figure 11.4 (right) *Syringa oblata* at the Fayuan Si, Beijing

but apparently occurred spontaneously in what was then called Manchuria. It is sweetly scented and very similar in general appearance to the common lilac, *S. vulgaris*, but is a much better foliage plant, being little affected by leaf diseases. In autumn the leaves assume yellow and purple tints which can be very attractive (fig. 11.3).

It seems likely that this is the plant which John Barrow (1804) noted as being conspicuous around the shores of the West Lake at Hangzhou in 1794 and identified as '*Syringa Vulgaris*'. Presumably, too, it was *S. oblata* which moved Juliet Bredon (1931), who lived in Beijing early this century, to write: 'In April the lilacs reign supreme. Against a background of walls and buildings toned by the hand of time to mellow hues, the haze of purple and white blossoms is most effective—exquisitely fragrant too, and the happy hunting ground of Humming Bird Moths and Swallow Tailed Butterflies who seem to know instinctively how well their glossy black wings show up against the flowers'.

These days it is not easy to find a background 'toned by the hand of time to mellow hues' against which the lilacs can be viewed. However, one can still have this experience in the southern part of the old Chinese area of Beijing at the Fayuan Si, long renowned for its lilacs which are all this species (fig. 11.4). In mid-April the courtyards of this charming old temple are a wonderful sight and, of course, are filled with the scent as well.

In Chinese gardens most plants appear to be seedlings and are variable in colour, panicle size, and floriferousness. The best of them are splendid (fig. 11.5), absolutely covered with flowers and far better than any plants of this species which I have seen cultivated elsewhere. There is clearly an opportunity here for the introduction to other countries of selected forms. Even the plants with white flowers (figs 11.6, 19.39) show some variability, though I suspect that most of these are vegetatively propagated. An excellent white form was collected by Meyer in 1915 and sent back to the United States, where it is still growing well (Fiala, 1988).

Figure 11.5 (above) *Syringa oblata*, lavender form, in Zhongshan Park, Beijing

Figure 11.6 (left) *Syringa oblata*, white form, in the Daguan Yuan, Beijing

Green and Chang (1995) consider *S. oblata* to consist of two subspecies: subsp. *oblata*, from northern China, and subsp. *dilatata*, a morphologically distinct form found in the north-east of the country and in Korea. Most of the plants seen in gardens appear to belong to subsp. *oblata*, but since there is much variation some of them may be forms of subsp. *dilatata*.

S. pekinensis —*see* S. reticulata

S. × persica —*see* S. laciniata

S. pubescens 小叶丁香 (Xiaoye Dingxiang) *Small-leaf Lilac*

This lilac is at present considered to consist of the subspecies *pubescens*, *patula*, *julianae*, and *microphylla* (Green & Chang, 1995). According to Bretschneider (1898), what is now known as *S. pubescens* subsp. *pubescens* was first collected in the lower regions of the mountains near Beijing in 1831 by Bunge and Kirilov and described by Turczaninov in 1840. It was introduced to Europe from the Beijing mountains by Bretschneider in 1881 (Bean, 1980). It forms a shrub 2.0–2.5 m tall and has smaller leaves and flowers than *S. villosa*, which is found in the same region. Li (1959) records it as being cultivated in Chinese gardens.

Joseph Hers, who went to China in 1919 and again in 1922–23, said that what is now known as subsp. *microphylla* was much cultivated in Henan and that the flowers were used as a substitute for tea (Fiala, 1988). Li (1959) also records its cultivation in China. It has pinkish lavender flowers and gets its Chinese name, *Siji Dingxiang* (Four Seasons Lilac), from its habit of flowering intermittently throughout the summer and again in autumn.

Plants with off-white flowers which I saw in a street planting in Chengde (fig. 11.7) appeared to me to be a form of *S. pubescens*.

S. reticulata subsp. pekinensis 北京丁香 (Beijing Dingxiang) *Beijing Lilac*

This plant, formerly known as *S. pekinensis*, is another lilac which Bretschneider (1898) recorded in the Beijing mountains, where it had been discovered in the middle of the 18th century by d'Incarville. Unlike the other species described so far, it forms a small tree, sometimes growing to 7 m or more. It is native to the north of the country and produces cream, privet-scented flowers in terminal panicles in late spring or early summer. Until recently it does not seem to have been cultivated in China. However, there are trees of what appears to be this lilac in the garden of the Imperial Summer Villa at Chengde, so I have included it (fig. 11.8).

S. villosa 红丁香 (Hong Dingxiang) *Pink Lilac*

S. villosa is a large shrub or small tree, native to northern China, which produces terminal panicles of pinkish flowers in late spring on the current year's growth. It is easily distinguished from *S. oblata* by these characters and by its anthers projecting from the tube of the flower. I have included

it here because Li (1959) says that it is the most beautiful of the lilacs cultivated in China and is especially popular in gardens around Beijing. However he has probably mistaken *S. oblata* for this species, which does not seem to be grown except in botanical gardens.

S. villosa was first described by Vahl in 1805 from specimens collected by d'Incarville in the mountains near Beijing half a century earlier (Bretschneider, 1880). Bretschneider (1898) said that the local people called it '*ta ting hiang*' (= *da dingxiang*, big lilac). It was he who eventually introduced it to cultivation elsewhere, sending seeds some time between 1879 and 1882 to Kew, the Arnold Arboretum, Paris and St Petersburg (Lauener, 1996). Subsequently, in North America, it was used extensively in hybridising to produce a race of late-blooming lilacs.

Figure 11.7 (below) **A form of *Syringa pubescens* in Chengde**

Figure 11.8 (right) ***Syringa reticulata* subsp. pekinensis in the garden of the Imperial Summer Villa, Chengde**

CHAPTER 12

PEONIES

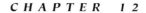

Of all the flowers I have seen, I doubt that any are
as extravagant and sumptuous as those of the tree
peonies. And while those of the Chinese herbaceous
kinds are smaller, they have a quality and refine-
ment which immediately claims the attention. As a
child I was greatly impressed by both kinds, though
many years were to pass before I learned something
of their fascinating history, a history which goes
back well into the 1st millennium BC. Amongst the
earliest Western writers to provide any details about
this are Cibot (1778b, 8; 1786c) and Hoffman
(1849). However, more extensive and accurate
surveys have been published subsequently by Li
(1959) and Needham (1986), whose works should
be consulted by anyone wishing to go into the topic
in greater detail than I have below.

The tree peonies (*Paeonia suffruticosa*), as almost
everyone must know, are shrubby plants which shed
their leaves in autumn, while the herbaceous kinds
(*P. lactiflora*) die back to ground level annually.
In China they are known as *Mudan* and *Shaoyao*,
respectively. While *Mudan* is usually translated as
'male red', the character *dan* also suggests medicinal
use, as in *danpi*, the name given to the root bark
of the plant. It is unclear what *Shaoyao*, a more

Figure 12.1 A tree peony in the Qiuxia Pu, Jiading

ancient name, originally meant, although the character *yao*, as used here, has long been interpreted as meaning 'medicine'. Anyway, whatever the true history of these names may be, there is no doubt that both plants came to be mentioned in the ancient literature on account of their medicinal use.

The name *Shaoyao* occurs earlier than *Mudan* but, although it is traditionally accepted that it applies to the herbaceous peony, it would be unwise to be dogmatic about it.

Figure 12.2 **A single herbaceous peony at the Tanzhe Si, south-west of Beijing**

Figure 12.4 **A double herbaceous peony in the Daguan Yuan, Beijing**

A famous song in the *Shi Jing* (Book of Odes), datable to the 8th or 7th century BC, tells how young men and women presented each other with *Shaoyao*, and a tradition developed that the words meant a 'binding herb'. This, of course, cannot be taken as an indication that the plant was cultivated at the time, though it seems certain that it was used for medicinal and perhaps magical purposes. The first known use of the name *Mudan* occurs in a text generally attributed to the 4th century BC, which says that it grew in southern Shaanxi and Shanxi and that the red ones were especially fine. Again, since these remarks are in a context of useful plants and minerals, the plants were probably being admired for their medicinal rather than their ornamental properties (Needham, 1986).

Both names also occur in the early pharmacopoeias, but it seems that there was some confusion over the species to which they were applied. According to Needham (1986), it was not until the Sui dynasty that the names they have borne ever since became firmly attached to the two groups. Then, when the peace and prosperity of the Tang period set in, the cultivation of ornamental plants began to flourish and gardeners started to collect together all the varieties that could be found and to develop new ones, the double kinds finding particular favour. Not only did people grow them but they began to write about them, so a remarkable amount of literature concerning their early development and cultivation has accumulated. Further reference to this is made below, where the two species are dealt with separately.

PAEONIA Paeoniaceae

P. lactiflora 芍药 (Shaoyao)

CHINESE HERBACEOUS PEONY

P. lactiflora is native to a large area of Mongolia, northern China, and eastern Siberia. On his travels north of Beijing in 1866 Armand David remarked on its abundance and its large pink or white, fragrant flowers (Fox, 1949). In China it has been known from antiquity onward as a medicinal plant, a parallel with *P. officinalis* in Europe, and something of its early history has been mentioned above.

P. lactiflora has been in cultivation for much longer than *P. suffruticosa*, but it is not clear when it began to be grown for purely ornamental purposes. However, in the course of the centuries hundreds of varieties arose, both singles and doubles in a range of colours, and during the Song dynasty the *Shaoyao* became almost as popular as the *Mudan*

(Li, 1959). However, it never quite reached the same pinnacle of esteem. In the time of the Empress Wu, about 690 AD, it was known merely as the 'dynastic ancestor of flowers', whereas the tree peony had become the 'king of flowers' (Needham, 1986). Similarly, Hoffman (1849) says that it was known as 'the king's minister'. Nevertheless, there is something about the informal shapes, the translucence, and the colours of the Chinese herbaceous peonies which sets them apart as aristocrats amongst the garden plants.

During the Song the place most renowned for herbaceous peonies was Yangzhou, which was called Guanglin in those days. Li (1959) quotes an old saying from the floral annals: '*Mu dan* in Luoyang, *shao yao* in Guanglin'. Situated where the Grand Canal crosses the Yangtze, Yangzhou was then the richest and most prosperous city in China, and the *Shaoyao* was a favourite garden plant amongst all classes. In the 12th century the garden of the Zhu family in that city is reported to have contained almost 60,000 plants and to have received streams of visitors in the season (Li, 1959).

Wang Guan's *Yangzhou Shaoyao Pu* (Treatise on the Herbaceous Peonies of Yangzhou) appeared in 1075 and several other monographs were produced a few years later, each describing over 30 kinds. It is also recorded that artists were commissioned to make paintings of the varieties. After the Song there was a long gap until towards the end of the Ming when further monographs appeared (Needham, 1986). Apparently Yangzhou was still famous for the plant at this time, as Li (1959) quotes a 16th century writer as saying that there were more than 30 varieties in cultivation there, both singles and doubles, but that only the roots of single-

Figure 12.3 An anemone-centred herbaceous peony in the Daguan Yuan, Beijing

flowered forms were used in medicine. There are also mentions in encyclopaedic works such as the *Qunfang Pu* of 1630, which records white, red, purple, and yellow varieties, and says that there are singles, doubles, and 'piled-up' doubles, examples of all of which are still grown today (figs 12.2–12.7). As with the tree peonies, the so-called 'yellows' were considered the rarest and most valuable, though it seems likely they were merely yellowish towards their centres or had pale yellow petaloid stamens. Both Li (1959) and Needham (1986) record that the old literature also included instructions for cultivation. Amongst other things it was recommended that the roots be dug, cleaned, and divided every so often and replanted in well-manured soil, but that this should not be done too frequently, a recommendation that still holds true today.

The Chinese methods of cultivation, propagation, and so on were also described by Cibot (1786c), the first European writer to give us a detailed account of the Chinese

Figure 12.5 (left) A silver-edged double herbaceous peony in the Daguan Yuan, Beijing

Figure 12.6 (below left) A pink 'piled up' double herbaceous peony in the Daguan Yuan, Beijing

Figure 12.7 (below) A red 'piled up' double herbaceous peony in the Daguan Yuan, Beijing

Figure 12.8 A bed of white herbaceous peonies photographed in the 1920s at Sengwangfu, on the outskirts of Beijing. Note the soil banked up around the bed in the traditional manner. From *Gardens of China* by Osvald Sirén (1949).

herbaceous peonies. He noted that they would grow in all soils but that in order to exhibit their full beauty they should be given rich, moist soil and a situation protected from strong winds. He described how the Chinese dug sunken beds, using the soil removed to build up borders around spaces in which the plants were grouped two by two, four by four, six by six, or in even greater numbers (fig. 12.8). While he seems to have recognised the skills of the Chinese cultivators of this plant, he was at pains to remark, for the edification of Europe, on details, refinements, and subtleties which, as he put it, astonish by their futility. For example, he drew attention to what he regarded as the idolatrous stupidity of insisting that regard must be paid to the phases of the moon when planting the offshoots, and noted the significance that Chinese gardeners gave to the choosing of auspicious days for carrying out other operations. Apparently, too, it was the custom to expose the mother root before digging and dividing it, marking each of the offshoots to be removed so that they could be replanted with exactly the same orientation in regard to the points of the compass. Otherwise, it was believed, the flow of sap would be interrupted and they would die. Also, offshoots to be sent away were placed in a piece of bamboo surrounded by the soil in which they had grown.

In spite of his views concerning some of these practices, Cibot (1786c), was greatly impressed by the beauty of the flowers and reported that there were more than 40 kinds, including singles, semi-doubles, and doubles of various colours, some with a strong disagreeable odour and others with

an exquisite fragrance superior to that of the most expensive perfumes. He went on to describe how the lateral buds were removed to produce the best flowers, and how some growers enclosed the plants in a kind of tent made of matting to protect them from dust, wind, and heavy rain. It appears that Yangzhou was still famous for herbaceous peonies in Cibot's day, as he described how plants already in bud were annually brought on barges from there to Beijing to decorate the emperor's apartments.

The *Shaoyao* has remained a favourite flower in China, where plants are found in most gardens and parks in the areas which suit it. And although it is not a success in regions with mild winters, roots are shipped south annually as is done with the tree peonies. Herbaceous peonies have also been grown in China for cut flowers (fig. 2.15), a use to which in many ways they are better suited than as garden ornaments. Cibot (1786c) mentioned that in Beijing the plants were not sold, only the flowers, and that burning the bases of the stems made them last longer in the vase. Fortune (1863) later noted that the flowers were much sought by the ladies of Tianjin and saw fields planted with them along the Grand Canal a little to the west of that city. And Tun (1965), writing some 30 or 40 years later of the 4th month in Beijing, said that herbaceous peonies were produced at Fengtai, a little to the south-west of the city, the place where a few years later Frank Meyer was able to buy plants of the 'Meyer' lemon and *Syringa meyeri*. Tun noted that these peonies were cut while still in bud and sold throughout the city, varieties such as 'Concubine Yang' and 'Muddle-headed White' being available everywhere. He also noted: 'Unlike the tree peony these are exceedingly regular so that even though heat be applied to warm them, they cannot change their time and open before their proper season. Among flowers they are indeed the ones that resist the demands of man'. It seems that the florists of Beijing, who had become so adept at bringing tree peonies and other plants into bloom for the Chinese New Year, were not able to impose their will on the herbaceous kinds.

Although they were not usually planted in the elaborate beds and terraces often constructed for the tree peonies, it is clear that they were considered to be of special value from the prominent positions in which they are depicted in paintings of gardens. For instance a double white cultivar is shown in its own special enclosure in a 16th century painting of The Garden of Solitary Enjoyment (Harrist, 1993, fig. 6). Herbaceous peonies are frequently depicted in Chinese art and those seen in paintings include double whites, pinks, and reds (e.g. Bartholemew, 1985a, cat. no. 8b; Weidner et al., 1988, cat. no. 9). However they have not been given the enormous symbolic significance accorded the tree peonies.

Those wishing to see herbaceous peonies in China will have no difficulty as there are plantings in many places, including almost all those mentioned below with regard to tree peonies. There is a large collection in the garden of the Imperial Summer Villa at Chengde, and in Yangzhou they are present in various gardens, including that at the Slender West Lake (fig. 12.9). The reds and rich cerises are especially popular, though soft pinks and whites are common, and flowers with a yellowish tinge towards their centres are also seen. In old temple gardens one sometimes sees pale pink singles which seem not to differ from the wild types (fig. 12.10). There are also cultivars with purple foliage (fig. 12.9) and, as Cibot noted, some varieties have a ravishing scent, while that of others is less entrancing.

The Chinese herbaceous peony was introduced to Japan in medieval times and has been popular there ever since. As they have done with the tree peonies, the Japanese have developed cultivars to suit their own taste, particularly the so-called 'anemone-centred' forms which have come to be called 'Japanese' types. The flowers are much admired in Japan and herbaceous peony gardens there attract large crowds in late spring.

The wild form of *P. lactiflora* was introduced to the West from eastern Siberia in the late 18th century (*Edwards's Botanical Register* 1815, t. 42), but it seems that the Chinese garden forms did not begin arriving until early in the 19th (Sabine, 1817), although there is some evidence that one or two may have been imported earlier (Gorer, 1970). The cultivated kinds were presumably obtained from Guangzhou, where the plants were annually brought from further north. Their popularity as garden plants in Europe was enhanced by the enthusiasm of the French breeders, who were offering numerous named varieties from the mid-19th century on. They were followed by the Kelways in England, who by 1884 were offering 250 varieties, Krelage and Zoon in Holland, who listed 500 varieties in 1892, and later by breeders in North America (Wister, 1962). Eventually the cultivars derived from *P. lactiflora* were hybridised with other herbaceous species and further variation has been introduced. However, it is the old Chinese types which retain my affection.

I suspect that very few Chinese herbaceous peonies have reached the West since the early part of last century. The only recent importation I know of is of a collection of over 50 named cultivars sent to the University of British Columbia by the Nanjing Botanical Garden. Hence it may well be that a survey of Chinese gardens and nurseries would still prove rewarding.

P. suffruticosa 牡丹 (Mudan) *Male Red*

TREE PEONY, MOUTAN

Probably no flower in China has ever surpassed the tree peony in winning and holding public esteem. It is looked upon as the flower of the yang principle, that of brightness and masculinity, qualities suggested by its name, *Mudan*. It is regarded as the 'king of flowers', a symbol of honour, riches, and aristocracy, as well as an emblem of love, affection, and feminine beauty, sometimes representing spring amongst the flowers of the four seasons (Bickford, 1996; Koehn, 1952; Williams, 1975). It is one of most frequently depicted subjects in Chinese painting and is used as a decorative motif on porcelain and in many other ways.

As a symbol of riches and honour the tree peony is sometimes called just that, *Fugui*, a name that is used in many puns. For instance a picture of crabapple, *Haitang*, and

Figure 12.9 (left) A bed of herbaceous peonies at the Slender West Lake, Yangzhou, edged with *Ophiopogon japonicus* and *Oxalis articulata*. Note the purple-leaved cultivar.

Figure 12.10 (above) A single herbaceous peony at the Baiyunguan, Beijing

peony suggests *mantang fugui*, literally 'full main-room riches and honour', meaning 'riches and honour to the whole family'. The addition of *Magnolia denudata* to this combination, as depicted in figure 2.7, suggests the phrase *yutang fugui*, which means 'wealth and rank in the jade hall', as pointed out earlier.

While there there seems so far to have been no argument about the origin of the Chinese herbaceous peonies, from exactly what wild type or types the cultivated tree peonies have been derived is unclear, particularly since *P. suffruticosa* is a name based on a cultivated plant, a double form sent to England. Whether existing wild peonies of this type should all be regarded as forms of *P. suffruticosa* or as several separate species, as is done by Chinese botanists (Osti, 1994; Page, 1997), remains to be seen.

As mentioned above, it appears that the early use of the tree peony was purely medicinal. The demand for the root bark ensured that it became known and grown in many Chinese provinces at an early date (Haw, 1985b). Its use in medicine persists to this day and Osti (1994) reports seeing thousands of hectares of plants in Anhui cultivated mainly for pharmaceutical purposes. He says that a variety called 'Fendianbai'('Phoenix White') (fig. 12.11), thought to be a cultivar of what the Chinese regard as *P. ostii*, is the one that is principally used as it is very vigorous. He also notes that the collection of roots for medicinal purposes over thousands of years threatens to eliminate tree peonies from the wild.

The time at which the tree peony was first cultivated as an ornamental plant is unclear, as references to it before the Tang are rare. It seems that people began to transplant it to gardens and foster new varieties in the 4th century AD in Zhejiang. Famous paintings of tree peonies were made in the 6th century and by the end of the 7th enthusiasm for the plant had spread to the Tang capital, Changan (present-day Xian), which became a great centre for its culture. In the 10th and 11th centuries it was supplanted in the east by Luoyang in Henan and in the west by Tianpeng in Sichuan. Later other places became important, such as Chenzhou in Henan, Caozhou (now Heze) in Shandong, Bozhou in Anhui, and centres near Shanghai (Needham, 1986). Heze appears to be the main centre now for the production of plants but undoubtedly there are others.

Stephen Haw (1985b), who visited Heze in 1982, reports that, as well as being grown for their flowers and roots, the tree peonies of Heze are used to make another saleable product, mudan wine. This sweet and fragrant liqueur distilled

Figure 12.11 (above) Tree peony 'Phoenix White' in the Beijing Botanical Garden

Figure 12.12 (left) Tree peony 'Yao Huang' in Jingshan Park, Beijing

from the fermented petals is produced on a commercial scale as a local specialty.

The first details of the development of the garden varieties and the extraordinary enthusiasm of the people for them come from the Tang, a period of peace and prosperity which was propitious for gardeners. A publication of about 1085 records the tradition that the Empress Wu banished the tree peonies from Changan to Luoyang somewhere around 690 AD because they were late in coming out (Needham, 1986). It has also been suggested that the Empress Wu's actions are responsible for the precocious blooming of a particular peony variety to this day (Morris, 1983). Presumably this is

the same Empress Wu whose encounter with the flower fairies features in the 19th century novel *Flowers in the Mirror*, mentioned in Chapter 2. Anyway her banishment of the tree peonies, if it took place at all, does not seem to have been long-lasting, as by the 8th century they were being grown by all classes in the capital and there was a plantation of tens of thousands in the emperor's garden at nearby Li Shan. Needham (1986) quotes an author who, about 860, described how the nobility and gentry had been making excursions for the past 30 years to admire the flowers each spring.

It was during this period that double varieties were mentioned for the first time, and the individual horticulturalists who produced them were immortalised in the literature. A situation arose not unlike the tulipomania of 17th century Holland, enormous prices being asked and obtained for the newest and finest varieties. When they were in bloom their owners screened them with wattle fences and protected them from the sun with awnings.

By the 10th and 11th centuries Luoyang had become the chief centre for their culture, and the double varieties, often described as 'hundred-petalled' or 'thousand-petalled', were the most admired and the most expensive. It seems that in general the best varieties were those of the Wei and Yao families, 'Wei Hua' ('Wei Family Flower') and 'Yao Huang' ('Yao Family Yellow') (fig. 12.12) being particularly famous. We can read about many such varieties in the *Qunfang Pu* (Assembly of Perfumes) of 1630, which lists them in their order of introduction. 'Yao Huang' was the culmination of a series each given the family name of a gardener or patron, which ran through the Tang, Five Dynasties, and Song periods. Mr Yao, for instance, was a horticulturalist, a man of the people, while Mr Wei was a Prime Minister around the middle of the 10th century. 'Wei Hua' was a double pale pink said to have more than 700 petals (Needham,1986).

During the following eight centuries no less than 20 important works on tree peonies appeared, together with a wealth of short essays, tracts, memoranda, and poems (Needham, 1986), and they became a favourite subject with painters, embroiderers, and decorators of porcelain. The apogee of enthusiasm for them appears to have occurred in the 10th century. In the *Xuanhe Huapu*, a catalogue of paintings in the imperial collection compiled around 1120, the name *Mudan* occurs in the titles of three times as many works as does *Mei* (*Prunus mume*).

The compilers of the *Xuanhe Huapu* commented that in depicting both tree and herbaceous peonies, often along with the phoenix or peacock, the painter brought out rich and aristocratic qualities. The peony, with its large, brilliantly coloured blooms and deep green foliage, was associated with voluptuous women, robust sensuality, opulence, and aristocratic splendour. In contrast, depictions of pine, bamboo, plum, and chrysanthemum suggested simplicity and seclusion. In this regard it is interesting to note that in works painted towards the end of the 11th century *Mei* overtakes *Mudan*, indicating a shift in taste. There seems to have been some rejection of aristocratic opulence in the later Song (Bickford, 1996). Even so the immediate appeal of tree peonies has ensured that they have remained favourite garden plants.

The first monograph about them was by a monk, Zhong Xiu, who in 986 produced his *Yuezhong Mudanhua Pin* (Grades of the Tree Peonies of Yue). He described 32 of the most beautiful varieties in southern Zhejiang, the area where wild varieties had first been recorded. What was probably the next treatise, describing 52 varieties, has been lost, but this was followed by the outstanding book concerning them, Ouyang Xiu's *Luoyang Mudan Ji* (Account of the Tree Peonies of Luoyang) written in 1034 (Li, 1959; Needham, 1986). Ouyang Xiu described how in the flowering season homes and temples were decorated with them, and how in old temples and deserted mansions people put up curtains and screens on which to display the flowers. He also said that

flowers of the finest varieties, 'Yao Huang' and 'Wei Hua', were sent from Luoyang to the emperor at Kaifeng, packed in bamboo cages filled with cabbage leaves so that they would not shake, the stems sealed with wax to keep them fresh. As well as this he listed 24 varieties grown in the city in his time and described methods of cultivation, pruning, watering, grafting, seed raising, and control of pests.

Ouyang Xiu recorded how in early spring people brought cuttings (presumably rooted pieces) collected from plants in the hills for sale in Luoyang, where the purchasers planted them in their gardens for grafting, which had to be carried out, it was believed, between about 23 September and 23 October. He said that to make the graft the stock should be cut off about 13–18 cm above ground, and that the joint should be sealed and wrapped with mud, with the whole thing surrounded with loose soil and protected with a hood of rush leaves open to the south for ventilation. It appears that professional grafters were employed to carry out the procedure, and that they were paid the following spring when the success of their efforts could be seen. Single scions of sought-after varieties brought huge prices when they first appeared, the price falling as their cultivation spread. Even so the people of Luoyang were deeply attached to their peonies and were loath to allow the spread of new or rare cultivars such as 'Yao Huang'. Ouyang Xiu records that, if privileged patricians or high officials came looking for this plant, they were likely to be given a graft which had been killed by dipping it in hot water.

New varieties continued to be produced, and in his *Luoyang Huamu Ji* (Records of the Flowers and Trees of Luoyang) of 1082 Zhou Shihou listed as many as 109 different cultivars, of which 59 were double and 50 semi-double. The Chinese preference for very large doubles had obviously already had a marked effect on the selection of forms for cultivation (Haw, 1985b). By the early 17th century at least 266 kinds were known, and a work of 1809 described in detail the cultivation methods for 103 of the best varieties (Needham, 1986). A stroll around tree peony plantings today shows that many of the old cultivars still exist and seedlings are still being raised (Haw, 1985b).

While some varieties may have arisen as mutations, it seems likely that most of the old cultivars were raised from seed, though there is no evidence that cross-pollination was carried out. It was said that seeds should be collected about the 6th month when the pods begin to split. Then they should be stored in moist soil after drying in the air for a few days, and finally planted in the autumn, after being tested in water, the empty ones being discarded. It was thought that, if the seeds were collected after the pods were fully dried, few would germinate (Li, 1959). This may well have been so, as special conditions are required to break their dormancy fully. If kept moist while temperatures are still high enough, the seeds start producing roots. However, they do not subsequently produce shoots until a cold period has occurred (Wister, 1962). My own experience confirms that if seeds are harvested late or are kept dry and then planted in autumn, winter, or spring, they will not produce roots until conditions become warmer and then do not send up shoots until another winter has passed.

As well as producing new varieties by raising seeds or by the propagation of mutations, it seems to have been believed, as mentioned in Chapter 1, that variants could be produced by grafting onto such unlikely stocks as parsnips, radishes, ashes, and various nuts (Cibot, 1778b, 8). Li (1959) states that many new varieties were developed by grafting on the herbaceous peony, another improbable state of affairs. Perhaps he merely meant that new varieties were increased in this manner. While this may well have been so, it is unlikely that a claim of the successful grafting of a tree peony onto a plant of *Toona sinensis*, so that its blossoms could be admired from an upstairs window (Li, 1959), has any basis in fact.

Another curious claim was made by Duan Chengshi, writing about 860 in his *Youyang Zazu* (Miscellany of Youyang). He described a bush at a monastery which early in the 9th century used to put forth 1200 blossoms in various shades of red, purple, yellow, and white (Needham, 1986). This sounds fanciful but, since the Chinese have long been masters of grafting several varieties of a plant onto the one rootstock, perhaps it was true; or perhaps what the writer took for a single plant was a clump of several growing together.

It was also asserted that the colours could be changed by treating the soil with chemicals and plant extracts (Cibot, 1778b, 8; Li, 1959). The methods were usually kept secret by their originators, but one author records that watering a white peony with an extract of *Lithospermum officinale* would make the flowers purple, by using *Crocus sativus* a red colour would be achieved, and that with powdered root of *Atractylis ovata* the petals would become striped with a yellow band whatever the original colour of the flower. Good-

ness knows what truth there is in any of this.

There can be no doubt, however, that the Chinese were competent propagators of tree peony plants. Cibot (1778b, 8) described the Chinese methods of seed raising, division, and grafting. His descriptions of the rigmarole involved in grafting stem on stem and stem on root make fascinating reading, but he does not seem to have been acquainted with the procedure of grafting onto the roots of herbaceous peonies. Perhaps this was because it had largely been superseded, as Xue Fengxiang's *Mudan Bashu* (Eight Epistles on the Tree Peony) of about 1610 says that the art of grafting special forms onto the roots of wild forms rather than onto those of the herbaceous peony was mastered only about 1600 and gave much better results (Needham, 1986). However, this seems an odd statement as the method described by Ouyang Xiu in 1034 seems likely to have involved grafting onto wild tree peonies, particularly since the graft was made well above ground level.

Wherever the truth may lie in all of this, it is clear that grafting onto the roots of herbaceous peonies was carried out with a high rate of success near Shanghai at the time of Robert Fortune's visits. Fortune (1850g, 1852) visited the tree peony nurseries a little to the west of the city where the stock used was a herbaceous variety with small, single flowers. He saw how, at the beginning of October, short pieces of the tip-growth of the current season were cleft-grafted onto pieces of root the thickness of a finger. The union was then tied with raffia and sealed with clay before the grafted pieces were planted out of doors in the shade.

As regards the cultivation of the plants, Cibot (1778b, 8) went into considerable detail about soils, the preparation of the beds, and control of diseases and pests. He also mentioned pruning, noting that this was done at the end of autumn, only the strongest branches with flower buds being left, a procedure which ensured the production of the best flowers. He commented that the plants in the emperor's garden were not properly cared for, their flowers being smaller and less brilliant than those of plants grown in the south. Also he recorded that plants were sent to the emperor every year from 'Hou-kouang' at the end of autumn and brought into flower by the florists in December and January. Where 'Hou-kouang' was and what its present name is I have as yet been unable to ascertain.

Fortune (1852) was struck by the attention lavished on tree peonies by their owners, describing a specimen on the

Figure 12.13 (above right) **Tree peony bed in the Liu Yuan, Suzhou**

Figure 12.14 (right) **Tree peony bed in the Shizilin, Suzhou**

Figure 12.15 (above) **Peony terrace, Beijing, from *Chinese Gardens* by Dorothy Graham (1938)**

outskirts of Shanghai which produced 300–400 flowers annually. He recorded that 'during its entire flowering period the plant was protected from the sun with shading, while in front of this stupendous example there was a chair on which the many visitors could sit to enjoy the spectacle of its blooms'. No doubt because of the esteem in which they were held and the money expended in obtaining the plants, tree peonies were often planted in special beds edged with carved stone or picturesque rocks. These can still be seen in many old gardens today (figs 12.13, 12.14). Those who could afford it sometimes built special terraces on which to display their plants, one row rising above another and shaded by awnings erected to protect the blooms from the sun. Juliet Bredon (1931) was impressed by these structures, as was Osbert Sitwell (1935, 1949), who saw them during his visit to Beijing. There is a photograph of one (fig. 12.15) in Dorothy Graham's (1938) *Chinese Gardens* and an example still exists at the Summer Palace.

Sitwell felt that the peonies in China had attained 'a hitherto undreamed of perfection'. He was particularly impressed by those which he saw at the Summer Palace. And Dorothy Graham of more than 200 kinds planted in the marble terraces at the 'Tung Hsiao Ssu', a temple in Beijing, mentioning varieties with white, deep red, velvety black, and green flowers, as well as a pink with double and single blooms on the same plant, and another with flowers 'the palest of yellow like a faint gleam of candlelight'. The green one was probably the famous variety 'Doulü' ('Pea Green'), the flowers of which are tinged with green as they open, although they soon fade to white, and the variety with both single and double flowers may have been 'Bingzhaohongshi' ('Ice-covered Red Rock') (fig. 12.16).

Figure 12.16 A tree peony producing both single and double flowers in the garden of the Song Qingling Museum, Beijing

This mention of 'the palest of yellow' leads me to remark that the existence of truly yellow tree peonies had been inferred from the frequent mention from at least as early as 1034 of 'yellow' varieties such as 'Yao Huang'. But, even though Chinese artists in Guangzhou produced paintings of a rich yellow and a coppery bronze tree peony for John Reeves early last century (Gorer, 1970; Synge, 1953), no-one has recorded actually seeing such things in China. And while a Chinese stamp has been issued in recent times depicting 'Yao Huang' as a double buttercup-yellow, a cultivar still in existence and bearing this name is actually a double white, tinged with cream (fig. 12.12). In short there is no reliable evidence that the Chinese ever developed truly yellow cultivars. Hence I tend to go along with the view of Clarke Abel (1819) who recorded that, since the plants were not in flower at the time of his visit, 'I could not therefore judge, from my own observation, of the fidelity of those Chinese drawings which represent this plant with yellow flowers, and was unable to meet with any person at Canton who had either seen or believed it to exist'. Joseph Sabine (1826) was of like mind, saying that 'the existence of a Yellow Moutan, is

altogether disbelieved by those best capable of forming a judgement on the subject'. This view was repeated by Robert Fortune (1847a), who reported that the variety the Chinese call 'the yellow' was white with a slight tinge of yellow near the centre. The first truly yellow hybrids appear to be those raised in Europe and North America this century by crossing *P. lutea* with cultivars of *P. suffruticosa*.

Tree peony plants were exported from China to Japan from the 8th century on, but they did not reach Europe until much later. The first mention of the plant by a European is apparently that made by Martinus Martini in his *Novus Atlas Sinensis* of 1655 (Bretschneider, 1880) and, according to Wister (1962), another first-hand report was made by a member of the Dutch East India Company's embassy to China in 1656, who described them as being like roses but without thorns and twice as large, mostly white with a little purple, but also yellow and red. As with the herbaceous kinds, the first detailed account of the tree peonies to reach Europe was that of Cibot (1778b, 8), who, amongst other things, said that there was reason to believe that they eclipsed the glory of tulips and carnations, and gave details of their

180

history as well as descriptions of the kinds available. It appears that such descriptions and the representations of the flowers in Chinese drawings excited in Sir Joseph Banks and others an ardent desire to import plants.

It became possible for Europeans to obtain plants in Banks's time thanks to the Chinese practice of sending plants annually from the north to Guangzhou where, although the climate is too warm for them to become established, there was, and still is, a demand for this most esteemed of Chinese flowers. The plants are potted up on arrival and, owing to the difference in temperature, soon come into bloom, usually at the time of the Chinese New Year (fig. 0.3). Fortune (1852) noted that in his day the price depended on the number of buds and that the plants were discarded after flowering, as no doubt is still done.

The early history of the introduction of tree peonies to Britain is chronicled by Sabine (1826), who recorded that the first plant to reach Britain was imported by Banks in 1789. Others followed and some of them grew into wonderful specimens bearing hundreds of flowers. Nurseries on the continent also imported plants in the first half of last century and raised seedlings, but the importations from Guangzhou all seem to have been pale pink or off-white with purple markings. It was suggested that early attempts to introduce the pure white and more richly coloured varieties failed perhaps because the Chinese did not want foreigners to have them (Sabine, 1826; Wister, 1962). However, it seems that the plants sent to Guangzhou year after year came from Hunan and western Jianxi, where the growers did not have the same collection of varieties as those in other parts of the country (Fortune, 1880). Ultimately the growing of tree peonies in Europe received its greatest impetus from the efforts of Fortune, who succeeded in obtaining some of the finest varieties from a nursery near Shanghai. Amongst those he brought back to Britain in 1846 were whites, lilacs, purples, 'the yellow', reds, and one 'nearly black' (Fortune, 1847a, 1852). These were given European names such as 'Bijou de Chusan', 'Glory of Shanghai', 'Lord Macartney' and 'Pride of Hongkong'. A scarlet one was named 'Robert Fortune' in honour of the collector. These newly introduced varieties were quickly propagated and by the 1860s were listed by both British and continental nurseries.

The 1860s, 70s, and 80s were the years of the greatest popularity of the Chinese varieties in Europe. Seedlings were raised and during the 1860s Krelage, in Holland, was able to offer 190 varieties. In the 1870s van Houtte, in Belgium, listed 168 varieties and, in Germany, Spaeth claimed to have 350. As late as the 1890s, Paillet, near Paris, listed 337 kinds (Wister, 1962). After that the numbers declined. As the old growers died it appears that the desire to have large collections died with them. Probably this was at least partly because the huge fully double flowers of many varieties were so heavy that the stems could not properly support them, so that they were hidden under the foliage. Even so a few of the better varieties, for example the pink 'Reine Elizabeth', the enormous purple 'Souvenir de Ducher', and Fortune's pure white 'Bijou de Chusan', have continued to be offered by nurseries here and there, at least up to recent times.

As mentioned above, it appears that tree peonies were brought to Japan soon after they became popular in China. They were taken up there with the same enthusiasm and over the centuries the Japanese have produced a huge number of varieties of their own. Since trade with Japan opened up in the second half of last century, Japanese nurseries have exported huge numbers of plants, so that most tree peonies seen in Western gardens and offered by nurseries these days are Japanese varieties. Even so, as mentioned above, some of the Chinese varieties imported long ago still persist, as do cultivars raised from them.

On the whole the Chinese varieties seem longer lived, less susceptible to die-back, and capable of forming larger and more compact bushes than the Japanese ones. It is certainly time for Western gardeners to re-assess them. According to Wister (1962) there is no record of the importation of any Chinese varieties from Fortune's time up to the publication of the *Peony Manual* in 1928. I suspect that this has continued to be true until very recent times. Osti (1994) reports the importation, presumably to Italy, of more than 1000 'Phoenix White' and 'Phoenix Pink'. And in recent years the Cricket Hill Garden of Thomaston, Connecticut, has brought in what is probably the largest collection of Chinese cultivars outside China, including the 'green' and some of the 'yellows', though apparently it was only very recently that anyone was prepared to sell them a plant of 'Yao Huang' (Aleksinas, 1995).

Apart from those mentioned above, I am aware of only one or two small-scale introductions and it seems doubtful that there have been many others. Apparently even as late as the 1990s some varieties were not permitted to be exported, a restriction interpreted as being a misguided attempt by the Chinese to protect their cultural heritage (Aleksinas, 1995).

Thus a survey of Chinese nurseries and gardens might well reveal further kinds worthy of wider cultivation, particularly as Chinese growers are said to be actively promoting their export now and to be offering over 480 varieties (Page, 1997).

Colours I have seen cover the same range as those exhibited by the plants collected together by the Cricket Hill Garden and include white, white tinged with cream, white tinged with green, pinks ranging from the palest to rich cerise, various mauves, purples (fig. 12.1), and reds of various shades including a very dark blackish-red (fig. 12.17). In a few cultivars the petals are edged with white (fig. 12.18) or a very pale tint of the main colour, and in one or two the flowers are striped in the manner of some of the old roses and carnations. One of these is called 'Er Qiao' ('Two Beauties'), being named for two famous beauties of the Three Kingdoms period (fig. 12.19). As well as the singles, semi-doubles, and doubles, there is at least one cultivar, as mentioned earlier, which regularly produces both single and double flowers on the one plant (fig. 12.16). Also in most plantings there are specimens of a single white, which appears to be the 'Phoenix White' (fig. 12.11) mentioned by Osti (1994).

The tree peony remains the country's most famous flower and there are plants to be seen wherever one goes. Almost every garden and temple has at least a few plants, special displays of potted plants are mounted in many places at flowering time, and there are large collections in many parks and botanical gardens. There is large tree-peony garden at the Shanghai Botanical Garden, and in Beijing there are collections in Zhongshan Park, the Imperial Garden of the Forbidden City, Jingshan Park, the Beijing Botanical Garden, the garden surrounding the Wofo Si, the Fragrant Hills Park, and numerous other places. The 'Tung Hsiao Ssu' or 'Tsung Hsiao Ssu', also known as the Temple of Supreme Service, first built in 627 AD in the south-west part of the city, and once famous for its displays of tree peonies (Arlington & Lewisohn, 1967; Bredon, 1931; Graham, 1938), apparently no longer exists. However, there are good collections at the Tanzhe Si

Figure 12.17 (above left) A 'black' tree peony in the Imperial Garden of the Forbidden City, Beijing

Figure 12.18 (left) A tree peony with white-edged petals in the Beijing Botanical Garden

Figure 12.19 (above right) Tree peony 'Er Qiao', a bicolour

and the Jietai Si to the south-west of the city. Of all those I have seen, the old plants in the tree-peony garden in Jingshan Park are particularly charming (fig. 12.20). And at all these places visitors can witness the enthusiasm of the Chinese for the flowers. Many people photograph each other in front of them, others discuss the respective merits of the different varieties, and a few are still seen drawing or painting them.

Elsewhere there are numerous other sites where they can be seen. Luoyang, for instance, is still be famous for its tree peonies, there is a collection in the Xian Botanical Garden, and Heze, which has been a famed centre of cultivation for the last 300 years, has an annual peony festival. Also Osti (1994) describes a visit to the Mountain of 10,000 Flowers in Shaanxi near Yanan, some distance to the north of Xian, where thousands of plants have been cultivated 'from time immemorial'. These are more or less naturalised under a canopy of thujas (*Platycladus orientalis*), and he counted 16 new cultivars amongst them. He also notes that some of the pavilions there contain paintings depicting the legend of the origin of tree peonies from the union of two beautiful fairies with two sturdy peasants. Regrettably he gives no indication of the detail in which this interesting phenomenon is depicted.

Figure 12.20 Tree peonies, Jingshan Park, Beijing

CHAPTER 13

WISTERIA AND OTHER VINES

It might well be claimed that *Wisteria sinensis* is the showiest of all vines. It is found in almost every Chinese garden and, following its introduction to Britain early last century, its cultivation rapidly spread around the world. Much the same can be said of *Parthenocissus tricuspidata* which most people, in defiance of its origin in eastern Asia, call Virginia Creeper. In addition to these two, various other ornamental climbing plants are cultivated in China. For instance, travelling in the countryside north of Beijing in the 1860s, Freeman-Mitford (1900) observed that most of the cottages had a garden fenced in by a 'hedge of millet stalks, trailed over with gourds, convolvulus, and vines'. Hence I have included in this chapter a selection of the climbers which one is most likely to see in Chinese gardens, some of which are not indigenous but arrived long ago and became incorporated in the garden flora. Also I could not bring myself to exclude a few others which are not grown as ornamentals but are likely for other reasons to claim attention. And although some of the jasmines are shrubby, for convenience I have included them

Figure 13.1 **Wisteria in the Yu Yuan, Shanghai**

here along with the climbing types, just as I have kept all the roses together in Chapter 15.

Climbing plants are used in Chinese gardens in several traditional ways. Most frequently they are planted to grow over rocks or artificial rock piles. Otherwise, as in the West, they are grown on pergolas and arbours, across bridges, on walls, and over doorways. Frequently, too, they are allowed to clamber over trees or climb up the trunks as they would in nature. In many places, for instance, wisteria has been planted to cover ancient trees which have died (fig. 2.11). As well as this, some species, particularly those with fragrant flowers, are kept in containers and brought out when in bloom for use as room ornaments or garden decorations. A few, wisteria in particular, are also trained as penjing.

ACTINIDIA Actinidiaceae

Actinidia is a genus of woody vines native to eastern Asia. According to Huxley et al. (1992) several species are grown for their edible fruits in China, where they have been cultivated since at least 770 AD. Meyer (1911), on the other hand, said that the fruits of various species are collected and eaten where they grow. He recorded that the species with the largest fruits was *A. deliciosa* but that, as far as he could ascertain, it was not cultivated at all. However, these days both imported and home-grown fruits are on display in the shops.

A. deliciosa 美味猕猴桃 (Meiwei Mihoutao)
Delicious Monkey-peach

CHINESE GOOSEBERRY, KIWI FRUIT

According to Huxley et al. (1992) this deciduous vine with large, heart-shaped leaves was introduced to Britain by Robert Fortune in 1847. However, Spongberg (1990) says that it was from Ernest Wilson's collections of 1900–1902

Figure 13.2 Fruits of *Actinidia deliciosa* bought in Beijing

that it was first successfully introduced to cultivation in the West, and that stock from his introduction growing in England was taken to New Zealand. It arrived in that country in 1906 but it was not commercially exploited there on a large scale until after World War II (Huxley et al., 1992). As a result of a successful marketing campaign, involving the introduction of the name Kiwi Fruit for the plant previously known generally as the Chinese Gooseberry, it soon became familiar in the supermarkets and fruit shops of the world, including those of China itself. Now it is cultivated in many countries, including China, where, I am told, it is principally grown on Hainan Dao, the southernmost part of the country. A cultivar with large fruits is now marketed throughout the country (fig. 13.2).

AKEBIA Lardizabalaceae

A. quinata 木通 (Mutong) *Wood to Open Up*

A. quinata is considered to be native to China, Korea, and Japan. Its woody stems have conspicuous tubes running through them and formerly were used in folk medicine in the belief that they would open up or loosen various parts of the body. It is occasionally grown in gardens and is a fast-growing, semi-evergreen or deciduous climber with digitate leaves composed of five leaflets, each notched at the apex. The brownish purple, vanilla-scented flowers, which are relatively inconspicuous, are followed by purple, ovoid fruits 5–10 cm long. The white pulp contains numerous reddish-brown or black seeds and is edible.

In the old days in Beijing on New Year's Eve a family would set out in a courtyard a 'Table of Heaven and Earth' on which appropriate offerings were placed. Above this were hung figures of the Eight Immortals threaded together with the stems of *A. quinata* (Tun, 1965), which presumably had some significance for this occasion.

CALYSTEGIA Convolvulaceae

C. hederacea 打碗花 (Dawanhua) *Broken Bowl Flower,*
小旋花 (Xiaoxuanhua) *Small Twining Flower,*
兔耳草 (Tuercao) *Rabbit's Ear Plant*

This perennial climbing herb, which produces small pink flowers, is sometimes depicted in Chinese art, for example in an 18th century painting reproduced by Weidner et al. (1988, cat. no. 45). A double form was raised by the Horticultural Society of London from a small piece of root found amongst the roots of a peony sent by Fortune in 1844. The flowers were pale pink with no trace of stamens or pistil and it was given the name *C. pubescens* (Lindley, 1846). It is now known as *C. hederacea* 'Flore Pleno' (fig. 13.3). Fortune (1846) mentions seeing it in a mandarin's garden in Ningbo.

CAMPSIS Bignoniaceae

C. grandiflora 凌霄花 (Lingxiaohua)

Rising to Heaven Flower

TRUMPET CREEPER

This striking deciduous climber with pinnate leaves and loose panicles of large, bright orange flowers in summer (fig. 13.4) is recorded in the earliest pharmacopoeias as a medicinal plant, though the pollen is believed to be harmful, so it is not usually planted close to living quarters (Li, 1959). In Chinese gardens it is often seen growing over rocks or on the slender upright pieces of fossilised wood known as 'stone bamboo shoots'. It is also occasionally used for penjiing (Hu, 1982). It obviously appealed to Chinese artists as it appears in paintings, at least from the 17th century on (e.g. Capon & Pang, 1981, cat. no. 65; Weidner et al., 1988, cat. no. 34), and is recorded in the *Jieziyuan Huazhuan* of 1701 (Sirén, 1949). It is native to China and Japan and was introduced to Britain in 1800 (Chittenden, 1956), subsequently making its way around the world. The flowers, stems, leaves, and roots are used in Chinese medicine (Wang & Ma, 1995).

Figure 13.3 (top right)

Calystegia hederacea 'Flore Pleno', from *Edwards' Ornamental Flower Garden*, 1854

Figure 13.4 (right)

Campsis grandiflora

CITRULLUS Cucurbitaceae

C. lanatus 西瓜 (Xigua) *Western Melon*

WATERMELON

Crook (1935) discusses the known history of the watermelon in China where, if the silence of the records is any guide, it was not known before the 10th or 11th centuries AD. As the fact that it is called Western Melon suggests, it almost certainly arrived from that direction. It has become enormously popular, as much for its seeds as the flesh of its fruits. I have included it here on this account, as any visitor to the country cannot fail to notice the popularity of these seeds as a snack food (fig. 13.5). In the shops there is usually more than one kind available and people can be seen everywhere cracking them open between their teeth. Crook claims that eating the seeds is mentioned in a letter written about 1080, and that there may be earlier references. He noticed that many Chinese had grooves in their incisor teeth, worn away from opening these seeds since childhood, and suggested that the skulls of Chinese of earlier centuries might reveal some clues to the antiquity of the watermelon in China, and to the time of origin of the melon-seed habit. The continued popularity of watermelon and other similar seeds has ensured that these grooves can be seen in the teeth of many Chinese today.

The first foreigner known to have recorded the practice is the Abbé Huc, who, in the early part of last century, was astounded by the quantity consumed all over the country

and was moved to write '… for if there is in the world a disappointing dish … it is surely this' (Crook, 1935). This view was not and is not shared by the Chinese. For example, in describing a visit in the 1860s to a theatre in Beijing, where the programme included 'indecent farces', Freeman-Mitford (1900) tells us that the patrons were eating fried melon and gourd seeds. They still do on such occasions, though the entertainment provided these days is usually rather less exciting. Wilson (1913), too, noted that 'The seeds of the water-melon are esteemed a great delicacy. They are slightly roasted, and are consumed in enormous quantities: no banquet is complete without them, and over their gossip in tea-shops or restaurants, scholars and coolies alike regale themselves with these delicious morsels'. And nowadays, even despite the onslaught of Western-style junk foods, the number of watermelons which must be grown to meet the demand is almost unimaginable.

Fresh watermelons are also extremely popular. Tun (1965) records that in the old days varieties with white, yellow, and red flesh began to be sold on the streets of Beijing in the 6th month. He said they were cut up like 'lotus petals' (slices) or like a 'camel's back' (halved), and that they countered the effects of heat and helped to overcome the effects of intoxication. Watermelon remains enormously popular in China, but during my visits I have seen only the red-fleshed kinds.

CLEMATIS Ranunculaceae

Li (1959) says that several species are sometimes planted in gardens but they are not common. I have not come across any reliable evidence that clematis are at present grown in traditional Chinese gardens but, since the genus is so well known and since China is the home of such striking species as *C. armandii*, *C. florida*, *C. lanuginosa*, and *C. montana*, I felt that I should give it at least passing mention.

C. florida 铁线莲 (Tiexianlian) *Wiry Lotus*

According to Li (1959), this is the species most frequently planted in China. The fact that the name *Tiexianlian* is listed in ordinary Chinese dictionaries as meaning 'the cream clematis' suggests that at least it must be a familiar plant. In Western gardens it is known chiefly as a form introduced from Japan, which has cream flowers with purple stamens.

Figure 13.5

Watermelon

seeds

bought

from a

food store

in Beijing

Figure 13.6 A cream clematis in the Beijing Botanical Garden

The only clematis I have seen in a garden in China is a plant with cream flowers with cream stamens growing in the Beijing Botanical Garden (fig. 13.6). However, since I am not acquainted with the Chinese forms of *C. florida*, I am not convinced that this is a representative of the true species. The situation is further confused by a report of Armand David who, while journeying north of Beijing in July 1866, saw what he took to be *C. florida* growing wild in abundance and said that its large double flowers were either blue, purple, or pale slate-grey (Fox, 1949).

C. lanuginosa 毛叶铁线莲 (Maoye Tiexianlian)
Hairy-leaved Wiry Lotus

Whether *C. lanuginosa* has ever been grown in Chinese gardens I do not know, but I have included it here because of its importance as an ancestor of the large-flowered hybrids. It was introduced to England from China in 1850 by Robert Fortune, who in July of that year found a blue-flowered form growing wild near the Tiantong Si, not far from Ningbo (Lindley & Paxton, 1853).

CUCURBITA Cucurbitaceae

C. pepo 西葫芦 (Xihulu) *Western Bottle Gourd*

PUMPKIN, MARROW, SQUASH

The North American squashes and pumpkins belong to this species and are included here because the roasted seeds are eaten as a snack food. The large white seeds are seen on sale all over China, as noted by Meyer (1911). However, they seem to be less popular than the ubiquitous watermelon seeds. The seeds of other species of *Cucurbita* are perhaps used similarly.

Box (1902) records that in the country districts near Shanghai, on the evening of the 15th of the 1st or 8th lunar months, childless women went secretly by moonlight to the fields in search of a pumpkin or squash, believing that if they took one home a son would be born to them.

FICUS Moraceae

F. pumila 薜荔 (Bili)

CREEPING FIG

In its juvenile phase the stems of this Asian species cling to rocks, walls, and tree trunks in the manner of ivy. The adult phase has larger leaves on spreading branches on which the typical fig-shaped fruits are borne. These may be as much as 7 cm long and become purplish when ripe. It is commonly planted in Chinese gardens to clothe rocks and tree trunks (fig. 13.7). It was introduced to Britain about 1771 (*Curtis's Botanical Magazine* 1882, t. 6657) and is now found throughout the world.

IPOMOEA Convolvulaceae

While the indigenous *I. nil* has apparently been grown in China for centuries, several of the American species became popular after their introduction, so I have included them here as well. The seeds of some species are used medicinally, while those of others contain hallucinogenic substances.

I. batatas 番薯 (Fanshu) *Exotic Yam,* 红薯 (Hongshu) *Red Yam*

SWEET POTATO

G. T. Lay (1846) wrote of Fuzhou in November 1844: 'Sweet potatoes now in great abundance. They are much relished by the common people who find them ready cooked as they pass along the street'. As this remains the case today, with roasted sweet potato tubers sold everywhere on the streets in

the cooler months, I have included a brief mention of this plant here. Since it was widely distributed throughout the world in pre-Columbian times, no-one seems certain of its true home, though it is frequently suggested that it is native to tropical America (Merrill, 1954). The date for its introduction to China was cautiously set at 1594 by Goodrich (Ho, 1955), having reached Fujian by sea. However, Ho suggests that it may have arrived decades earlier overland via Yunnan.

I. coccinea 园叶鸟萝 (Yuanye Niaoluo) *Round-leaved Bird Vine,* 橙红鸟萝 (Chenghong Niaoluo) *Orange-red Bird Vine*

RED MORNING GLORY, STAR IPOMOEA

This annual species from the United States has ovate leaves which are entire or coarsely toothed, and small, long-tubed, scarlet flowers which are yellow inside at the base (fig. 13.8). Its Chinese names relate to both its similarities to and differences from *I. quamoclit*, which is described below.

I. nil 裂叶牵牛 (Lieye Qianniu) *Divided-leaf To-lead-an-ox)*

MORNING GLORY

This morning glory has white-throated, blue flowers about 5 cm in diameter and large, entire or 3-lobed leaves. It is an annual and is found throughout the tropics. Plants fitting this description appear from time to time in Chinese paintings, for instance in the *Song Hundred Flowers* in the Palace Museum, Beijing, in an album of garden flowers dated 1540 (Barnhardt,1983, cat. no. 26), and in various works of the 17th, 18th, and 19th centuries. It is also represented on a Ming blue and white vase of the 15th century and on various other pieces of porcelain (Bartholemew, 1985a; Krahl, 1987). According to Bartholemew (1985a), it is a symbol of marital bliss. Enthusiasm for this plant in China has never compared with that exhibited for it in Japan, where many spectacular variants have been raised.

The use of the name *Qianniu* (To Lead an Ox) relates to the tradition that the seeds once cured a farmer of his illness and he led his cattle to the fields to give thanks to the plant which saved his life.

Figure 13.7 (opposite page) *Ficus pumila* **growing over rocks at the Tianyi Ge, Ningbo**

Figure 13.8 (left) *Ipomoea coccinea*

I. quamoclit 鸟萝 (Niaoluo) *Bird Vine,* 锦屏封 (Jinpingfeng) *Brocade Screen,* 游龙草 (Youlongcao) *Writhing Dragon Plant*

CYPRESS VINE

I. quamoclit is an annual climbing herb introduced from tropical America. It has fern-like foliage composed of deeply pinnatisect leaves and small, star-shaped, scarlet flowers. Clarke Abel (1819) saw it in 1816 in the neighbourhood of Beijing trained on small frames of trellis and concluded that it was 'from its frequent culture, obviously a favourite'. Morrison (1822) records its flowering in the 7th moon in Guangzhou.

It gets the names *Niaoluo* (Bird Vine) and *Youlongcao* (Writhing Dragon Plant) because, with its twining growth, its appearance suggests birds in flight or a writhing dragon. The name *Jinpingfeng* (Brocade Screen) is an allusion to the general appearance of the plant, when its red flowers stand out against the green of the foliage.

JASMINUM Oleaceae

It seems probable that the best-known jasmines, *J. officinale*, Common Jasmine, and *J. sambac*, Arabian Jasmine, are native to the Middle East or perhaps even India, but they have been in cultivation for so long that it is impossible to be certain. Nevertheless it seems likely that they were taken both west and east by the Arabs. The first record of these plants in China appears to be that in Ji Han's *Nanfang Caomu Zhuang* (Account of the Plants and Trees of the Southern Regions) of 304 AD, in which the author states that they were enthusiastically cultivated in the south at that time, and the flowers were worn by women and girls in their hair. Also he quotes a writer of some 475 years earlier, who mentioned two fragrant flowers brought from foreign countries and noted that, unlike the orange, which was believed to change into *Poncirus trifoliata* when taken north, they did not change when taken to a different environment. While it is impossible to know for certain whether these plants were jasmines, it appears that the scented jasmines were introduced at some time prior to the end of the 3rd century AD, probably by the sea route already established between western Asia and southern China. For *J. officinale* Ji Han used the name *Yeximing*, a name presumably derived from the Persian 'Yasmin', like its European name. *J. sambac* he called *Moli*, a name believed to be derived from the Sanskrit 'Mallika' (Schafer, 1948; Li, 1959).

Jasmine is mentioned again by 9th century writers who also record that 'men of the western regions' gather the flowers and press them to make a very smooth and fragrant oil (Schafer, 1948). However there is no evidence that the flowers were used for this purpose in China as early as this, though eventually they were (Williams, 1975). The *Lingwai Daida* (A Reply [to questions about what lies] Beyond the Ling Range) of 1178 records the distilling of both sorts in Guangzhou and Guilin (Needham, 1986). And eventually they came to be included amongst the flowers used for scenting tea, though this may not have happened until the Qing dynasty (Wang & Ma, 1995).

Both species became great favourites in southern China, particularly from the 14th to the 18th centuries in Guangzhou, where the jasmine gardens extended many kilometres along the banks of the Pearl River. Li (1959) quotes accounts from the 17th and 18th centuries which describe how, before dawn, women gathered buds due to open that day. These were then purchased by flower merchants and taken back to the city, where they were made into headdresses, lampshades, and other ornamental articles and offered for sale.

It seems that in the evening, after such buds were placed in the hair, they began to open, becoming more fragrant with human warmth and lasting till dawn. It was believed that they had a cooling effect and that breathing the scented air was beneficial for the lungs. They were also incorporated in garlands for decorating buildings for religious ceremonies, for making perfumes and cosmetic oils considered to be beneficial to the skin and hair, and for scenting food, wine, and other drinks. According to Li (1959), it was said that at the height of the flowering season each summer 'The city was like snow in the night, and was fragrant everywhere'.

Although enthusiasm for these plants has waned, other foreign species have been introduced from time to time and several of the native yellow-flowered species, which are scentless, have been and still are grown in gardens. Species cultivated in China are described below.

J. floridum 探春花 (Tanchunhua) *Searching-for-spring Flower*

Rather than being considered a climber this species is usually a semi-evergreen shrub with arching branches. It has alternate leaves, each with 3–5 leaflets, and yellow flowers. It was collected in or near Beijing by Bunge in 1831 (*Curtis's Botanical Magazine* 1883, t. 6719).

J. mesnyi 云南黄素馨 (Yunnan Huang Suxin)
Yunnan Yellow Jasmine

PRIMROSE JASMINE

J. mesnyi, which was earlier known as *J. primulinum*, is very similar to *J. nudiflorum*, having opposite leaves composed of 3 leaflets, but is evergreen, more vigorous, and has larger flowers, most of which are double or semi-double (fig. 13.9). Wilson (1906) recorded that the same obtained even in wild plants which were common in hedgerows in southern Yunnan, from where he collected it for the nursery firm of Veitch. Whether these plants were truly wild and what the history of this plant is are both matters for conjecture. As it failed to produce seeds, it had to be introduced as plants. While it is the jasmine most frequently planted in Chinese gardens nowadays, it is not clear for how long this has been the case. However, it appears in an 18th century *Hundred Flowers* reproduced by Capon and Pang (1981, cat. no. 65), so it has probably been grown for well over 200 years.

J. nudiflorum 迎春花 (Yingchunhua)
Welcoming-spring Flower

WINTER JASMINE

This deciduous, somewhat straggly shrub, has opposite leaves composed of three leaflets. The yellow flowers, borne singly on the bare branches in the axils of the previous season's leaves, are 6-lobed and appear in winter and early spring. It is native to northern China and valuable for its hardiness and for flowering in mid-winter. It was found by Alexander von Bunge in Beijing in 1830–31 and later in nurseries and gardens in and around Shanghai, Suzhou, and Nanjing by Robert Fortune, who introduced it to England from Shanghai in 1844 (Lindley, 1846; *Edwards' Ornamental Flower Garden*, 1854). It was later seen by Fortune (1863) to be one of the few cold-hardy plants to be offered by nurserymen around Tianjin and Beijing. According to *Edwards's Ornamental Flower Garden* (1854), 'The Chinese often graft it on the more common kinds, about a foot from the ground, which improves its appearance'. Wang and Ma (1995) say that potted plants can be trained into a variety of shapes.

J. odoratissimum 浓香探春 (Nongxiang Tanchun)
Strongly-scented Searching-for-spring Flower

J. odoratissimum is an introduced species which has strongly scented, yellow flowers borne in panicles of up to 20. Unlike *J. officinale* and *J. sambac*, this species was a much later introduction, coming from Madeira by way of obvious Iberian channels (Needham, 1986).

J. officinale 素方花 (Sufanghua) *White Square Flower*

COMMON JASMINE

The common jasmine is a deciduous or semi-evergreen climber which can become up to 10 m tall. It has opposite leaves with 5–9 leaflets and the highly fragrant, white flowers, 2–2.5 cm wide, are borne in terminal clusters in summer and early autumn. Its Chinese name refers to its white flowers and the stems which are square in cross-section. As mentioned above, it is believed to have been introduced to southern China before the end of the 3rd century AD. Its subsequent popularity there is undoubtedly due to its strong and delicious scent.

J. officinale f. *grandiflorum* 素馨花 (Suxinhua)
White Fragrant Flower

SPANISH JASMINE

This is more vigorous than the typical form and larger in all its parts. The flowers, which may be up to 4 cm in diameter, are flushed with pink on the exterior (fig. 13.10). According to Li (1959) it was introduced quite early from India and is extensively cultivated in southern China.

Figure 13.9 *Jasminum mesnyi* in bloom on the Su Causeway, Hangzhou

Figure 13.10 *Jasminum officinale f. grandiflorum*

Fortune (1857) described a visit to a tea factory in Guangzhou and gave a list of the flowers used for scenting the tea, including those which he identified as *J. paniculatum*, which he said were generally mixed with those of *J. sambac* in the proportion of 1:3 by weight. He gave it a rating of 2 or 3 on his scale of 1–7 for the value which the Chinese gave to the various flowers used for scenting tea. However, I have so far found no reliable evidence that *J. paniculatum* is or was grown for this purpose in China and, as Fortune gave the Chinese name of this plant as 'Sieu-hing-hwa', it seems possible that the flowers he encountered were those of *J. officinale* f. *grandiflorum* or perhaps just those of the original form of that species.

J. sambac 茉莉花 (Molihua) *Moli Flower*

ARABIAN JASMINE

The Arabian Jasmine is an evergreen climber with broadly ovate, undivided leaves borne opposite each other on the stems or whorled in threes. The small white flowers are highly scented and produced throughout the warmer part of the year, usually in threes. They often become pinkish-purple as they fade.

Apparently introduced to southern China at about the same time as *J. officinale*, the Arabian Jasmine soon became the most popular of the species in cultivation there. Fortune (1847a) recorded that large quantities were cultivated in the fields near Fuzhou, where it was used to 'decorate the hair of the ladies and to garnish the tables of the wealthy'. Li (1959) says it was grown in special gardens in many cities and that

every day throughout the summer the unopened buds were gathered before sunrise and sold for perfuming tea and for making hair ornaments. It is still grown extensively for scenting tea, and the largest centre of production is around Fuzhou where, according to Li, well over 1,000,000 kg of flower buds are produced annually for this purpose. Fortune (1857) said that in this regard the Chinese rated it only behind the rose and flowering plum, and that it was the most favoured for scenting teas for the foreign market. This is certainly still the case.

Cibot (1778b, 4) described it in great detail as it occurred in his time in Beijing, where the flowers were highly valued for perfuming tea, syrups, sweetmeats, and clothes. It was grown there as a pot plant as the climate was too cold for it to be left out of doors. When in flower, these plants were used to decorate and scent the emperor's apartments. Cibot described how it was propagated by cuttings, the bases of which were pushed through holes in pieces of half-rotted boards, which were then floated in basins of water in the shade. The pieces of board were broken away after roots had appeared and the cuttings were then planted in soil.

J. sambac is still grown occasionally for ornament as a pot plant, and produced in greenhouses in areas otherwise too cold for it. It is usually seen in China as the double form known in the West as 'Grand Duke of Tuscany' (fig. 13.11). Presumably this was the plant described as 'a double flowering jasmine' by John Barrow (1804), who saw it in nurseries in Guangzhou in 1794. Ji Han wrote that the *Moli* flowers resemble white roses and that their fragrance exceeds that of the *Yehximing* (Li, 1979). This might indicate that the form known to him had double flowers, as he seems to have taken the resemblance to white roses as a mark of distinction between the two kinds. Also, I have not found any early Chinese painting of a single form, though paintings of the double form are common from the early Song on (e.g. Sirén, 1973, vol. 3, pl. 139; Suzuki, 1982, vol. 4, JP1-005), and Chen (1953) illustrates it with double flowers in his *Illustrated Manual of Chinese Trees and Shrubs*. However Weidner et al. (1988, cat. no.34) reproduce a 17th century *Hundred Flowers* which includes a single form of this species, and a drawing from the *Zhiwu Mingshi Tukao* (Investigations on the Identity of Plant Names) of 1848 (fig. 13.12) is similar. These flowers have only 6–8 petals, as did those seen in Beijing by Cibot (1778b, 4), whereas those of the double variety which I have examined have proved to be a hose-in-hose form with a total of 14–20. While these observations might suggest that the single form was introduced to China much later than the double, much further research would be needed to clarify the position.

LAGENARIA Cucurbitaceae

L. siceraria 葫芦 (Hulu)

CALABASH, BOTTLE GOURD

L. siceraria is a vigorous annual climber with white flowers which open in the evening and continue to grow during the night, reaching up to 10 cm in diameter, and often remaining open till the following midday. The fruits are of various curious shapes, those of some varieties being waisted, resembling bottles. They can be used as a vegetable when young and the seeds are sometimes eaten as well. Meyer (1911) noted that these were 'boiled in salted water and eaten when cold as appetizing delicacies by the rural classes in China'. The dried fruits were used as floats tied to the backs of children by the boat people in Guangzhou and were also used as containers, particularly for medicines. The sign-boards of shops usually involved images or pictures of the articles sold, hence those of drug and wine shops were often in the shape of a bottle-gourd. Also these gourds or representations of them were used as charms to dissipate or ward off pernicious influences, particularly on the morning of the Dragon Boat Festival (Williams, 1975; Bartholemew, 1985a).

The bottle gourd is a Daoist symbol of magic and an emblem of purity. By virtue of its numerous seeds it is a symbol of fertility and also, perhaps because of its durability when dried, of longevity. It is seen tied by its waist to the long staff of the God of Longevity, as one of the symbols of his supernatural powers. Also Li Tieguai, one of the Eight Immortals, has a bottle gourd as his emblem. In Chinese mythology there are numerous instances where immortals uncorked their gourds in the height of battle and vanquished their foes by sucking them into these vessels (Bartholemew, 1985a; Gulland, 1928; Koehn, 1952). No doubt it is on account of all these associations that vases in the shape of bottle gourds have been popular for a very long time. An illustration of a good example from the Yuan dynasty is given by Bartholemew (1985a, cat. no. 35). It also appears in paintings from time to time and is one of the plants depicted in the *Jieziyuan Huazhuan* (1975).

Princess Der Ling (1911) records that the Empress Dowager grew bottle gourds at the Summer Palace outside Beijing, taking her ladies-in-waiting daily to see how they

Figure 13.11 (below)
The double form of
Jasminum sambac
Figure 13.12 (right)
Jasminum sambac,
from the *Zhiwu*
Mingshi Tukao **of 1848**

were progressing as they ripened. She would tie ribbons around those with the smallest waists so as not to lose sight of them and then, when they were harvested, she would scrape the outer skin with a bamboo knife, wipe them with a wet cloth, and leave them to dry. After a few days, when they had turned brown, they were hung as ornaments in the palace. In one room alone there were over 10,000 gourds of different shapes. These observations leave us in no doubt that the Dowager Empress firmly believed that the efficacy of auspicious objects and symbols was enhanced by repetition. It was a duty of the court ladies to wipe these gourds periodically with a cloth to keep them shiny, and to prepare any new ones. Princess Der Ling remarked that 'None of us cared very much for this work excepting Her Majesty'.

LONICERA Caprifoliaceae

L. japonica 金银花 (Jinyinhua) *Gold-and-silver Flower,* 金银藤 (Jinyinteng) *Gold-and-silver Vine*

JAPANESE HONEYSUCKLE

The Japanese Honeysuckle is an evergreen vine native to eastern Asia which is often planted to cover walls, fences, arbours, and rocks in Chinese gardens. A somewhat weedy species, it appears to have attracted attention on account of its

strongly scented, white flowers which change to yellow as they fade (fig. 13.13). In China it exhibits, not unexpectedly, some variation. A form with purplish stems and flowers pink on the outside was sent to England by William Kerr early last century and came to be regarded as a distinct variety, *L. japonica* var. *repens*, though one wonders whether it deserves more than cultivar status.

Fortune (1863) noted *L. japonica* as one of the few plants not requiring protection in winter among those offered by nurserymen in northern China, most of the others being of southern origin. Besides its use as an ornamental plant, its flowers and leaves are used for making a cooling essence for summer drinks (Li, 1959). The flowers have also been used for flavouring tea. Princess Der Ling (1911) describes how the Dowager Empress was brought tea in a white jade cup with a gold lid and saucer, accompanied by two similar cups, one with honeysuckle flowers and the other with rose petals. She took the golden cover off the cup containing the honeysuckle flowers and, using gold chopsticks, placed some of them in the tea. She considered this gave it a very delicate flavour. Princess Der Ling was offered some and considered it to be the most delicious tea she had ever tasted. For those less fortunate, it seems that, according to the *Jiuhuang Bencao*, the roots and young leaves could be eaten in times of food shortage (Bretschneider, 1881).

Like most of the popular Chinese garden plants it turns up from time to time in paintings, particularly those of the 'hundred flowers' genre. An example from the 17th century is reproduced by Weidner et al. (1988, cat. no. 34), and it shows the typical form with both white and yellow flowers and without any pink pigmentation.

Figure 13.13 *Lonicera japonica*

Figure 13.14 *Parthenocissus tricuspidata* growing over rocks at the Jingxinzhai, Behai Park, Beijing

PARTHENOCISSUS Vitaceae

P. tricuspidata 爬山虎 (Pashanhu)
Mountain-climbing Tiger

VIRGINIA CREEPER

How, as a child, was I to know that the Virginia Creeper, which grew on our kitchen chimney and was greatly admired when flaunting its autumn colours, was actually a native of China and Japan? Only much later did the reason for this anomaly become clear to me, when I discovered that the name, which had originally been applied to *P. quinquefolia* from North America, was later also given, understandably perhaps, to its closely similar Asian relative.

Be that as it may, *P. tricuspidata* is seen in a great many Chinese gardens, almost invariably planted to grow over rocks, either on its own (fig. 13.14) or with other vines such as *Campsis grandiflora* or *Wisteria sinensis*. The leaves are very variable, sometimes being composed of 3 leaflets, at others merely being 3-lobed.

P. tricuspidata was given its Chinese name because it is considered to be as good as a tiger at climbing high mountains. When one sees it overwhelming buildings in the West the appropriateness of this choice is immediately apparent.

TELOSMA Asclepiadaceae

T. cordata 夜香花 (Yexianghua) *Night-scented Flower,*
夜来香 (Yelaixiang) *Night-coming Scent*

When I first read Cibot's (1778b, 9) detailed description of '*Le Yê-hiang-hoa*', carefully trained by the Beijing nursery-men on bamboo stakes and rings to display its pale-green, night-scented flowers, I wondered what on earth it could be. Fortunately Cibot's description is detailed and accurate, so I was able to confirm my preliminary guess at its identity by turning to the illustration (fig. 13.15) in *Edwards's Botanical Register* (1819, t.412, as *Pergularia odoratissima*). *T. cordata* is a tropical plant which found favour with Chinese garden-ers, but to what country or countries it is native is unclear. Macmillan (1991), who calls it 'Tonkin Creeper', gives China and Sumatra, while Mabberley (1997a) says India to south-east Asia. Loureiro (1790) recorded its cultivation in Guangzhou, where Morrison (1822) listed it as flowering in the 5th moon (June).

Cibot said that its scent was sweet, suave, and so exqui-site that he knew of no other flower which would compare with it. He also said that it grew easily in the southern

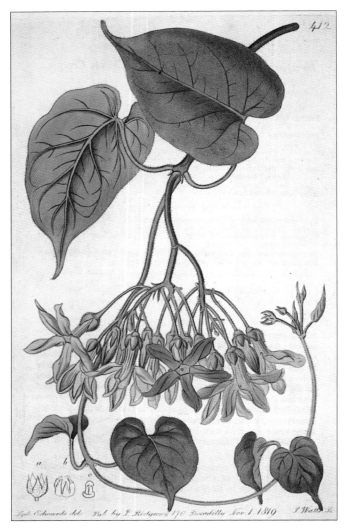

Figure 13.15 *Telosma cordata*, from *Edwards's Botanical Register* 1819, t. 412

provinces but that the gardeners of Beijing had to go to great pains to conserve it in their greenhouses in winter. As a result it was very expensive, a good pot costing 20 to 30 ounces of silver. He also recorded that the 'Vice-Roi du Tchekiang' sent it every year to put in the apartments of the emperor.

TRACHELOSPERMUM Apocynaceae

T. jasminoides 络石 (Luoshi) *Rock Net*

STAR JASMINE

This charming Chinese climber is present in almost every garden in those parts of the country which are not too cold for it. With its shiny, opposite leaves and its fragrant, white flowers produced abundantly in early summer, its popularity is easy to understand (fig. 13.16). In Chinese gardens it

is usually grown, as are so many vines, over rocks or on tree trunks. It gets its Chinese name because it covers rocks in the manner of a rope net. It was introduced to Britain by Fortune from Shanghai in 1844 (Lindley, 1846), and has now deservedly become a favourite garden plant in many parts of the world.

VITIS Vitaceae

V. vinifera 葡萄 (Putao)

GRAPE

The grape and lucerne (alfalfa) are recorded as having been brought back to China from central Asia about 128 BC by Zhang Qian, an envoy of the Emperor Wu, and are the first foreign plants for which such documentation exists (Laufer, 1967). However, there are now so many varieties of grape in China that it seems probable that subsequent introductions were made, and presumably new varieties were raised locally as well. Both green and black varieties are recorded earlier than 527 AD, and during the Song a seedless variety is mentioned. Laufer (1967) records that the Chinese soon began drying grapes to make raisins and notes their skill in storing the fresh fruit to keep it on the market almost continuously. Likewise Williams (1975) recorded that there were many varieties in China and that the fruit could be kept in good condition for several months by keeping it underground, a procedure earlier noted by Fortune (1863) during his visit to Tianjin in 1861.

Cibot (1780a) described the various types of raisin available in Beijing. He said that those of 'Ha-Mi', a small tributary kingdom south-west of Beijing with a very dry climate, were the most famous. There were two kinds, one used in medicine and the other for the table. Cibot (1778b, 14) had noted earlier that the grapes of 'Hoai-lai-hien' (presumably Huailai Xian, a little to the south-west of Beijing) were gigantic, bigger than any in France, and in huge bunches, and available from April right up to the end of August. Unusually large grapes are still seen in shops and markets in China today, but whether they are the same variety as that described by Cibot I do not know.

Almost a century after Cibot recorded his observations, Sampson (1869a) reported what was said to be a method for producing grapes of large size. It involved planting a vine beside a tree of the Chinese Date (*Ziziphus jujuba*), boring a hole through the trunk of the tree in spring, and pulling a branch of the grape through it. Then, when the natural growth of the wood had completely closed the opening, the grape was cut off from its own root. It was claimed that it then combined with the foster tree and produced grapes as large as those of the Chinese Date. This curious procedure is yet another example of the Chinese penchant for claiming success with unlikely combinations of stock and scion.

In spite of the early introduction and successful cultivation of grapes, the Chinese were slow to adopt the Iranian custom of making and drinking wine, and the method was not introduced until the Tang (Schafer, 1963). Wine-making subsequently seems to have been carried out principally in the northern province of Shanxi, whence the product was carried all over the country. Even now, however, wine made from grapes has not become as popular as it has long been in the West, failing so far to oust traditional rice wine. Nevertheless locally made wines appear on the shelves of almost every provedore today at a most reasonable price. I have found the whites of the Great Wall and Dynasty brands, while not perhaps masterpieces of the oenologist's art, a great comfort at the end of a hard day's sightseeing. Various joint ventures with foreign winemakers are now under way, so undoubtedly the quality will greatly improve and the number of brands and types will rapidly increase, as no doubt will the prices.

Vineyards are seen chiefly from the Yangtze valley north and the plants are trained in a manner different from that usually followed in the West (fig. 1.3). They are allowed to develop very long main leaders and are grown over arbours or flat trellis-work supported about 3 m above ground, as described by Fortune (1863). In the north, at the approach of the first frosts, the vines are taken down and, after the laterals have been pruned back and the leaders coiled up if necessary, they are buried to protect them from the severe cold of winter. Whether this is really necessary in all areas where it is practised I do not know, but it is still done, the plants being unearthed again in spring and replaced on the

Figure 13.16 *Trachelospermum jasminoides*

supports. On the other hand, in the more recently planted vineyards of the joint ventures, the vines are grown according to the usual Western practice.

As well as being grown on a commercial scale, grape vines are trained over arbours in various gardens in China, as they are in other parts of the world. No doubt they are appreciated in the same way for their appearance, the shade they provide, and for the fruit they produce. There is a good example in the garden of Gongwangfu in Beijing. I was interested to see, at the beginning of November 1996, that the vines there had been taken down, pruned and coiled up ready for burial, just as happens on a larger scale (fig. 13.17). Also it appears that, in the past at least, grape vines have been dwarfed for ornamental purposes (Livingstone, 1822).

The grape has also been depicted frequently in paintings and used as a decorative motif or border pattern (Williams, 1975). For example, Bartholemew (1985a, cat. no. 2) illustrates a Tang bronze mirror with a raised decoration of grapes and animals, Krahl (1987) records its portrayal on porcelain, and it appears in a 16th century painting reproduced by Kao (1988, no. 50.5). According to Koehn (1952), its long trailing habit has been interpreted as indicating continuity.

Figure 13.17 Grape vines prepared for winter burial at Gongwangfu, Beijing

WISTERIA Leguminosae

Wisteria is a small genus with representatives in eastern Asia and eastern North America. It seems probable that only one species, *W. sinensis*, is native to China and this is the species seen in gardens throughout the country. At some stage the Japanese species *W. brachybotrys* and *W. floribunda* have been introduced and are occasionally planted. It is not known when they were first brought to China. However, just as in Japan the Chinese species is rarely grown, so it is in China with the Japanese ones.

W. brachybotrys 白花紫藤 (Baihua Ziteng)
White-flowered Wisteria

SILKY WISTERIA

I have seen this Japanese species only once in China, grown as a large penjing specimen in the Shanghai Botanic Garden. This example was of the white-flowered cultivar known in Japan as 'Shiro Kapitan' and long cultivated in the West under the name *W. venusta*. It twines in an anticlockwise direction, as does *W. sinensis*, but is distinguished from it by its persistently pubescent foliage, its larger individual flowers, and its shorter and broader racemes.

W. floribunda 多花紫藤 (Duohua Ziteng)
Many-flowered Wisteria

JAPANESE WISTERIA

Except in botanic gardens, I have not seen *W. floribunda* in China other than trained as penjing. It twines in a clockwise fashion and is thus easily distinguished from *W. brachybotrys* and *W. sinensis*, from which it also differs by having smaller flowers. The few plants I saw in China had mauve flowers and appeared to be seedlings. I have not encountered any of the named cultivars which are such a feature of Japanese gardens and which are also to be seen in Korea.

Wisteria sinensis 紫藤 (Ziteng) *Purple Vine*

CHINESE WISTERIA

One of the most spectacular garden plants, this wisteria (figs 2.11, 3.6, 13.1, 13.18, 13.19, 13.20) has been grown in Chinese gardens for a very long time, being admired both for its blossoms and its dragon-like, gnarled and twisted trunks. Even so it does not seem to have attracted the attention of the artists, poets, and writers in quite the same way that many other plants have, so it is probably not possible to trace its history accurately. Nevertheless Uehara (1961) says

Figure 13.18 *Wisteria sinensis* 'Alba' growing over a rock pile in a garden beside the Slender West Lake, Yangzhou. *Yucca gloriosa* can be seen on the far left and *Ophiopogon japonicus* in the foreground.

that it is often mentioned in old Chinese poems, Wang (1988) includes it amongst plants cultivated in the Six Dynasties Period, and Schafer (1965) quotes a reference indicating that it was a favourite in Changan courtyards in the Tang. So it seems that it has been cultivated for at least 1500 years and today there is hardly a garden without it. Except for the white forms, most of the cultivated plants appear to be seedlings, show considerable variation, and look no different from plants growing wild.

There appear to be two white cultivars in China at present. One, of which there are specimens in the Yu Yuan, Shanghai, and beside the Slender West Lake, Yangzhou, agrees in all particulars with the one introduced by Fortune and known in the West as 'Alba' (fig. 13.18). The other, which I have seen in the Zhuozheng Yuan, Suzhou, at the Huanglong Si, Hangzhou, and in the Guyi Yuan, Shanghai, is very similar but more strongly scented. It appears to be identical with the cultivar 'Jako' sold by Japanese nurseries. In the West these cultivars exhibit greater cold-hardiness than the mauve form introduced to England from Guangzhou and presumably they originated further north.

Not only is *W. sinensis* grown for ornament, but the seeds, stems, and flowers are all used in Chinese medicine. Also it is listed and illustrated in the *Jiuhuang Bencao*, where it is stated that the flowers can be eaten in times of food shortage. Both Du Cane (1908) and Wyman (1949) mention the use of the flowers as food in China and, as I have earlier recorded (Valder, 1995), the parboiled flowers are eaten fried or made into sweet cakes. The making of these cakes in Beijing has also been mentioned by Tun (1965), who referred to them as *tengluo bing* (wisteria cakes), *tengluo* being another name for the plant.

In spite of its popularity over the centuries, Chinese illustrations of *W. sinensis*, recent times apart, seem to be rare. The earliest I have found so far is the woodcut in the *Jiuhuang Bencao* (fig. 2.17). This is followed by a painting of about 1530 of a wisteria climbing over a dead tree (Barnhart, 1983, cat. no. 23), then from the 17th century on it has been depicted more frequently. According to Bartholemew (1985a) it was a popular subject with certain 20th century artists, who took delight in using their calligraphic skills to depict the twining branches.

Its use as decorative motif is attested by the exhibition at both the Imperial Summer Villa, Chengde, and the Eastern Qing Tombs, 125 km east of Beijing, of gowns belonging to the Dowager Empress Cixi, which are embroidered with wisteria blossoms. She seems to have been fond of the plant, as it is also depicted on a set of porcelain bowls which she commissioned (Bartholemew, 1985a, cat. no. 16). And Mrs Archibald Little (1901), who lived in Beijing early this century, wrote: 'One of the most fairylike pieces of embroidery I have ever seen was mosquito-curtains worked all over with clusters of wistaria for either the Emperor or Empress, and somehow or other bought, before being used, out of the Imperial Palace by a European collector'.

Figure 13.19 *Wisteria sinensis* trained through a rock in the garden of the Song Qingling Museum, Beijing, with a potted pomegranate in the foreground

which was about 1.8 m high and had its branches trained horizontally in a symmetrical fashion, so that when it was in bloom it resembled a floral fountain. And early this century, in a courtyard near the Audience Hall at the Summer Palace outside Beijing, there were two immense baskets 4–5 m tall, built of natural logs, over which wisteria grew. These were great favourites of the Dowager Empress and she took delight in showing them to people when they were in bloom (Der Ling, 1911). Alas, no trace of these structures remains and I have not come across anything similar elsewhere.

The first known mention of this plant by a European was made by Domenicus Parennin, a French Jesuit missionary who went to China in 1698 and died in Bejing in 1741. In a letter he mentions a climbing plant, 'teng lo' (*tengluo*), with beautiful violet flowers hanging down in large bunches (Bretschneider, 1898). In and around Beijing wisteria is still occasionally referred to by this name, as in the case of the *tengluo bing* mentioned above. However, attention was not drawn to the plant in Europe until it was seen early in the 19th century in the garden of a merchant in Guangzhou, some 1900 km south, further south even than its known natural range in the wild. It was introduced from there to England in 1816, presumably due to the efforts of John Reeves, who wrote in his retirement of the plants he obtained '… it was only by engaging to purchase them that I got Wisteria propagated for sale …' (Le Rougetel, 1982).

Some years later, in 1844, Robert Fortune saw numerous plants, both cultivated and growing wild, in the Yangtze delta region. He also encountered a white form growing in a garden in Ningbo, from which layers were made and sent to the Horticultural Society's garden at Chiswick (Fortune, 1847a). He was also able to obtain a white form from a nursery in Suzhou, but whether it was the same as the one from Ningbo and what happened to it seems not to have been recorded.

Mrs Little's use of the spelling 'wistaria' prompts me to point out that, even though the genus was named after Caspar Wistar by Thomas Nuttall in 1818, he spelled it *Wisteria* and subsequently this has become accepted internationally in botanical circles.

The plant is usually seen growing in Chinese gardens trained in one or other of several traditional ways, either as a shrub, on a pergola or portico, on bridges, on rocks at the waters edge, over a rock-pile (figs 13.1, 13.18), or even through an ornamental rock (fig. 13.19). Sometimes, too, it is allowed to climb in a natural fashion over a living tree, and in old gardens and temple grounds it is frequently planted to grow over ancient trees which have died (fig. 2.11). It is also a popular subject for penjing. Robert Fortune (1851) was greatly impressed by a specimen he saw in Shanghai,

Unless it has occurred in recent times, I know of no evidence that *W. sinensis* has been imported to Europe again. Ernest Wilson (1916) earlier expressed the same view and pointed out that almost all the plants in cultivation outside China had descended from the original introductions. The only plant of independent origin of which he knew was one on a house in Boston raised from seed received from Shanghai in 1887.

Plants growing in Chinese gardens have continued to impress European visitors and residents. Juliet Bredon (1931) devoted quite some space to the wisterias in Beijing, including amongst her remarks that 'their pendant flower clusters, hiding all traces of the stem, give the distant effect of mauve smoke rising among the trees' and that 'No one knows better than the Chinese gardener how and where wisteria should be planted to show off the long purple trails of blossom to best advantage'. Osbert Sitwell (1935, 1949) was similarly impressed by Chinese methods of training and cultivating this plant. One of the most famous specimens grew at the Heilongtan (Black Dragon Pool) outside Beijing on the way to the Dajue Si. It seems foreigners were allowed to bathe in the pool over which it hung (Hubbard, 1923). George Kates (1967), who went to Beijing in 1933 and stayed seven years, wrote:

> The glory of this place in the spring was the oldest, the largest, and the most writhing wisteria vine that I ever beheld. A stout framework, so extended so as to cover part of the pool, supported it; and when the flowers were in bloom, and the air alive with the urgent sound of bees, one might swim below this trellis as lavender blossoms dropped onto the green water. China gave, neither measuring nor caring, so many pleasures like these!

Dorothy Graham (1938) also became effusive when describing this particular vine. I tried to find it in 1996 but, although the locality retains the name, the people consulted there knew nothing of either a pool or a wisteria. Perhaps I was expecting too much, as some 60 years earlier Osbert Sitwell (1949) observed that most places in and around Beijing were rotting away, 'even Dragon Pool and Jade Fountain', and predicted that soon little would be left.

Apart from Ernest Wilson (1913), who noted that plants growing wild near Yichang varied very much in shade of colour, no-one seems to have commented on the variability of this plant. However, variants in China differ from one another almost as much as do the named cultivars of *W. floribunda* in Japan. There is thus considerable scope for making further introductions. The forms growing in and around Beijing, for instance, must surely exhibit greater cold-hardiness than the earlier introductions from areas to the south. Not only this but amongst them are individuals with very long racemes, in most of which the standard petal bears a predominantly white rather than a yellow blotch (fig. 13.20). These long racemed forms are much more effective on pergolas than is the original introduction from Guangzhou. I obtained seeds from Beijing in 1994 and succeeded in raising about a dozen plants, but they have yet to flower.

Figure 13.20 *W. sinensis*, northern type, in Beijing

CHAPTER 14

AZALEAS

Azaleas belong to the genus *Rhododendron*, of which China has more species than any other country. While many of these are astonishingly handsome and their introduction to other parts of the world has caused considerable excitement, in China itself only a few have found their way into traditional gardens and almost all of these are the types we know as azaleas.

Azaleas and rhododendrons are generally known in Chinese as *Dujuan*, which also means 'cuckoo'. Apparently this name arose from a legend in which red azalea flowers sprang up from blood spat out by the constantly calling cuckoo (Li, 1959). Two scarlet lines mark the corners of the cuckoo's beak and, in another version of this legend, it is said that it spits out blood as it sings, staining the azalea red and hastening spring (Wang & Ma, 1995). In the 9th century Li Deyu refers to the azalea as *Dujuan* and a poem, of perhaps a century later, associates it with the seasonal singing of the bird at its time of flowering in Sichuan (Schafer, 1965).

Apart from this, azaleas apparently bring to the Chinese mind few of the associations of the type conjured up by so many other garden plants,

Figure 14.1 A cultivar of *Rhododendron simsii* in the Yi Yuan, Suzhou

though from time to time they are seen as symbols of good fortune, fire, blood, or the so-called fair sex (Schafer, 1965, 1967; Williams, 1975). Likewise, in spite of their popularity as garden plants, they have rarely been the subjects of paintings. The earliest I have come across is dated 1540 (Barnhart, 1983, cat. no. 26) and occasional examples can be found in works of the 17th, 18th, and 19th centuries.

The first references to the cultivation of azaleas in China date from the first half of the 9th century AD, when wild species were brought into the gardens of the Yellow River valley from the south (Schafer, 1965, 1967). Amongst the writers of the time they were praised by Li Deyu, who grew them in his garden near Luoyang, and Meng Guan thought they surpassed all the southern flowers. But, according to Schafer, not everyone found them entrancing. Han Yu thought they were interesting when massed but not to be compared with the flowering trees of his northern home. Whether azaleas had been grown before this in southern China we do not know. However, it is clear that from that time on they became popular garden plants. For example, after the orchids, the azalea was the favourite plant of Shen Fu (1983), a native of Suzhou whose *Six Records of a Floating Life* is believed to have been written in 1809. In this book he mentions that, although it has no fragrance to speak of, its colours are long-lasting and it is easy to prune. However, he complained that his wife would not let him prune his plants heavily, so it was difficult for him to train them properly.

There is an interesting mention of *Duyuan* in Zhang Zhichun's *Yongchang Erfang Ji* (Notes on Two Ornamentals at Yongchang) of about 1495, where 20 kinds are recorded, including 'yellows, purples and greens' (Needham, 1986). There has apparently been no other detailed treatment of the genus in the Chinese literature until the present century. As far as I know this book has not been translated into any Western language, but it would be interesting to know whether the 20 kinds were all azaleas, as Yongchang (now Baoshan) lies in western Yunnan between the Salween and the Mekong, a region rich in rhododendron species. In general, however, the conservatism of Chinese gardeners is such that, apart from azaleas, other rhododendrons are almost universally ignored except in botanic gardens.

There seems to have been a weakening of this resolve only in areas rich in wild species. In Sichuan and Yunnan, for instance, interest in the genus has been stimulated by botanists studying the Chinese flora and by collectors from abroad. As a result one may now occasionally come across collections of species in gardens and temple grounds. Also I have seen *R. decorum* used for penjing in Yunnan and large plants of *R. arboreum* subsp. *delavayi*, apparently dug from the wild, grown in containers. But if species such as these are ever planted in the ground this must be unusual. For what it is worth, I have not seen it except in botanic gardens. But then, except in municipal plantings, azaleas are usually kept in containers and brought out only when in bloom (pre-title pages and figs 14.2, 14.11), as are so many of the plants used to ornament gardens. Also they are often dwarfed and trained as penjing.

An early European mention of the Chinese azaleas is made by Clarke Abel (1819), who saw them at the Fa Tee Gardens in late 1816 or early 1817: 'The *Azalias* exhibited no striking variety of colour, being either white or red, of different degrees of intensity; but were certainly the most beautiful plants I met with in China'. After visiting these gardens almost 30 years later, Robert Fortune (1844) described the azaleas as 'most splendid'. He said that the principal kinds grown were 'Indica, Indica alba, phoenicea, lateritia, variegata and the yellow *Azalea sinensis*'. These were probably *R. simsii*, a white form of *R. mucronatum*, a brick red form of *R. simsii*, something like the cultivar 'Alphonse Anderson', and *R. molle*, respectively. Fortune (1852) also visited the nurseries near Shanghai which specialised in azaleas, from which he sent to England some cultivars with white flowers striped with red or purple, generally considered to be forms of *R. simsii*, together with forms of what became named *R. × obtusum*, said to have come from Suzhou. In writing about the Shanghai azaleas, Fortune (1852) said that layering was the common method of propagation, but that grafting and striking from cuttings were also carried out successfully. He also reported that, during the hot summer months, both young and old plants were shaded from the midday sun. An interesting, non-horticultural use of azaleas in China is recorded by Morris (1983), who says that the roots are dried, sanded, and lacquered to become stands for plants, the crown being cut across to form a flat surface on top.

While Chinese azaleas may have been taken to Japan long ago, they were not introduced to other parts of the world

Figure 14.2 (opposite page) **Various azaleas in the Zhuozheng Yuan, Suzhou**

until the first half of the 19th century. In Europe they were used to breed hundreds of cultivars, the principal group being the so-called Belgian Indian Hybrids. While in areas with cold winters the Belgian hybrids are used mainly as pot plants for indoor decoration, they, together with forms with similar parentage and the original Chinese introductions, have proved enormously successful out of doors in milder parts of the world, such as the southern United States, Australia, and New Zealand. I have seen forms of *R. simsii* and *R. × pulchrum* growing well not only in southern China but in Indonesia and north Queensland. On the other hand, in the colder areas it is principally the Japanese azaleas and their derivatives which are grown in the open.

The azaleas seen in Chinese gardens nowadays are almost all plants which have been developed in cultivation, some of which have no known wild antecedents. Thus to assign them to species is not an easy task and probably not even sensible. All the same they have been referred to one or other of *R. indicum*, *R. molle* , *R. mucronatum* , *R. × obtusum*, *R. × pulchrum* and *R. simsii*, so I shall follow suit in describing them below together with *R. farrerae*.

RHODODENDRON Ericaceae

R. farrerae 丁香杜鹃 (Dingxiang Dujuan) *Lilac Azalea*

Although it seems unlikely that this early-blooming deciduous azalea is grown in Chinese gardens, I have included it

Figure 14.3 *Rhododendron farrerae* on a hillside in the New Territories, Hong Kong

because of its association with the Chinese New Year as a symbol of good fortune in the south-east of the country, where it is native. Metcalf (1942) mentions its presence in the markets of Guangzhou at this season, but does not say whether he saw cut branches or plants dug from the hillsides. It is a very beautiful species with pink or mauve-pink flowers produced before the leaves. I have seen it on Victoria Peak, Hong Kong, and in the New Territories (fig. 14.3). It has proved difficult to grow and is rarely seen outside its native habitat.

R. indicum 印度杜鹃 (Yindu Dujuan) *Indian Azalea*

This is a Japanese species Linnaeus named *indicum*, as it had been introduced to Holland from the East Indies somewhere around 1680, though doubtless it had been brought there originally from Nagasaki (Wilson & Rehder, 1921). While the wild plant usually has red flowers, it has given rise to many cultivars in a wide range of colours and forms. When *R. indicum* was first brought to China I do not know, but cultivars of this species were seen last century near Shanghai by Fortune, who sent back a form with frilled petals which became known as 'Crispiflorum' (Galle, 1985; Wilson & Rehder, 1921). It seems probable that the azaleas identified in China as *A. variegata* belong here, as Fortune (1852) said they flowered fully three weeks or a month later than the type he called *A. indica* (*R. simsii*).

Metcalf (1942), commenting on flowers available for the Chinese New Year in Guangzhou, says that a few years before the time of his writing the only kinds available there were the locally native *R. farrerae* and *R. simsii*, but that more recently Japanese varieties had been introduced and that many beautiful kinds were available during the holiday season. Of course these may not have been cultivars of *R. indicum*. That they were in flower at this time, whether forced or otherwise, suggests that they may have been Kurume azaleas.

R. molle subsp. molle 羊踯躅 (Yangzhizhu) *Goat Staggers*

This common deciduous shrub has been known as *Yangzhizhu* for centuries because when goats eat it they become paralysed, stagger, and fall. Thus it was recognised early that the plants are poisonous. In fact it is probable that all rhododendrons are poisonous, but in general grazing animals avoid them. The word *yang* also means 'sheep', so

Figure 14.4 (right) **Rhododendron molle**

Figure 14.5 (below right) **Rhododendron**

mucronatum in the Liu Yuan, Suzhou

perhaps they are included too. However, sheep are more discerning when it comes to diet than are goats, so are less likely to be affected. On account of its toxic properties the Chinese subspecies of *R. molle* has long been exploited as a narcotic drug and pesticide (Needham, 1986; Schafer, 1965; Young & Chong, 1980). Sometimes it is called *Naoyanghua* (Trouble-goat Flower), or simply *Huang Dujuan* (Yellow Azalea). It is broadly distributed in the Yangtze valley and south to Fujian and Guangdong.

Although pharmacologists of earlier ages had found it of interest, no-one, it seems, accorded it aesthetic merit until Tang times, when it is referred to as growing splendidly in a garden in a poem of the 9th century (Schafer, 1965). It is certainly very handsome with its scented yellow flowers which appear just before and with the leaves (fig. 14.4). Nevertheless for some reason it remains rare in Chinese gardens. I have not seen it in any of those I have visited. Ernest Wilson also reported that he had not seen it in Chinese gardens during his visits early this century, but that long ago it must have been grown in those of Guangzhou, Hangzhou, and Suzhou (Wilson & Rehder, 1921). However, Robert Fortune (1844) mentioned seeing it at the Fa Tee Gardens.

While the Japanese variant of this species, subsp. *japonicum*, is well-known abroad and has given rise to the popular Mollis Azaleas, subsp. *molle* is rarely seen outside China. Forms from the south of its range would probably be useful for breeding deciduous azaleas suited to areas with hot summers and mild winters. Collecting seeds, however, may not be easy, as I have been told that the capsules are usually picked before they ripen for use in Chinese medicine.

R. mucronatum var. *mucronatum* 白花杜鹃

(Baihua Dujuan) *White-flowered Azalea*

Most of the large-flowered white azaleas commonly seen in Chinese gardens (fig. 14.5) tend to be placed in this variety of *R. mucronatum*, which does not seem to be known in the wild in China and has probably been derived from var.

ripense, a native of Japan. A white form called 'Indica Alba' or 'Ledifolia Alba' and a double mauve called *Azalea indica* var. *plena* were introduced to England from China by Joseph Poole in 1819, and in the middle of the century a double white was introduced by Fortune under the name *A. narcissiflorum* (Galle, 1985; Lee, 1978; Wilson & Rehder, 1921).

Figure 14.6 **Kurume azaleas in the Zhuozheng Yuan, Suzhou**

R. × obtusum 石岩杜鹃 *Shiyan Dujuan* (Rocky Crag Azalea)

Amongst the azaleas sent back in the middle of last century by Robert Fortune were three with small leaves and flowers (Wilson & Rehder, 1921). Fortune found two of these during a visit to two nurseries near Shanghai, known to the foreign residents as the 'Pou-shan Gardens', and was greatly impressed by their neat habit and brilliant flowers (Fortune, 1852). Another with white flowers was sent from Hong Kong, but whether Fortune collected it there or on the mainland is not known (Wilson & Rehder, 1921). The first of these azaleas to be named, a single cerise, was called *Azalea obtusa* by Lindley, later being transferred to *Rhododendron*. The second, with 'hose-in-hose' flowers, was named *Azalea amoena*, and the white one *A. ramentacea*. These have subsequently come to be regarded as cultivars of *R. × obtusum* (Davidian, 1995).

Fortune is recorded as saying that *Azalea amoena* had been brought to Shanghai from 'the far-famed city of Soochow-foo' and that further than that its origin was unknown. He speculated that it probably came from a country further north than any of its race in China or from a higher elevation, and that it was likely to prove hardy in Britain (Lindley & Paxton, 1853). It now seems certain that these showy, small-flowered plants are Japanese, although it is not known when azaleas of this type were first brought to China. However it is clear that they fall within the range of variation of

the Kurume group, thought to have been derived from *R. kaempferi* and *R. kiusianum*. These two species hybridise naturally in Kyushu where their ranges overlap but, since both are usually deciduous and the Kurume azaleas are usually evergreen, it seems possible that other species such as *R. indicum* have been involved. Anyway, if a botanical designation is wanted for this group of cultivars, it seems sensible, for the time being at least, to group them all in *R. × obtusum*. There are now hundreds of cultivars in a wide range of colours and floral forms.

The Kurume azaleas did not begin to reach the West from Japan until early this century, and the story of their introduction is given in detail by Lee (1978). However, similar cultivars must have reached China much earlier, as Fortune's observations and collections make clear. Whether other azaleas of this type were present in China at the time of his visits I do not know, but dozens of cultivars are now grown in many parts of the country (fig. 14.6). They are especially popular for training as penjing and exhibitions of these plants are mounted at flowering time. As well as this, large plants are brought out when in bloom and placed on stands in gardens along with cultivars of *R. simsii*, *R. indicum*, and *R. × pulchrum*. Also I have seen them in street plantings, hundreds of plants being grown in this situation in Ningbo, for instance. Whether any cultivars have been raised in China I do not know.

R. × pulchrum 锦绣杜鹃 (Jinxiu Dujuan) *Splendid Azalea*

The plants placed here are amongst the azaleas most frequently seen in Chinese gardens. It has been suggested that they have arisen from crosses between *R. mucronatum* and *R. simsii* or *R. scabrum*, but what evidence there is for this I do not know. There are no known wild equivalents and it seems that they have arisen in Chinese gardens. They form large vigorous shrubs and those most commonly seen have magenta flowers (Pre-title pages and fig. 14.7). An example of this type was introduced to England in 1824 and subsequently named *R. phoeniceum*. In Japan there is a widely grown cultivar known as 'Omurasaki', a name which means 'Big Mauve', and a form introduced to Australia has become one of the most popular azaleas there with the name 'Magnifica'. In the United States the plant known as 'Formosa' is similar and other variants have been called 'Maxwellii' and 'Smithii' (Galle, 1985; Lee, 1978; Wilson & Rehder, 1921).

Almost as common in China are forms with pink flowers, white at the edges, and with a darker blotch (Pre-title pages and fig. 14.8). These presumably are chimaeras, as occasionally they produce branches of the magenta form. The plants grown in Australia as 'Alphonse Anderson' and in the United States as 'George Lindley Taber' are of this type. It

may well be that the 'variegata' which Fortune (1844) saw in the Fa Tee Gardens, apparently in bloom at the same time as *R. × pulchrum*, was something along these lines too. It seems unlikely that the 'Variegata' now considered to be a cultivar of *R. indicum* would be in bloom so early. These forms of *R. × pulchrum* cannot stand severe frost and apparently they are grown without protection only from the Yangtze valley south.

Figure 14.7 (right) *Rhododendron × pulchrum*, **magenta form and** Figure 14.8 (below) *Rhododendron × pulchrum*, **pink form, in Hangzhou**

Figure 14.9 (left) A large-flowered white azalea at Beigushan, Zhenjiang

Figure 14.10 (below left) *Rhododendron simsii* on Victoria Peak, Hong Kong

I have also seen in China azaleas with large white flowers (fig. 14.9) which appear to belong here rather than in *R. mucronatum*. They are very similar to or possibly identical with a plant grown in Australia as 'Alba Magna'. This has scented white flowers spotted with green, cannot stand severe frost, and does well in areas with mild winters.

R. simsii 杜鹃 (Dujuan) *Cuckoo*

This species is found abundantly from the Yangtze valley south and extends into Burma and Thailand. In the wild its flowers are almost invariably some shade of vermilion (fig. 14.10), and its alternative Chinese name *Yingshanhong* (Reflect Mountain Red) indicates its abundance on many hillsides. Along with *R. molle* it is known to have been brought into cultivation in the 9th century AD (Schafer, 1965, 1967) in the Yellow River valley region and called *Dujuan*, the name used by Li Deyu and his contemporary, Bai Juyi. Li Deyu also refers to it as *Shanliu* (Mountain Pomegranate) and Bai Juyi as *Hong Zhizhu* (Red Staggers). Zhang Ji wrote of the *Zhizhu* flower reddening the head of a mountain stream. The use of these names suggests that it was recognised that it had poisonous properties of the type exhibited by *R. molle* or at least that the plants were closely related.

Because of a poem which mentions an azalea cultivated 'close to flagstones', it is assumed that this species was kept in gardens in the first half of the 9th century. Apparently it had been brought north from Jiangxi, where perhaps it had been cultivated earlier. That it created an impact is made clear by the 9th century poets who wrote of the cuckoo's blood colouring its branches or dripping onto it to form the flowers, and compared it with fire, red stars falling from the sky, or the rosy cheeks and ruddy jewels of exciting women (Schafer, 1965). Azaleas can certainly be said to have got off to a good start.

R. simsii was seen by many of the early European visitors to southern China, although it was usually identified as *R. indicum*. Metcalf (1942) records its popularity as an ornament for the Chinese New Year in Guangzhou but, in spite of its abundance in the wild, the ordinary red form is not frequently grown in Chinese gardens nowadays. However, various cultivated azaleas are considered to belong to this species, including the white varieties striped with purple or red which Robert Fortune introduced from Shanghai and which provided such a boost to the breeding of the so-called Indian Azaleas (Galle, 1985; Lee, 1978; Wilson & Rehder, 1921). There are many plants of this type in Chinese gardens, but where their affinities really lie is unclear. Early this century George Forrest recorded examples cultivated near Dali as having flowers 'rose to white' (Millais, 1917). Forms differing in their generally smaller and paler flowers have been distinguished as var. *mesembrinum* (Chamberlain & Rae,

Figure 14.11 An azalea of the *Rhododendron simsii* type in a pavilion at the Yi Pu, Suzhou

1990). In 1980 I saw similar plants in a wide range of colours in Kunming, and again in Suzhou in 1996. There are also many doubles which flower at the same time and which presumably have been derived from *R. simsii* (fig.14.11). For all I know these may be Belgian Indian kinds which have found their way to China. Anyway these types and others which bloom about the same time are usually called *Dahua Zazhong Dujuan* (Large-flowered Hybrid Azaleas).

CHAPTER 15

ROSES

The general name for plants of the genus *Rosa* in China is *Qiangwei*. Numerous garden varieties exist but their history does not seem to have been as well documented as that of many other flowers. Even so an emperor of about 140 BC is said to have judged a rose to be more beautiful than the smile of his favourite concubine (Le Rougetel, 1988) and, according to the *Zhongguo Huajing* (China Floral Encyclopaedia), roses were grown in the gardens of the Emperor Liang in the period 502–547 AD (Ogisu, 1996). During the Song dynasty rose cultivation flourished and Hazel Le Rougetel (1988) quotes a Song poem which mentions a rose which blooms repeatedly throughout the season. This type of rose has long been known as *Yuejihua*, which literally translated means 'Month Season Flower', a name thought to allude to its being in bloom every month of the year.

An examination of the literature has shown that, as well as being grown for their appearance, roses have had more practical uses in China. For instance, it is recorded in the *Jiuhuang Bencao* that the leaves can be eaten in times of food shortage

Figure 15.1 A cultivar of the *Rosa multiflora* type at the Beitang, Beijing

(Bretschneider, 1881). Also, roses have been employed in various ways in Chinese medicine and the petals have been used in the preparation of scents and flavourings.

In spite of their general appeal, during the late Ming the view was expressed by Wen Zhengheng, as mentioned in Chapter 2, that climbing roses were not really flowers for people of taste (Clunas, 1996). Clunas's translation of Wen's remarks on the subject is as follows:

> I once saw in someone's garden a screen made from bamboo, with five-coloured climbing roses trained across it. The Bank's roses were trained on a wooden frame, called a 'rose arbour'. When they flowered he would sit beneath it—what difference is there between this and dining in the market place? However neither of them can be planted without a framework, so perhaps they should be planted round the women's quarters for the servant girls to pluck—this is just about acceptable.

While Wen's views may not have been widely held, the fact remains that, in spite of their popularity, roses do not seem to have become involved in the Chinese world of art, literature, and symbolism to the extent that many other garden plants have. There is only one Chinese book of which we know devoted entirely to these plants, a fact which perhaps says something about the esteem in which roses were held by Chinese scholars. This book is the *Yuejihua Pu* (Treatise on the Monthly Rose) by an 18th century writer who used the bureaucratic pseudonym Pinghuaguan Zhu (Master of the Flower-Criticism Office). According to Needham (1986) it is a work of value, giving much information on grafting, pruning, climatic and edaphic requirements, and control of pests. Le Rougetel (1988) gives a précis of part of a translation of it, which deals with the romantic names given to various cultivars. Cantoniensis (1867), incidentally, recorded that roses in China were grafted by making an incision in the stock and inserting a young scion beneath the bark, a procedure akin to budding.

Le Rougetel (1988) and Phillips and Rix (1993) have given interesting accounts of roses in China and have summarised the history of their introduction to the outside world. From their observations in China, Phillips and Rix conclude that many of the Chinese garden roses, including the repeat-blooming types, very likely had their origin in the west of the country. It was the arrival of repeat-flowering Chinese roses in Europe in the late 18th and early 19th cen-

turies which revolutionised rose breeding, making possible the hybridising which led to the production of the widely-grown modern roses. The early use of the Chinese roses in breeding has been documented by Hurst (1941).

Genetic studies have shown that the repeat-blooming characteristic is brought about by a recessive gene which governs the continuous production of flowering shoots throughout the year (Hurst, 1941). And while the Chinese roses have also contributed various scents, colours, and growth habits due to other genes, this repeat-blooming gene, first exploited by Chinese gardeners over a thousand years ago, has wrought the greatest improvement on the world's best-loved flowering plant.

Just as the Chinese roses were taken to the outside world, more recently the modern roses incorporating their genes have been brought to China. For the present-day visitor this makes it more complicated than ever in deciding what is what, particularly since they are all called *Yuejihua*, the name originally given to the Chinese roses from which they are descended. Fortunately the modern roses do not seem as yet to have gained a foothold in the traditional gardens though, with their enormous flowers and brilliant colours, they are often used as pot plants and in plantings elsewhere. It is to be hoped that they do not oust their indigenous ancestors, for the old Chinese garden roses are plants of considerable character and charm. The principal types are dealt with below, although I have omitted *R. rubus*, which Li (1959) says has been popular in Chinese gardens since the Song, since I have not seen it in any garden nor seen it mentioned by anyone else. The situation is further confounded by the fact that in modern dictionaries the name *Tumi*, which he gives for this plant, is one of the names applied to *Rubus rosifolius*.

An interesting place in which to see Chinese roses is the rose garden in the park surrounding the Temple of Heaven in Beijing. Also, exhibitions of potted and cut roses are held from time to time in the adjoining nursery area. No doubt there are many other places where similar displays may be seen. The shrubby types are planted either in pots or in the ground, and the climbing types are grown on walls, pergolas, and arbours, or over rock piles, as are so many other vines in China. And while many of the roses existing in Chinese gardens exhibit evidence of hybridity, Western botanists have chosen to include them under a variety of species names. Even so, there are many roses in Chinese gardens which do not fit comfortably in any of the categories so far provided. No doubt the situation will be clarified when their taxonomy is re-examined using modern techniques.

ROSA Rosaceae

R. banksiae 木香花 (Muxianghua) *Woody Perfumed Flower*

BANKSIA ROSE

This vigorous climber is usually thornless and grows into a very large plant, covered in spring with bunches of small white or yellow flowers. It is a great favourite in China for growing on walls and over arbours, porches, and doorways. The form most frequently seen is the violet-scented double white (fig. 15.2), but in spite of its popularity it does not seem to have received much attention from the painters and poets. However, what appears to be this plant is represented in a southern Song painting (Suzuki, 1982, vol. 2, S10–013) and in an 18th century *Hundred Flowers* reproduced by Capon and Pang (1981, cat. no. 65), so presumably it is a very old cultivar. Perhaps the 'white tiny roses twisted into an intricate trellis' mentioned by Dorothy Graham (1938) as being used for perfuming rooms were this species.

The double white was sent by William Kerr to Kew from Guangzhou in 1807 and named for the wife of the director, Sir Joseph Banks. The double yellow (fig. 15.3), which occurs less frequently in Chinese gardens and which so far I have not seen in any painting, arrived in England in 1824, having been collected by John Parks (Cox, 1986). The single yellow arrived later, reaching England via Italy in 1871. Alice Coats (1963) says it was noted in Beijing in 1816 by Dr Clarke Abel. I have not seen any single form in Chinese gardens, though the single white, from which garden forms have arisen, is not uncommon in the wild. Chinese fishermen are said to use the bark of its roots to dye their nets (Harkness, 1978). It was not known in Britain until 1905 when an old plant, believed to have been brought from China in 1796, was found in Scotland (Coats, 1963; Pizzetti & Cocker, 1975).

R. × beanii

R. beanii forms a spreading shrub with loose clusters of small double flowers. The petals are pale pink, the inner ones rather narrow and ragged. It is thought to be a hybrid of garden origin and was introduced to England from China in 1844 (Chittenden, 1956). I have not succeeded in finding a Chinese name for this rose and do not know whether it still

Figure 15.2 *Rosa banksiae* var. *banksiae* in the Wangshi Yuan, Suzhou

Figure 15.3 *Rosa banksiae* 'Lutea' in the Sheng Yuan, Shaoxing

Figure 15.4 A cultivar of *Rosa chinensis* in the Zhuozheng Yuan, Suzhou. 'One Thousand Lights',
a cultivar recently imported to Australia, is very similar to this.

exists in China. *R. anemoniflora* is considered by some to be a synonym of *R.* × *beanii*, but this name is used by Chinese botanists to refer to a rose with single pink flowers known as *Yinfen Qiangwei* (Silver Powder Rose), a name which alludes to the whitish undersurfaces of the leaves.

R. chinensis 月季 (Yueji) *Month Season,* 月季花 (Yuejihua) *Month Season Flower*

CHINA ROSE, MONTHLY ROSE

Presumably the Chinese name applied to the cultivars of *R. chinensis* indicates that they are in bloom in every month. They are certainly the most famous of all the Chinese roses on account of this ever-blooming habit, and they became known in the West as Monthly or China roses. They are evergreen, except in areas with cold winters, usually have

double flowers, and have leaves with 3–5 leaflets. The wild single form, which is native to western China, was thought until recently to bear only pink or reddish flowers. However, it has been recorded recently in over ten locations in Sichuan by Mikinori Ogisu (1996), who has reported individuals with flowers which are white, white changing to crimson, and pale yellow deepening to orange-yellow. His study of the Chinese literature has also revealed something of the history of the garden varieties, although their relationship to the wild forms is still somewhat unclear. Most of the cultivated varieties known today are double with pink or red flowers and usually are scented (figs 15.4, 15.5).

It appears that during the Song these perpetual flowering roses were extensively cultivated, 41 named varieties being known in Luoyang alone (Ogisu, 1996). Enthusiasm for roses continued into the Ming, and 300 years ago the Chinese led the world in rose growing, both in terms of the number of cultivars and techniques of cultivation. Even so, roses of this type rarely appear in Chinese paintings, though occasional examples can be found from the 17th century on (e.g. Barnhart, 1983, cat. no. 33; Capon & Pang, 1981, cat. no. 65; Weidner et al., 1988, cat. nos 17, 24, 40) and instructions for painting them are included in the *Jieziyuan Huazhuan* (1975). The China rose also appears as a decorative motif on porcelain (Li, 1959, pl. 1; Krahl, 1987).

It seems probable that many of the old cultivars have been lost, including perhaps the one called 'Five Coloured', against which Wen Zhengheng railed, which Fortune (1847) found in a garden at Ningbo, and which produced red, white, and striped flowers at the same time. Even so, various forms are planted throughout China and are usually seen as arching shrubs up to 2 m tall with stems which are thornless or bear only a few thorns. All those I have seen are double except for two upright types which are widely planted. One of these has single, pale pink flowers (fig. 15.6) and the other, which is similar, has carmine flowers with white centres (fig. 15.7). There are also dwarf repeat-flowering roses which are assigned to *R. chinensis*, including

that called 'Minima' in the West. This produces single pale pink flowers which are usually only about 3 cm in diameter (*Curtis's Botanical Magazine* 1815, t. 1762). Double pink forms are also known, and at the Temple of Heaven in Beijing I saw a tiny, deep scarlet rose of this type.

Figure 15.5 (top left) **A cultivar of *Rosa chinensis* at Wuxi**

Figure 15.6 (top right) **A single pink form of *Rosa chinenis* in the Daguan Yuan, Beijing**

Figure 15.7 (right) **A single bicoloured form of *Rosa chinensis* in the Daguan Yuan, Beijing**

It is said that a Chinese rose, called 'Koushin' in Japanese, had already been brought to Japan by the Heian period (794–1192 AD). Although 'Koushin' means 'every other month', the name must be treated with some care as it may refer not only to *R. chinensis* but in more general terms to roses from China. The oldest definite record of *R. chinensis* in Japan occurs in a diary of the Kamakura period (1192–1333 AD), and it is depicted in a scroll of 1309 (Ogisu, 1996).

The China rose is believed to have reached Europe in the 15th century, possibly via Iran (Gorer, 1970). According to Hurst (1941) we may safely conclude that it was in cultivation in Italy in the early 16th century, as it appears in a painting by Bronzino from that time. It may have been this rose which Montaigne saw in flower in Ferrara in November 1578 and was told that it flowered all year round. It appears that a pink China rose was grown by Philip Miller at the Chelsea Physic Garden as early as 1752, but no more is heard of the species in Britain until the arrival in 1789 of two new varieties which were given the names 'Slater's Crimson' and 'Parson's Pink' (Coats, 1963). These two introductions and two tea roses (cultivars of *R.* × *odorata*), also from China, were used to develop all the main groups of 19th century roses.

R. × *fortuniana* 大花白木香 (Dahuabai Muxiang)
Woody Perfumed Rose with Big White Flower

This rare rose came to Europe from China in the middle of last century. It is a large, scrambling shrub with small, sparse prickles and double white flowers, 5–10 cm across, borne singly. It shows similarities to *R. banksiae* and it has been suggested that it is a hybrid of this species and *R. laevigata*. Whether it still exists in China I do not know. It was introduced by Robert Fortune for the Horticultural Society of London and given its name by Lindley and Paxton (1853, Vol. 2), who recognised that it is probably a hybrid. However they (1853, Vol. 3) went on to use the same name for 'Fortune's Yellow', a plant which is quite different. Hence it may be that some clarification is needed as to which plant the name should properly be appled.

R. multiflora 野薔薇 (Ye Qiangwei) *Wild Rose*

This climber, which is native to China, Korea, and Japan, produces large clusters of small white or pink flowers. It was seen wild in western China on several occasions by Wilson (1913)

and cultivated forms have long been grown in Chinese gardens. The variety *cathayensis* has single pink flowers and is known in China as *Fentuan Qiangwei* ('Powdered-ball Rose'). Another variety is known as *Shizimei* ('Ten Elder and Younger Sisters') (Needham, 1986) or *Qizimei* ('Seven Elder and Younger Sisters') (Li, 1959). This has double, pinkish purple flowers which fade to white, the names presumably having been given to it because of the number of different colours appearing in the flower cluster at the one time. It was sent to Britain by Charles Greville in 1815 and became known as 'Seven Sisters Rose' or 'Grevillei'. A cultivar with double pink flowers had already been introduced to Britain in 1804 and given the name 'Carnea' (Coats, 1963). There are other roses of this type in China, including a handsome, semi-double white (fig. 15.1), which is popular in Beijing and is perhaps a hybrid. It is possible that the single white rose depicted in the 10th-century *Jade Hall Peony* (fig. 2.7) is this species, and variants seen in paintings of the 17th century and later include white and pink singles, a semi-double pink, and the 'Seven Sisters Rose' (Barnhart, 1983, cat. no. 33; Capon & Pang, 1981, cat.no. 65; Weidner et al., 1988, cat. nos 30, 34).

The forms of *R. multiflora* are not repeat-flowering, are deciduous, and often have leaves with 7 or more leaflets and stipules with fimbriated margins, characteristics which separate them easily from the China roses and the tea roses. However, many of the roses in Chinese gardens appear to be hybrids between the multiflora types and the China roses. These have larger flowers borne in smaller clusters, fewer thorns, and fewer leaflets per leaf, but retain the fimbriated stipules (figs 15.8, 15.9, 19.52). Although they bloom only in late spring, they are very showy and are often seen clothing banks or growing on rock piles. Phillips and Rix (1993) noted various examples in western China, and I have seen several kinds in gardens in the Yangtze delta region and in and around Beijing.

Outside China *R. multiflora* has been used to breed climbers and ramblers, and a dwarf form was an ancestor of the first of the polyanthas. *R. multiflora* was superseded as a parent by *R. wichuriana* which was subsequently introduced from Japan. However, it remains an important stock on which to bud or graft other varieties.

Williams (1975) records the superstition that the wild rose should not be grown in a private residence owing to its thorny nature, because it may thus cause dissension in the family.

Figure 15.8 A rose, apparently a hybrid of *Rosa multiflora*, at the Beitang, Beijing

Figure 15.9 A rose, apparently a hybrid of *Rosa multiflora*, in the Xi Yuan, Nanjing

Figure 15.10 A pale cream rose of the tea-rose type on a wall in Hangzhou

R. × *odorata* 香水月季 (Xiangshui Yueji)

Scented Monthly Rose

TEA ROSE

The Chinese garden roses included under this name are thought to have arisen from a hybrid or hybrids between *R. chinensis* and *R. gigantea*, a vigorous climber from Upper Burma and Yunnan with large, scented, white or cream flowers. As well as there being pale pinks, some tea roses have white or yellow flowers, colours usually not mentioned in relation to the cultivars of *R. chinensis*. The double yellow roses depicted in Chinese paintings probably belong here (Sze, 1977; Weidner et al., 1988, cat. no. 34). The earliest examples I have found so far date from the 17th century, but Hurst (1941) says that there are 10th century paintings of pale pink roses which appear to be identical with 'Hume's Blush Tea-scented China'. This appears to be the first rose of this type to reach Europe, and was sent from Guangzhou to Sir Abraham Hume in 1808 (Coats, 1963). A yellow variety, 'Parks' Yellow Tea-scented China', followed in 1824. Although these varieties were less cold-hardy than the China roses and were subsequently lost, they played an important part in the development of the modern roses. Recently these two lost tea roses are said to have been found again, but whether they are exactly the same as those originally introduced is uncertain.

Another interesting rose which can perhaps be included in *R.* × *odorata* is 'Fortune's Yellow', which was found by Fortune (1846, 1847a) in a garden at Ningbo. It is a climber with yellow flowers flushed with red. Also Meyer placed a free-blooming, pale pink pillar rose which he collected in 1908 in this species (Cunningham, 1984). Roses of this type are still seen in China, particularly in the areas with mild winters. Phillips and Rix (1993) found a pink and a yellow in Yunnan, and I saw a pale cream with glossy leaves in Hangzhou in 1996 (fig. 15.10).

How the cultivars of *R.* × *odorata* came to be called 'tea' roses is a puzzle. Perhaps the first introductions to England, if they differed from those now bearing their names, were tea-scented, but this characteristic has not been transmitted to their descendants. Coats (1963) suggests that it was the odour of the *fresh* bruised tea leaf that the scent was supposed to resemble. It has also been proposed that, having been brought on tea clippers, the plants smelled of tea on arrival. On the other hand, perhaps the name arose merely because the first examples to reach Europe were brought on tea clippers. Another reason suggested for the name is that the plants were bought from the Fa Tee Gardens in Guangzhou. A more plausible theory was that proposed recently by Martyn Rix: that the tea which was imported was scented (Phillips & Rix, 1993). Undoubtedly much of it was, and at times rose petals were used for the purpose. For instance, Fortune (1857) listed roses amongst the eight different flowers which he saw being used in a tea factory in Guangzhou, and later authors also mention the use of rose petals for scenting tea (Graham, 1938; Li, 1959). Thus there is considerable evidence in favour of Martyn Rix's theory. It appears that the principal rose employed was *R. rugosa* (Li, 1959), and further details concerning this are given below.

R. roxburghii 缫丝花 (Saosihua)
Silk-reeling Flower

CHESTNUT ROSE

R. roxburghii is a stiff shrub which produces its double pink flowers, darker towards their centres, over a long period in midsummer (fig. 15.11). It was taken to England in 1823 from Calcutta, having earlier been introduced to the Calcutta Botanic Garden from Guangzhou by its superintendent, Dr Roxburgh (*Curtis's Botanical Magazine* 1881, t. 6548). Whether this double cultivar still exists in China I do not know. The wild single form is found in western China (Phillips & Rix, 1993; Wilson, 1913). It gets its Chinese name because it blooms at the time of the silk reeling. In the West it became known as the Chestnut Rose on account of its spiny hips.

R. rugosa 玫瑰 (Meigui)

This shrub rose has large, purplish-pink, single flowers, conspicuous red hips, and lustrous, rugose foliage which turns orange in autumn. It is native to Japan, northern China, Korea, and neighbouring parts of eastern Russia. It is very variable with many varieties in cultivation, most of them having originated in China. A double purple (fig.15.12) and a double white (fig. 15.13) are the kinds usually planted there at the present time.

As mentioned earlier, the petals are used for scenting tea (Graham, 1938; Li, 1959). Fortune (1857), commenting on

3490

Figure 15.11 *R. roxburghii*, from *Curtis's Botanical Magazine* 1836, t. 3490

the use of various flowers for this purpose, indicated that a rose he said was called 'Tsing moi-qui-hwa', a name which suggests that it was probably *R. rugosa*, topped the list in the esteem of the Chinese, narrowly beating the flowering plum (*P. mume*). Likewise it seems probable that the rose petals, which Princess Der Ling (1911) described being brought in a jade cup to the Dowager Empress in case she wished to place them in her tea, were of this species.

Li (1959) says that the petals of the darker-coloured flowers are picked when newly opened and made into a paste with a small amount of water. This is strained, sweetened with sugar, and used in very small quantities to impart its sweetness and fragrance to foods. Presumably this is the 'delicious jam' which Bredon and Mitrophanov (1927) say was made from the roses sold by street hawkers in the 4th lunar month in Beijing, the same month in which Tun (1965) records the making of 'mei-kuei ping' (rose cakes) in that city. A scented liquid known as the 'Dew of Roses' is also made from the petals for preparing a cool summer drink.

Graham (1938), Hubbard (1923), and Needham (1986) record that the flowers were also used for flavouring wine.

Whether any of this still happens I do not know, but Hubbard (1923) gives an evocative description of the peasants harvesting the petals from both cultivated and wild plants at the beginning of June around the 'Miao Feng Shan' (Mountain Peak Temple) in the Western Hills, and sending them down in donkey-loads to Beijing.

Figure 15.12 **A cultivar of *Rosa rugosa* in the Daguan Yuan, Beijing**

Figure 15.13 A cultivar of *Rosa rugosa* in the Daguan Yuan, Beijing

Figure 15.14 *Rosa xanthina* in Zhongshan Park, Beijing

R. xanthina 黄刺玫 (Huang Cimei) *Yellow Thorny Rose*

With its fern-like foliage, arching branches, and profusion of semi-double golden flowers, this rose is a notable feature of Beijing gardens in spring (figs 3.1, 15.14). Although commonly cultivated in northern China, it does not seem to have been brought into cultivation elsewhere until 1906, when Meyer sent it from Beijing to the United States (Cunningham, 1984). It was first described from this semi-double form, but the wild form, *Danban Huang Cimei* (Single Yellow Thorny Rose), which is usually paler, is also occasionally cultivated in and around Beijing (fig. 15.15). It has been named forma *spontanea*. This was also sent to the United States by Meyer who found it in 1907 growing in dry, rocky localities in Shandong and collected seed (Cunningham, 1984). Meyer (1916) also saw it in Shanxi. What look to be both single and double forms appear in Qing dynasty paintings (e.g. Capon & Pang, 1981, cat. no. 65; Suzuki, 1982 vol. 3, JM1–075).

Figure 15.15 *Rosa xanthina* f. *spontanea* at Badachu, near Beijing

CHAPTER 16

THE LOTUS AND OTHER AQUATIC PLANTS

The Chinese use only a small range of aquatic plants in their gardens. Of these the lotus (*Nelumbo nucifera*) is by far the most important. It is cultivated in the ponds and lakes of most Chinese gardens and in smaller gardens and courtyards, where it is, or at least was, planted in large earthenware or porcelain containers specially made for the purpose. Clarke Abel (1819) observed it grown in this manner in the environs of Beijing in 1816, describing it as being prized above all other flowers and raised 'in capacious vases of water, containing gold and silver fish, supported on stands a few feet from the ground. These were surrounded by steps of different elevation, supporting other plants mingled with artificial rocks, representing hilly country and covered with diminutive houses, pagodas, and gardens'. Thus, while not necessarily going to such lengths, even those with very little space could accommodate it and have brought to mind all the thoughts of exaltation which this flower evokes. The fact that seed pods, flowers, and buds

Figure 16.1 A double cultivar of *Nelumbo nucifera*

Figure 16.2 A variegated cultivar of *Acorus gramineus*

seems to have been as a charm to ward off evil influences, particularly on the morning of the 5th day of the 5th month, the day of the Dragon Boat Festival, when it was tied in bunches with artemisia and hung over doors and gates (Gulland, 1928; Tun, 1965). This practice arose from the ancient belief that the artemisia leaves looked like a tiger and the calamus leaves like a sword. In general appearance its leaves resemble those of a large flag-iris. No doubt because of its reputation, it is occasionally seen in paintings and is represented in the *Jieziyuan Huazhuan* (1975). The roots are included among the famine foods listed in the *Jiuhuang Bencao* (Bretschneider, 1881).

A. gramineus 石菖蒲 (Shi Changpu) *Rock Acorus*

A. gramineus is smaller in all its parts than *A. calamus* and its leaves do not have prominent midribs (fig. 16.2). It gets its Chinese name on account of its growing in shallow water on or amongst rocks. There are several known cultivars including dwarf and variegated forms. According to Li (1956a) the smaller types are used for penjing and tray gardens. The plants can be dwarfed by removing all the leaves once or twice a year.

ELEOCHARIS Cyperaceae

E. dulcis 荸荠 (Biqi)

WATER CHESTNUT

The Chinese water chestnut is the corm of *E. dulcis*, which is a sedge. It is grown in shallow water in paddy fields in rotation with rice, the fields being drained before harvest. The round corms, which resemble those of gladioli, are about 4 cm in diameter and consist of crisp white flesh covered by a thin layer of blackish brown skin. Three of these corms are

are present at the same time denotes the three stages of existence—past, present, and future. The many seeds in each seed head suggest abundant progeny and, because its rhizomes are firmly rooted in the mud and its flowers and leaves are numerous, it is a symbol of steadfastness and prosperity in the family. And the pronunciation of the Chinese names for the plant, *Lian* and *He*, is the same as it is for other characters denoting continuity and peace, respectively. Hence the plant brings with it these connotations as well (Koehn, 1952).

In addition to the lotus, I have included here a few other aquatic plants, such as the waterlilies (*Nymphaea* species and hybrids), which may be encountered in gardens, together with one or two which probably would not be but which are interesting food plants or have played an important part in Chinese culture. Others, for example *Marsilea quadrifolia*, a fern with floating leaves each of which looks like a four-leaved clover, appear in paintings, but since I have not seen them in any garden they have been omitted.

ACORUS Acoraceae

A. calamus 菖蒲 (Changpu) *Fortunate Bullrush*

SWEET FLAG

Known generally as Sweet Flag in the West, *A. calamus* gets its Chinese name because of its supposed magical qualities. To see the plant in bloom was regarded as an omen of good fortune and it was believed to prolong life and promote intelligence when taken as an infusion. Its chief use, however,

amongst the various objects depicted in figure 2.15. When peeled they may be eaten fresh or cooked. They are recognisable as the crunchy vegetable in many Chinese dishes.

Cibot (1778b, 5) commented on the water chestnut, '*Pi-tsi*', in Beijing, saying that it was abundant in the gardens of the emperor and much more agreeable to eat than the water caltrop, '*Lin-kiu*'. He recorded that it could be eaten fresh, dried and made into flour, or baked with sugar or honey, and that, as well as being good for feeding poultry, it was eaten by the poor in years of scarcity. He said that it was cultivated by throwing the seeds into deep water at the end of autumn, and that sites adjacent to banks facing south and exposed to the sun were the most suitable.

EURYALE Nymphaeaceae

E. ferox 芡实 (Qianshi), 鸡头米 (Jitoumi) *Chicken-head Rice*

E. ferox is a giant waterlily from Asia with leaves which may be 1 m or more in diameter. These leaves are sparsely spiny on their upper surfaces but their lower surfaces, which become purple as they mature, are covered in radiating, spongy, densely prickly veins. The petioles, flower stalks, and calyx are prickly as well. The relatively small flowers, many of which fail to open, have numerous purple or lilac petals. They are followed by prickly, many-seeded fruits. *E. ferox* is very similar to the species of *Victoria* native to South America but the margins of its leaves are flat, not upturned as in that genus.

The words making up the botanical name of this plant allude to its thorny appearance. Euryale was one of the Gorgons, intimidating women who had serpents instead of hair on their heads, and 'ferox' is a Latin adjective meaning 'fierce'. The Chinese name *Jitoumi* refers to the appearance of the flower bud and fruit. Tun (1965) records that by the middle of the 7th month pedlars on the streets of Beijing were shouting 'Old chicken heads! Just out of the river'. He said that along with water chestnuts they were grown in the imperial waterways. The roots and seeds are listed as famine foods in the *Jiuhuang Bencao* (Bretschneider, 1881).

Cibot (1778b, 6) was the first European to give an accurate description of this striking plant as it occurs in China. He recorded the names '*Le Lien-kien* ou *Ki-teou*' for it, saying that it had been known and cultivated in China since the Zhou dynasty. He mentioned that flour was made from the

seeds and that the root was used in medicine. Professor Zhang Zhiming of the Beijing Botanical Garden tells me that the rhizome, young leaves, and flower buds are eaten as a vegetable, the seeds are still used to make a flour, and the whole plant is used in medicine. Nowadays it is probably grown for ornament only in botanic gardens, as it is at the Beijing Botanical Garden.

NELUMBO Nelumbonaceae

N. nucifera 荷花 (Hehua) *He Flower*

LOTUS

Considering its long history in China, it is not surprising that there are several names for the lotus, the most frequently used, in addition to *Hehua*, being *Lian* and *Furong*. The history of the names for the plant is given by Zou and Zhao (1997), who also trace its history in China. The lotus was well known to the ancient Chinese, and special characters employed in early times for every part of the plant indicate its long and close association with the people.

The stems, leaves, flowers, seed-heads, seeds, and rhizomes have all been used in Chinese medicine (Cibot, 1778b, 1; Wang & Ma, 1995; Zou & Zhao, 1997) and, although they were regarded as famine foods by the author of the *Jiuhuang Bencao* (Bretschneider, 1881), the rhizome has long been a popular comestible and the seeds are much esteemed as a delicacy. In places where bodies of water are common, particularly the Yangtze valley, the lotus is planted extensively for the production of the rhizomes and seeds which are shipped all over the country. The harvesting of these products has traditionally been one of the more pleasant tasks assigned to farmers' children and has inspired many folk songs (Li, 1959). The leaves, both fresh and dried, are used, too, particularly for wrapping foods to be steamed. Also Cibot (1778b, 1) reported that, in his day, heavy smokers mixed the dried leaves with tobacco to reduce its strength.

The introduction of Buddhism to China brought with it the significance that the lotus had acquired in India over a long period, dating back earlier than the worship of Brahma, who is sometimes represented seated on a lotus flower arising from the navel of Vishnu who floats on his back on the ocean (Burkill, 1946; Watanabe, 1990; Williams, 1975). In China, as in India long before, it began to figure in temple architecture and decoration. Images of the Buddha and other deities are invariably placed on stands representing lotus

blossoms. The lotus is also the special symbol of the God of Mercy, Guanyin, and of He Xiangu, one of the Eight Immortals. To the Buddhists it symbolises the perfect man, rising from the mud, reaching up to the light and there revealing a stainless purity and integrity, undefiled by the darkness it has traversed.

Although the tenets of Buddhism were somewhat at odds with the beliefs of the Confucian scholars, they showed little hesitation in adopting the lotus as the symbol of the perfect or princely man. For example a renowned scholar, Zhou Dunyi, in the 11th century wrote a short essay praising the virtues of the lotus which was memorised by every schoolboy in former times. Amongst other things he wrote: 'In my opinion the chrysanthemum is the flower of retirement and leisure; the peony the flower of rank and wealth. But the lotus is the flower of purity and integrity' (Li, 1959). Another translation of Zhou Dunyi's essay is given by Lai (1977), along with translations of numerous poems about the lotus and reproductions of paintings of it. It has long been recognised as the principal symbol of summer, figuring as such in paintings representing the four seasons or the twelve months (Williams, 1975). It is often depicted, too, in its dying phase, with seed heads and hanging, brown leaves. The occasional stems which bear two

Figure 16.3 *Assembled Auspicious Objects*, hanging scroll by Giuseppe Castiglione, 1715, showing double white and pink lotuses, a double lotus seed head, millet (*Setaria italica*), and arrowhead (*Sagittaria sagittifolia*)

seed heads are considered to be particularly auspicious. Such a stem appears in a painting of auspicious objects by the Jesuit missionary Giuseppe Castiglione dated 1715, now in the National Palace Museum, Taipei (fig. 16.3).

Much can be added to what has already been said concerning the symbolism attributed to the lotus. It is often depicted along with a box (*he*) and a *ruyi* to form the rebus *hehe ruyi* (harmony and everything to your wish), an auspicious phrase used at weddings. The lotus and the box are attributes of the twin genii of harmony and mirth, immortals who bestow blessings on marriages. The alternative name *Lian* is also involved in such matters, as the character for 'continuous' is pronounced the same way. Thus a picture of a boy holding a lotus suggests the birth of a succession of sons (Bartholemew, 1985b).

Tun (1965) records that in Beijing at Zhongyuan Jie, which is on the 15th day of the 7th month, from twilight onward boys went along the streets singing, each carrying a lantern made from lotus leaves, in which a candle made a beautiful glow through the green. In a footnote to his translation of Tun's book (Tun, 1965), Derk Bodde says that, coming always at full moon, this was one of the most picturesque of all the Chinese celebrations.

From a purely aesthetic point of view, stands of this famous plant in ponds and lakes are amongst the most impressive sights in China in the summer months. Juliet Bredon (1931) gives a lyrical description of it blooming in the palace moat on the north side of Zhongshan Park in Beijing, to which visitors thronged for 'lotus viewing', as they still do in many places throughout the country. The expanse of green made up of the huge circular leaves with the immense translucent blooms rising above it has long been the subject of universal admiration (Frontispiece, figs 16.1, 16.4). Its stately habit and the great beauty and fragrance of the flowers have inspired many generations of poets and artists. Likewise the sound of rain falling on the leaves and the sight of the glistening droplets remaining in their centres is the subject of appreciation and comment. And the seed heads, *lian peng*, which develop after flowering have a beauty of their own (figs 16.3, 16.5). They are broadly obconical, with each seed (more correctly, each individual fruit) being embedded in a separate cavity on the upper side. These heads, which are

Figure 16.4 (right) **A white cultivar of** *Nelumbo nucifera*

Figure 16.5 (below) **Seed head of** *Nelumbo nucifera*

spongy, eventually break off the stem and float until decay releases the seeds, which then sink to the bottom. Also it is not surprising, on account of the resemblance, that the name given to a shower nozzle is *lianpengtou* (lotus-pod head).

N. nucifera is interesting in that it can be included among the plants which have seeds capable of surviving for very long periods. It seems that peasants in southern Liaoning were in the habit of digging down through the loess soil of their farms to an ancient lake bed to extract peat for fuel and that this peat contained lotus seeds, which were often eaten by children. About 1913 Japanese botanists obtained some of these seeds and showed them to be capable of germination. The palaeobotanist Ichiro Ohga (1923, 1926) investigated the phenomenon and concluded that the lake must have dried up about 400 years previously and that the seeds must be at least this age. Later carbon dating indicates that they are at least 1000 years old. So the children of the villagers had been eating a very long-keeping food.

According to Zou and Zhao (1997), Dr Sun Yat-sen took four of these seeds to Japan in 1919 and presented them to Mr Takashi Tanaka as a token of friendship. When plants raised from these seeds bloomed in 1930 a clone was named 'Sun Wen', Dr Sun Yat-sen's original name. Other such seeds have been found subsequently in northern China (Zou & Zhao, 1997). Although lotus seeds have an impervious seed coat and will not usually germinate unless this is chipped or treated with acid, it is thought that they are most likely to survive for long periods when buried in a moderately moist soil deficient in oxygen.

Although the first mention of the lotus in China appears to be the one in the *Shi Jing* (Book of Odes), no definite record of its cultivation occurs until the Han, when it was planted by Zhao Di, in the 1st century AD (Wang & Ma, 1995). In the Six Dynasties period there is mention of varieties with 'one hundred petals' and 'one thousand petals', and a book of the time lists eleven different kinds (Zou & Zhao, 1997). A white form is known to have been in Luoyang in the 9th century, as Li Deyu wrote a poem praising the white lotuses at the Donglin Monastery (Lai, 1977), though what may well have been a double white had already been described a century before (Schafer, 1965). In later writings white and pink single and double kinds, 'yellows', a dwarf form, and types with more than one flower per stem are recorded (Li, 1959; Zou & Zhao, 1997). Also, the Jesuit priests in Beijing in the 18th century were aware of several kinds, as

evident in Giuseppe Castiglione's painting of 1715, which shows a double pink and a double white (fig. 16.3). Cibot (1778b, 1), later in the century, described red, white, and striped singles and doubles.

In recent times there has been a resurgence of lotus breeding in China, in which, amongst other things, the yellow American lotus, *N. lutea*, has become involved. There are said to be more than 200 cultivars in China now, well over 100 of which are described and illustrated by Zou and Zhou (1997), who also give detailed information on propagation, cultivation, and the techniques of lotus breeding.

Lotuses are planted in ponds and lakes all over the country and collections of cultivars may be seen in many places, including the Zhuozheng Yuan, in Suzhou. And from time to time the Lotus Branch of the China Association of Flowers holds national lotus exhibitions in various parts of the country. For plantings on a grand scale in classical settings, those in Beijing at Behai Park and the Summer Palace are unforgettable sights at flowering time, and so, it is said, are those at the Imperial Summer Villa in Chengde, which were remarked upon over 200 years ago by Cibot (1778b, 1).

The lotus is believed to occur naturally in Asia from Iran to Japan and south to Australia. On the whole it is a plant of the warmer regions but forms from north-east Asia can tolerate remarkably low temperatures. However, it is the general practice in northern China to store the rhizomes over winter in cellars and to replant in early spring. By keeping the rhizomes in cool storage and planting them late, it is possible to delay blooming until the worst of the summer heat is past, a procedure which is recommended for some of the more delicate varieties (Zou & Zhao, 1997).

The lotus was a favourite of the Dowager Empress Cixi, who early this century took a great interest in the annual replanting in the lake at the Summer Palace. She would sit for hours on the Jade Girdle Bridge supervising the eunuchs at their work. At Behai she had herself photographed in an open boat amongst the lotuses (fig. 16.6), and at the same time, surrounded by cut blooms, she was also photographed as Guanyin, the Goddess of Mercy (Der Ling, 1911).

NYMPHAEA Nymphaeaceae

Although waterlilies (*Nymphaea* species and cultivars) are rarely mentioned in the Chinese horticultural literature, the leaves are frequently depicted floating on the surface of the

Figure 16.6 (right) The Dowager Empress Cixi amongst the lotuses, from *Two Years in the Forbidden City* **by Princess Der Ling, 1911**

Figure 16.7 (below right) Waterlily, detail from the *Song Hundred Flowers*

water in old paintings of the lotus (e.g. Kao, 1988, no. 99). This suggests that they may have accompanied it in garden plantings, and the positioning of a plant in bloom beside a lotus in the *Song Hundred Flowers* in the collection of the Palace Museum, Beijing, (fig. 16.7) suggests the same thing. Also Li (1979) considers that the plant called *Shuilian* (Sleeping Lotus) by Ji Han in his *Nanfang Caomu Zhuang* (Account of the Plants and Trees of the Southern Region), written in 304 AD, can be safely referred to a species of *Nymphaea*. Li's translation of Ji Han's entry runs: 'Among the beautiful flowers is the *shui lian* which resembles the lotus but has purple stems, soft and without bristles'.

The common native waterlily in China is the small-flowered *N. tetragona*, to which earlier commentators have referred Ji Han's plant. Li (1979), however, suggests that it is quite possible that this and other early records describe the red waterlily, *N. rubra*, cultivated in India as an ornamental plant since early times. He says that a work of the Six Dynasties Period records that a lotus-like plant with red flowers was planted at the palace in 274 AD, that it was moved about 50 years later but is now extinct, the pond having been filled in. There were many other records of a similar water plant during the Tang and Song, and the flower's habit of opening at night intrigued the early observers.

Whatever the truth of all this is, the waterlilies commonly encountered in Chinese gardens nowadays appear to be hybrids of the type popular in other countries, presumably having been introduced over the years (fig. 16.8). Metcalf (1942) recorded that water lilies sent from Zhangzhou, Fujian, were amongst the flowers sold in Guangzhou at the Chinese New Year.

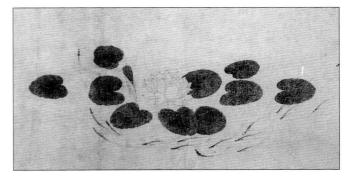

N. rubra 红睡莲 (Hong Shuilian) *Red Waterlily*

INDIAN RED WATERLILY

N. rubra has bronze leaves, up to 45 cm in diameter, and large red flowers, which open at night and are held well above the surface of the water. It is a native of India but, as mentioned above, it is possible that it was introduced to southern China as much as 1700 years ago.

N. tetragona 睡莲 (Shuilian) *Sleeping Lotus*

PYGMY WATERLILY

This species has leaves no more than 10 cm across and white flowers, which are usually only about 5 cm in diameter. The waterlily in bloom depicted in the *Song Hundred Flowers* (fig. 16.7) appears to be *N. tetragona*, and the leaves shown in many Chinese paintings look to be about the size of those of this species. While it may well be planted in garden ponds, I have not noticed it.

Figure 16.8 (left) **Waterlilies at the Wofo Si, Beijing**
Figure 16.9 (below left) *Trapa bicornis* (left) with *Cymbidium ensifolium* and *Magnolia liliflora*, detail from the *Song Hundred Flowers*

The roots and young leaves were among the famine foods listed in the *Jiuhuang Bencao* (Bretschneider, 1881), and Tun (1965) mentions that the young shoots were eaten in the 4th month in Beijing. Later in the season the stems bearing feathery seed heads are used in flower arrangements (Li, 1956a).

SAGITTARIA
Alismataceae

S. sagittifolia 慈姑 (Cigu)
Kind Aunt

ARROWHEAD

S. sagittifolia is chiefly cultivated for its tuberous roots which are eaten as a vegetable (Herklots, 1972; Usher, 1974), but it may occasionally appear in gardens. A cultivar with double flowers is known (Huxley et al., 1992).

PHRAGMITES Gramineae

P. australis 南方芦苇 (Nanfang Luwei) *Southern Luwei*

COMMON REED

This well-known plant is often referred to simply as *Luwei*. Liu (1982) lists it amongst the aquatic plants found in Suzhou gardens. Also it sometimes appears in paintings of waterside scenes (Barnhart 1983, cat. no. 28; She & Yuhas 1973, no. 44F). Depicted with two crabs it forms a motif known as *erjia chuanlu*, a pun meaning 'passing the civil service examinations with flying colours' (Bartholemew, 1985b, fig. 15).

The arrowhead gets its Chinese name because it produces numerous tubers and is thus seen as being 'kind'—generous enough to feed numerous 'nieces and nephews'. According to the *Jiuhuang Bencao*, the stems can also be eaten in times of food shortage (Bretschneider, 1881). It appears in the *Song Hundred Flowers* in the collection of the Palace Museum, Beijing, in Giuseppe Castiglione's painting of 1715 (fig. 16.3), in other paintings (Wang & Barnhart, 1990), and as a decorative motif on porcelain (Bartholemew, 1980, fig. 3; Krahl, 1987). According to Li (1956a) it was recommended by a 17th century writer as a subsidiary component in arrangements of more highly esteemed flowers.

TRAPA Trapaceae

T. bicornis 菱 (Ling), 乌菱 (Wu Ling) *Black Ling*

WATER CALTROP

T. bicornis is an aquatic perennial herb with two types of leaves. The submerged ones are long and threadlike and the floating ones, which form a rosette, are rhomboidal with petioles that are swollen in the middle. A rosette of these floating leaves is shown in the *Song Hundred Flowers* in the collection of the Palace Museum, Beijing (fig. 16.9). The plant is grown widely in still water in the warmer parts of the country for its fruits. These are shiny, black, two-horned nuts about 6 cm wide, which develop under the foliage in the centre of the plant. One of these was excavated from a tomb of 168 BC, which suggests that the water caltrop may have been cultivated at least as early as this (Bartholemew, 1980). Fortune (1852) observed women and boys harvesting the nuts in Lake Tai and recorded that he had seen three distinct varieties, so presumably there are several cultivars. This harvest at the end of summer is a happy occasion and has inspired numerous folk songs (Bartholemew, 1980).

The nuts are poisonous if eaten raw but are rendered harmless by prolonged boiling. One of the places where visitors will certainly encounter them is at the Yanyulou (Misty Rain Pavilion), a famous 16th century building much admired by the Qing emperors, which is built on an island in a lake at Jiaxing, about halfway between Shanghai and Hangzhou. Here they are sold by vendors who claim they are a product of the lake and who prepare them in elderly pressure-cookers. Vast quantities are sold to visitors, so that an excursion to this historic site is given a quality all its own by the crunch underfoot of the discarded shells.

Bartholemew (1980, fig. 3) reproduces a photograph of a Ming blue and white porcelain jar on which the plant is depicted, and points out that the shape of the fruit is not unlike that of a stylised representation of a bat. Since the bat symbolises good fortune the caltrop takes on this connotation as well. And as the character for intelligence is also pronounced *ling*, the plant becomes involved in various rebuses.

ZIZANIA Gramineae

Z. latifolia 茭儿菜 (Jiaoercai) *Jiaoer Vegetable*, 茭白 (Jiaobai) *White Jiao*, 茭笋 (Jiaosun) *Jiao Bamboo Shoot*

CHINESE WILD RICE

Z. latifolia is a perennial aquatic grass, related to rice, which is native to eastern Asia. It is cultivated not for its grain but for its stems, which become swollen as a result of infection by a smut fungus and resemble a stubby bamboo shoot about 17 cm long and 3 cm in diameter. These must be harvested at the right stage, since they become inedible as soon as the the black spores of the fungus begin developing in the swollen tissue. As the wild forms of *Z. latifolia* are somewhat resistant to the disease it seems that selected forms are grown which are susceptible. As Larkcom (1991) points out, this is a unique case in which disease-susceptible rather than disease-resistant plants are favoured for cultivation. The infected shoots are regarded as a delicacy and are used in a variety of dishes to absorb other flavours (Larkcom, 1991; Zee & Hui, 1990). Meyer (1916) recorded that they were eaten shredded or sliced, boiled in soups, or, when scalded, as a special salad. He noted that some Western residents had become so fond of this unusual vegetable they had it on their tables whenever it was procurable.

CHAPTER 17

CHRYSANTHEMUMS

The chrysanthemum was a particular favourite of the Confucian scholars and, together with the flowering plum, orchid, and bamboo, came to be regarded as one of the 'Four Gentlemen of Flowers' (Bartholemew, 1985a). Along with the tree peonies, it is probably the most generally admired of the Chinese ornamental plants. Robert Fortune (1844), for instance, came to this conclusion:

> The plants which stand next to dwarf trees in importance with the Chinese are certainly Chrysanthemums, which they manage extremely well—perhaps better than they do any other plant. So high do these plants stand in the favour of the Chinese gardener, that he will cultivate them extensively, even against the wishes of his employer; and, in many instances, rather leave his situation than give up the growth of his favourite flower. I was told the other day that the late Mr. Beale used to say he grew Chrysanthemums for no other purpose than to please his gardener, not having any particular taste for this flower himself.

The plants originally placed in the genus *Chrysanthemum* have been subject to various nomenclatural changes over the years. From the gardener's point of view the most confronting of these alterations was the removal some 30-odd

Figure 17.1 **A chrysanthemum at the Temple of Heaven, Beijing**

years ago of the perennial garden chrysanthemums to the genus *Dendranthema*. Mercifully this change has now been reversed, even though it requires the acceptance of other consequences (Brummitt, 1997; Trehane, 1995).

CHRYSANTHEMUM Compositae

C. coronarium 茼蒿 (Tonghao)

CROWN DAISY

This Mediterranean plant with golden flowers is an annual known as the Crown Daisy in the West. At some time it was introduced to the Orient and has become popular there as a vegetable, the leaves of young plants being served in various ways. In China it is often added to soups. There are several kinds varying in leaf size and the nature of the serrations. As one might expect, it has a characteristic aromatic taste. It is also used in Chinese medicine.

What appears to be this plant in bloom has been illustrated from time to time in paintings of the 'flowers of the seasons' and 'hundred flowers' types, for instance in the *Song Hundred Flowers* in the Palace Museum, Beijing, and another from the 18th century reproduced by Capon and Pang (1981, no. 65). Also the British Library has an accurate depiction in watercolours of this plant by an anonymous Chinese artist (Hulton & Smith, 1979, coloured plate 21). So perhaps it is, or has been, grown for ornament from time to time.

Although this plant is generally regarded as the 'edible chrysanthemum', owing to the nomenclatural changes mentioned above it should now be placed in the genus *Glebionis* (Brummitt, 1997; Trehane, 1995). However, since this is a recent decision, I have left it in *Chrysanthemum* here to avoid confusion until everyone gets used to the change.

C. indicum 野菊 (Ye Ju) *Wild Chrysanthemum*

C. indicum is a stoloniferous perennial which produces clusters of single yellow flowers, each only about 2.5 cm in diameter. The roots and young leaves are recorded in the *Jiuhuang Bencao* as famine foods (Bretschneider, 1881).

C. × morifolium 菊花 (Juhua) *Ju Flower*

CHRYSANTHEMUM

The origin of the garden chrysanthemums has been the subject of much speculation. Haw (1986a), for instance, says that there is little doubt that the cultivated chrysanthemum is a complex hybrid resulting from the interbreeding of several species, but considers that it must now be virtually impossible to ascertain which, if any, made the greatest contribution to its development.

The chrysanthemum is well chronicled in Chinese literature and both Li (1959) and Needham (1986) give excellent surveys. It is mentioned at least as far back as the 7th century BC, the appearance of the yellow flowers indicating the arrival of late autumn. It was subsequently mentioned in early pharmacopoeias as a safe and beneficial medicament, promoting longevity and immortality. Translations of both prose and poetry involving the chrysanthemum may be found in T. C. Lai's (1977) *Noble Fragrance*, along with reproductions of paintings of the plant.

The symbolism attached to the chrysanthemum is discussed in some detail by Koehn (1952) and Needham (1986). Firstly there was its association with autumn, always seen as a romantic and melancholy season, when in ancient times sentences of execution were carried out, when the yang gives way to the yin, and when winter is known to be approaching. Then the chrysanthemum came to symbolise the Confucian scholar, for just as it withstood the shortening days and the onset of cold weather, so he too must withstand disfavour and disapproval for standing firm in support of some unpopular principle or rebuking some departure from Confucian ethics. If he failed he might well retire from public life and go to live in the countryside far from the cares of the city. The poet Tao Yuanming (365–427 AD) did just this and, according to Needham (1986), it was he more than anyone who gave impetus to the cultivation and popularity of chrysanthemums. The phrase 'Tao Yuanming planted three rows (of chrysanthemums) to give gay colour to his eastern fence' became proverbial, and only in the light of it can some of the titles of later monographs on chrysanthemum culture be understood. And just as the chrysanthemum avoided 'the crowded market-place of spring', so the Confucian was content to live as a high-minded recluse. And, thirdly, there was the theme of longevity and immortality, associated with the drinking chrysanthemum wine on the 9th day of the 9th month, a custom popular during the Han dynasty. Just as the chrysanthemum bloomed late in the year, so might old age be blooming and even bloom on for ever. Apparently chrysanthemum wine was made by collecting leaves and twigs from plants in full bloom, mixing them with glutinous rice and allowing it to ferment, the wine being ready the

Figure 17.2 (right) **Part of a chrysanthemum display at the the Temple of Heaven, Beijing**

Figure 17.3 (above) **A chrysanthemum cultivar with green flowers in the Beijing Botanical Garden**

following autumn (Li, 1959). The association of the chrysanthemum with longevity is also thought to relate to the observations of the physician Ge Hong (284–364 AD), who described a chrysanthemum-covered valley in Henan, saying that the local people who drank from the river which flowed through this valley lived long lives (Wang & Ma, 1995). As well as all this many puns involving the Chinese names for the plant have been used to express good wishes (Koehn, 1952).

Such were some at least of the components of the cultural background of those who, from Tang times on, took up the culture and development of the chrysanthemum. Originally only yellows were known, but from the 8th century on white-flowered forms were often praised in poems, and later purple ones appeared. And as well as being grown in gardens and as pot plants, it is made clear that chrysanthemums were used in flower arrangements by Shen Fu (1983), whose detailed instructions about the matter have been summarised in Chapter 2.

The emergence of the various coloured forms and those with double flowers or petals of unusual shapes was apparently the result of selection (figs 1.1, 17.2, 17.3, 17.6 and page 386). Plants were raised from seed as well as by vegetative propagation, and this must have been the chief source of the new forms. However, the usual strange beliefs were widespread, including that the shallow planting of whole withered flowers was conducive to the production of variants. It was also believed they could be induced by grafting on unrelated

rootstocks and that the very small-flowered wild chrysanthemums could be transformed into large-flowered garden varieties by planting them in gardens with very fertile soil (Li, 1959; Needham, 1986). Nowadays there are cultivars in an astonishing range of forms and colours. There are even kinds with green flowers (fig. 17.3). In fact all colours except blue and violet are found.

The chrysanthemum has inspired a wealth of writings probably more voluminous than for any other Chinese garden plant. As with the peonies, the first monographs appeared in the Song, notably Liu Meng's *Ju Pu* (Treatise on Chrysanthemums) of 1104. Many others followed, describing not only the varieties but methods of cultivation, propagation, and pest control. In the 18th century there was even mention of varieties imported from abroad, probably from Japan (Needham, 1986). From these works it is clear that from about 20 cultivars in the late 10th century the number increased to more than 2000 by the early 19th. Tun (1965), in about 1900, could specify at random 133 varieties, and

Figure 17.4 The entrance to a chrysanthemum display in Behai Park, Beijing

knew of another 200. Many cultivars were given picturesque names such as 'Silver Bowl', 'Purple Crab's Claw', 'Heaven Full of Stars', 'Drunk with Wine made from Peaches of the Immortals' , 'White Crane Sleeping in the Snow' and 'Willow Streamers Drooping Gold' (Li, 1959; Tun, 1965; Williams, 1975). Li suggests that extensive cultivation in all parts of China during the last few hundred years may well have increased the number of kinds to several thousand. He points out, however, that many of them may have or have had only a regional existence.

How many of this vast number of cultivars survives today I do not know. As late as 1947 John Blofield attended a party given in a courtyard in Beijing in honour of the chrysanthemums (Blofield, 1961), so some varieties must have survived the Japanese occupation. And in spite of the vicissitudes of the years that followed there are in China today a huge number of cultivars, many recently raised, and the displays mounted all over the country every autumn attract large crowds of admirers (figs 17.2, 17.4, 17.5). Varieties similar to those first introduced to Europe are still seen along with others with much larger flowers and forms which suggests that Japanese varieties have been used in their breeding. The existence of such flowers in China earlier this century is suggested by Dorothy Graham (1938) when she described the arrival of autumn in one of the gardens she

had in Beijing: 'Then the tawny chrysanthemums came. The garden varieties, ragged and growing at will, were supplemented by the stately plants trained by the flower men— great blooms of *sang de boeuf* and umber, foaming gold and fragile mauve'.

Tun (1965) and Bredon and Mitrophanov (1927), whose writings focus on the latter part of the 19th century and the early part of the 20th, describe the practice at every coming of the 9th day of the 9th month in Beijing of the building by rich and noble families of 'chrysanthemum hills'. Several hundred plants, each in a porcelain pot, were arranged on a scaffolding so they looked like a hill. When the four sides were raised to a point they were called *juhua ta* (chrysanthemum pagodas or chrysanthemum towers). Tiered displays of potted plants still feature at chrysanthemum shows (fig. 17.2) and in places such as the foyers of large hotels. And as well as being used for the traditional autumn displays, potted chrysanthemums are also marketed for the Chinese New Year. According to Goody (1993) the yellow varieties, such as 'Yellow Prosperity', are much in demand at this time.

An interesting picture of chrysanthemum growing in China is given by Chen Haozi in his *Hua Jing* of 1688. He described not only the different varieties known to him but their propagation and culture, giving details about matters such as the application of fertilisers, disbudding to produce large flowers (figs 1.1, 17.2, 17.6), and the pinching of plants to make them branch and form large numbers of buds (Rivoire, 1928).

Another informative article was written by Cibot (1778b, 7). He examined Chinese books on the subject and, on the basis of these and his own observations, commented on the history and cultivation in China of what he called 'Le Kiu Hoa ou la Matricaire de Chine'. He chose to omit the fanciful and keep to the essentials, noting that chrysanthemums were planted in open positions in well prepared

Figure 17.5 (opposite page)
Chrysanthemums in Behai Park, Beijing
Figure 17.6 (left)
A chrysanthemum pruned to produce a single large flower

and manured soil and that they were propagated by seeds, cuttings, grafts, and suckers. He observed that weak branches were staked, that fertilisers were withheld as flowering time approached, that various procedures were adopted to control pests, and that in the emperor's garden the plants were shaded with mats at midday. He also described how enthusiasts left only one bud per stem and assisted the opening of the flowers with a toothpick when difficulties were encountered. He also recorded that the emperor's apartments were decorated with chrysanthemums from mid autumn until the end of winter.

Robert Fortune (1849d, 1852) also described chrysanthemum culture in China. He wrote that cuttings were struck every year from the young shoots, as in England, and that when these had produced roots they were placed in the pots where they were to grow and bloom. The potting mixture was composed of rich soil from the bottom of lakes mixed with old night-soil, and liquid manure was applied during the growing season. As the season progressed the plants were pinched and trained to produce the desired shape and number of flowers. He observed plants trained in the forms of animals, such as horses and deer, and as pagodas.

In more recent times Haw (1986a) has also described plants trained in fanciful shapes, including those of peacocks and dragons. No doubt such creations are still to be seen, and huge plants pinched to produce a mass of stems, each tied to a bamboo stake and allowed to produce only one flower, are very popular (figs 6.15, 17.7), as are cascade types trained either up or down (fig. 17.8). Goody (1993) says that this

practice of growing plants with hundreds of blooms goes back to the Song. One sees examples in the public displays but the potted plants available at the markets are usually either low growing with several flowers, as they generally are in the West, or kept to one stem bearing a single huge flower.

The Dowager Empress Cixi took a great interest in the growing of chrysanthemums at the Summer Palace. She and her ladies-in-waiting took part in the work themselves, planting the cuttings and later disbudding them so that each stem bore only a single flower (Der Ling, 1911). Princess Der Ling, who was obviously no horticulturist, was greatly surprised by the planting of cuttings, as they had no roots. She was also puzzled by Cixi's apparent ability to know what colour they were going to be long before they bloomed, labelling the plants accordingly, sometimes before any buds had formed. It seems she was always right and when Princess Der Ling asked her how she did it she was told that it was a secret.

It seems that the grafting of chrysanthemums on artemisia was also a common practice. In nursery gardens at Tianjin Fortune (1863) observed chrysanthemums grown as standards, grafted on stout stems of a species of *Artemisia*. He said that they grew with great vigour on this stock, and appeared 'rather curious objects to those who have seen them only on their own stems'. Many years later Meyer (1916) observed the same procedure in this city.

Another method of grafting on artemisia was described by someone identified only as 'K.' (1870). As soon as the graft was made, the artemisia stem was passed through the hole in the bottom of a flower pot kept about 15 cm above ground

level by standing it on four bricks. Earth was then placed in the pot, whereupon the plants gradually united, taking root at the same time in the pot. Then, as soon as the chrysanthemum commenced flowering, the stem of the artemisia was severed immediately below the pot and

Figure 17.7 **Part of a display for the Chinese New Year at Lieshi Lingyuan, Guangzhou. The chrysanthemums have been trained to produce a mass of stems each with a single flower. Also present are African marigolds, antirrhinums, nasturtiums, dahlias, and *Dianthus chinensis*.**

the plant taken away for display. The procedure was said to be carried out to 'strengthen' the chrysanthemum, which would otherwise produce a poor flower.

Williams (1975) recorded the practice in Beijing of grafting many different kinds of chrysanthemum on the one stock, a 'wild variety' which grew in profusion on the city wall, thus producing a grotesque plant with blooms of many colours. The plant shown in one of the paintings done early last century for John Reeves is an example of this practice (fig. 1.5), and in more recent times Stephen Haw (1986a) has photographed plants in Nanjing bearing flowers of at least five different types. It also seems possible, if not probable, that the 'wild variety', which Williams (1975) said grew on the city wall and was used as the stock, was an artemisia.

The chrysanthemum has been a popular subject for painting since the Tang and is present in every aspect of Chinese art (Bartholemew, 1985a). In early paintings only small single yellow or white varieties appeared (e.g. She & Yuhas, 1973, no. 3), but as time went by a greater variety of kinds was depicted. As well as its frequent appearance in paintings, it is widely used as a decorative motif in carving, on lacquer, on porcelain, and in many other ways. In the republican era which followed the fall of the Qing dynasty it was employed as what we would now call a 'logo' by the army on uniforms, in contradistinction to the sheaves of rice emblem worn by the navy, and came to be regarded as a national symbol (Williams, 1975).

Apart from its aesthetic and symbolic appeal the chrysanthemum is still credited with medicinal properties. While chrysanthemum wine is seldom made now, a tea infused from the petals is said to have a cooling effect in summer. A small white-flowered variety is grown in certain regions to produce the dried flowers used for this purpose, and these can now be obtained from suppliers in many parts of the world. Also some people collect and dry large quantities of the petals for stuffing pillows, which are pleasingly fragrant and also believed to have a cooling effect (Li, 1959).

Li says that chrysanthemums were taken to Japan from China in the 8th century, but according to Pizzetti and Cocker (1975) they had already reached there from Korea in 385 AD. No doubt plants were taken to Japan at various times and were used, perhaps in combination with indigenous forms, to produce further variants. If anything, chrysanthemums became even more popular in Japan than in China. The esteem for the plant there is evident from the adoption

Figure 17.8 **Cascade chrysanthemums in the Daguan Yuan, Beijing**

of a 16-petalled chrysanthemum as the crest of the Japanese imperial family.

The earliest record of the chrysanthemum in Europe, made in 1688, says that six varieties were grown in Holland, though nothing is known of their provenance. It seems nothing further was heard of these. Likewise, a small yellow-flowered kind grown in the Chelsea Physic Garden in 1764 became lost. However, the plant seems to have become permanently established in Europe following the importations of M. Blancard of Marseille in 1789. After this there were numerous introductions, many of them due to the efforts of John Reeves during his residence in Guangzhou. Nearly 70 varieties were introduced between 1820 and 1830 and European breeders soon began raising seedlings of their own. The arrival in England in 1846 of two small-flowered varieties, sent by Fortune from Zhoushan, led to the development of the pompon types, and later in the century Japanese varieties began to arrive and become involved in the further development of the plant (Coats, 1968; Payne, 1890; Sabine, 1822). Chrysanthemums quickly became popular around the world and remained so for a long time. Nowadays in the West they are comparatively little used as garden plants, though they are grown extensively in pots and for cut flowers.

The autumn-flowering habit of chrysanthemums, incidentally, results from their response to the lengthening of the nights as autumn approaches, which triggers off the development of flower buds. An understanding of this phenomenon has enabled the chrysanthemum growers of today to produce flowers all year round by adjusting the light regimes to which the plants are exposed.

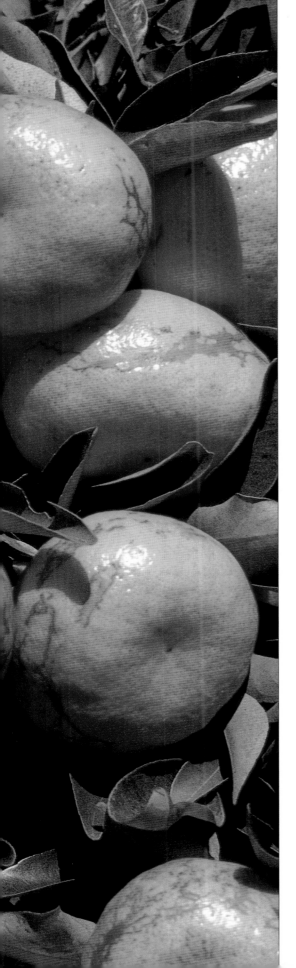

CHAPTER 18

CITRUS FRUITS

Under this heading I have included, together with *Poncirus trifoliata*, those plants which we think of as citrus fruits and which at times have been placed in the genera × *Citrofortunella*, *Citrus*, and *Fortunella*. In classifying them the position is complicated by the fact that, having arisen in cultivation, apparently by interspecific or even perhaps intergeneric hybridisation, many of them cannot be reliably referred to any known wild species. However, it is now believed that the commercial edible citrus fruits at present referred to the genus *Citrus* have been derived by selection and hybridisation from just three known wild species (*C. maxima*, *C. medica*, and *C. reticulata*) and two or more unknown plants. On this basis Mabberley (1997b) has published a classification to which readers wishing to be up to date should refer. Mabberley (1998) also has pointed out that in the light of modern taxonomic studies there appears to be no reason for keeping × *Citrofortunella* and *Fortunella* separate from *Citrus*.

Figure 18.1 **The mandarin orange (*Citrus reticulata*)**

The citrus fruits in China today have resulted from perhaps 3000 years of endeavour by Chinese gardeners and horticulturists. Ancient documents record that oranges and pomelos were sent as tribute to the King of Zhou, and a passage in *Yanzi Chunqiu* (Master Yan's Spring and Autumn Annals), a collection of stories about the statesman Yan Ying who lived in the 6th century BC, alludes to the belief about the orange and the thorny lime (*Poncirus trifoliata*) mentioned previously. It also includes a tale about the etiquette of peeling an orange at a princely court (Needham, 1986). These mentions, and numerous others dating from the 4th and 3rd centuries BC, make it safe to assume that citrus fruits were grown commercially in China from early in the 1st millennium BC, perhaps even earlier.

Needham (1986) records a Chinese writer of about 280 AD as saying that the number of sorts of trees in the orange and pomelo group was very large, that it included the mandarins and sour oranges, and that the best ones came from what is now Jiangxi. Some 900 years later Han Yanzhi wrote that the best ones, *Ju Gan*, with few or no seeds, came from the Nishan district of Zhejiang: 'They are thin-rinded and with a particularly delicious flavour. The fibres do not adhere to the segment walls, and in eating [the pulp] no refuse is left in the mouth' (Needham, 1986). Undoubtedly this was a superior cultivar of the mandarin.

It was Han Yanzhi who wrote the first book in any language devoted entirely to these fruits, the *Ju Lu* (Orange Record) of 1178 AD. In it he mentions 28 kinds of citrus but, since in categorising them his use of the words *gan*, *ju*, *cheng* and *luan* tends to cross what are now regarded as specific boundaries, it is difficult to be absolutely certain what they were. However, it seems that mandarins (*Citrus reticulata*), sweet and sour oranges (*C.* × *aurantium*), pomelos (*C. maxima*), kumquats (*C. japonica*), the trifoliate orange (*Poncirus trifoliata*), and perhaps the calamondin (*C.* × *microcarpa*) were included. As well as describing the 28 kinds, he discussed planting, cultivation, manuring, irrigation, seed raising, grafting, harvesting, storage, preservation, and medicinal use (Hagerty, 1923). Altogether a very remarkable achievement.

It is clear from the *Ju Lu* and subsequent works that in quite early times the Chinese understood that all these plants were related. Further, after the appearance of some Indian and southern forms, including the lemon (*C.* × *limon)*, the citron (*C. medica*), the lime (*C.* × *aurantiifolia*), and the

Rangpur lime (*C.* × *limonia* of some authors), these were quickly recognised as belonging to the same category. Other writings give further information about many varieties of the various types of citrus as well as about cultural practices and matters such as grafting and resistance to cold (Needham, 1986). And I have already drawn attention in Chapter 1 to the remarkable attempts to control citrus pests biologically with the citrus ant.

Citrus fruits in China are usually propagated by budding or grafting onto seedling stocks or onto the trifoliate orange. However, Wilson (1913) described a procedure in western China of propagating mandarins by notching the shoots which arise from the base of the tree, fixing earth around the cuts, and making a framework of bamboo or broken earthenware to keep the soil in place. The shoots were severed from the parent tree after roots had formed. As well as this, Livingstone (1822) recorded the air layering of branches loaded with flowers to produce dwarf plants with abundant fruit, stating that the layer was not separated until the fruit was nearly ripe. Fortune (1849b), on the other hand, recorded that all plants of the orange tribe which fruit in a small state were grafted, so no doubt the procedures varied.

While citrus fruits are only occasionally planted in the ground in Chinese gardens, potted plants are grown in huge numbers and sold for indoor decoration, particularly for the New Year. Covered with golden fruits, they are highly ornamental and are symbols of good fortune and prosperity. The growing of potted citrus for ornament seems to have had a long history, since in the *Ju Lu* Han Yanzhi mentions the *Jin Ju* (Golden Orange), perhaps a kumquat of some sort, saying that the fruits are inedible, that it is no more than a foot in height, and that it is only suitable for pots on balcony railings (Hagerty, 1923). Swingle and Reece (1967) claim this plant to be the wild kumquat, for which they give the name *Fortunella hindsii*, but this can be only a speculative identification.

While visiting a tea factory in Guangzhou, Fortune (1857) observed the use of citrus and other flowers for scenting tea. He saw a large heap of 'orange' flowers which a man was sifting to get out the stamens and other small parts. This was done so that the flowers might easily be sifted out as soon as the scenting had been accomplished. He also said that for this purpose the flowers must be fully expanded to bring out the scent, whereas the flowers of jasmine may be used in bud as they will expand and emit their fragrance during the time they are mixed with the tea. What the 'orange' used for this purpose was he did not say. It may perhaps have been the sour orange (*Citrus × aurantium*) or the pomelo (*C. maxima*), the flowers of both of which are exceptionally fragrant.

The various citrus cultivars, of course, are grown principally for their fruits, which are marketed all over the country. At the time of his explorations, Meyer (1911) recorded that there were said to be more than 80 kinds of edible oranges growing in the south-east of the country. And it is probably unnecessary to draw attention to the impact the Chinese citrus fruits have had on the rest of the world.

Attempts to put the citrus industry in China on a scientific basis seem to have begun in Guangzhou at the Canton Christian College, which had been established by the Presbyterians and which became the Lingnan University in 1921. Much work was done there in collaboration with the United States Department of Agriculture, in particular with Dr Walter T. Swingle, to whom we are indebted for revealing much of the ancient knowledge about citrus fruits to the rest of the world. After the departure of American Board of Trustees in 1952, the faculties of the Lingnan University were moved to another site and given the new name of South China Agricultural College. Then, during the Cultural Revolution, both Christian missionaries and the citrus industry associated with them were, to say the least, viewed with disfavour. In fact Mao Zedong called for the destruction of all the orange trees in China. As a result, in many areas of Guangdong the trees were chopped down and the land was used for growing rice. The citrus scientists on the staff of the Agricultural College were sent to the fields to work, while peasants took their place. A visitor to the college during this period could not find anyone who knew anything about oranges (Cooper, 1989).

Mao died in 1974 and was succeeded by Deng Xiaoping, who took a completely different view of the citrus industry. He proclaimed that orange culture and citrus research were of central importance to the communist concept of self-reliance and self-sufficiency. The replanting commenced and, beginning with 370,000 acres of oranges in 1978, which the Red Guards had somehow overlooked, China in 1988 reported that it had 1,300,000 acres planted and was aiming for 2,000,000 by the year 2000 (Cooper, 1989).

The various citrus fruits which I have included in this chapter are most easily identified from the attributes of their fruits and leaves. The petioles are more or less winged, and the size and shape of the wing is characteristic for certain kinds, as are the shape, size, and colour of the fruits and the number of segments they contain.

As well as being appreciated for their appearance, flavour, and scent, citrus fruits have also been used in traditional Chinese medicine, the dried fruits, peel, and seeds all being employed. According to Williams (1975), tincture of orange peel is greatly esteemed as a popular remedy, and Bredon and Mitrophanov (1927) record that, in the 5th month in Beijing early this century, people sipped water-ices flavoured with 'driving away heat' ingredients such as dried orange peel.

× CITROFORTUNELLA
see *Citrus × microcarpa*

CITRUS Rutaceae

C. × aurantiifolia 酸柠檬 (Suan Ningmeng) *Sour Lemon*

LIME

This small tree of straggling habit has short, sharp spines and narrowly winged petioles. The fruits are like small, smooth-skinned lemons and contain about 10 segments. They usually remain green in the tropical climates to which the plant is adapted, but sometimes become yellow when ripe. *C. × aurantiifolia* is thought to be a hybrid of *C. maxima*, its other parent or parents being unknown (Mabberley, 1997b). It cannot tolerate frost and is probably rare, even in southern China.

The lime was known to the Arabs from early times and was introduced to Europe about the 13th century (Purseglove, 1968). It was later introduced to the West Indies and other parts of the New World by the Spaniards.

C. × aurantium

ORANGE

Mabberley (1997b) groups the sweet and sour oranges, the tangelos, and the grapefruits together in *C. × aurantium*, since all are believed to be derived from hybrids between *C. maxima* and *C. reticulata*. Within *C. × aurantium* those oranges with more features of *C. maxima* make up the Sour Orange Group, and those with more of *C. reticulata* the Sweet Orange Group. The grapefruit is thought to have arisen as a result of hybridisation between the pomelo and the sweet orange (Bailey & Bailey, 1976; Mabberley, 1997).

1. Sour Orange 酸橙 (Suan Cheng) *Sour Orange*

This vigorous tree has been cultivated in China and southeast Asia for a very long time. It is hardier than the sweet orange and can stand several degrees of frost. The young shoots bear conspicuous spines and the petioles are broadly winged, the wings tapering towards the stem. The flowers, like those of the pomelo, are more strongly scented than those of the sweet orange, and the large, sour fruits are subglobose, depressed at the apex, and have rough, thick peel, 10–12 segments, many seeds, and a hollow core when ripe.

In China the sour orange is grown in orchards in Zhejiang, Fujian, and Guangdong (Cooper, 1989), and is occasionally used as an ornamental, either in a pot or planted in the ground (fig. 18.2). The plants called *Chengzi* and *Zhu Luan* in the *Ju Lu* are thought to be of this type (Needham, 1986). Although the fruits of the *Chengzi* were sour, Han Yanzhi recorded that people liked handling them because of their penetrating and fragrant odour, and used them for scenting clothes, seasoning vegetables, and for preserving in honey. And he said that people cut small fruits of the *Zhu Luan* in half, dried them in the sun, and passed them off as those of *Poncirus trifoliata*, which was highly valued as a medicine but sometimes hard to obtain in the south (Hagerty, 1923). In Han Yanzhi's time such a shortage may well have resulted from the fall to the Jin Tartars of the northern part of the country, where *P. trifoliata* was chiefly grown. He also described how the flowers of the *Zhu Luan*, which were larger and more fragrant than those of other kinds of orange, were distilled to make incense.

The sour orange arrived in Europe towards the end of the 1st century BC, presumably via the Roman–Indian trade route, was in orchard production in southern Europe by the early 4th century AD, and was subsequently brought to southern Spain by the Arabs (Needham, 1986). Its long association with this region is responsible, as is obvious, for the alternative name Seville Orange. Sour though the fruits may be, they ultimately became important for making marmalade. And it was the Spanish who first took this orange to the New World, where it became naturalised in Florida and other areas favourable to it.

2. Sweet Orange (*C. sinensis*) 甜橙 (Tian Cheng) *Sweet Orange*

The sweet orange forms a medium-sized tree easily distinguished from the sour orange by its narrowly winged petioles

and its globose fruits (fig. 18.3), which are sweet when ripe, have 10–14 segments, and do not develop hollow cores. The names given to citrus fruits in the medieval Chinese literature suggest that knowledge of the sweet orange dates back to the 2nd or 3rd century BC. Cooper (1989) is of the opinion that it probably originated in Yunnan and later became naturalised there and in adjacent parts of Burma.

Prior to the cultural revolution it seems that most of the commercial production was in Guangdong, but cultivation is now being attempted as far north as the Yangtze valley and parts of Sichuan, most of the plants being grafts on the cold-hardy *Poncirus trifoliata* (Cooper, 1989). Sweet oranges are at present second in popularity in China only to mandarins. Occasionally, too, they are grown for ornament.

The cultivation of oranges spread from China to the East Indies, Ceylon, India, and eastern Africa, probably some centuries before Vasco Da Gama rounded the Cape (Tolkowski, 1937). Needham (1986) says that, although tradition has it that the sweet orange was introduced to the West via Portuguese trade in the East Indies after Vasco da Gama, it is now

Figure 18.2 (opposite page) A sour orange (centre) in a courtyard of the Tiantong Si, near Ningbo, with a pomegranate (left) and *Buxus microphylla* and *Nandina domestica* (right)

Figure 18.3 (above) A sweet orange cultivar

thought to have reached Europe a good deal earlier, around 1470, by way of Genoese contacts through the Levant. However, it was not until the Portuguese introduced a better variety from India in the first half of the 16th century that it found favour with the public. Then, not only in Portugal itself but all over the Mediterranean region people took to growing 'Portugal oranges' commercially. In the late 16th and early 17th centuries both the Spanish and Portuguese made further introductions which became known as 'China oranges', though the cultivar said to have been brought to Lisbon from Goa by Don Francisco Mascarenhas in 1635 appears to have been the best (Tolkowsky, 1937). All this no doubt had much to do with the enthusiasm for orangeries at this time in those parts of Europe where the trees could not be grown in the open air.

Columbus is reported to have taken seeds of oranges from the Canaries to Haiti in 1493, though it is unknown whether these were those of sweet or sour oranges, or both. However, it seems logical to assume that sweet oranges were taken early to Central and South America by the Spanish and Portuguese and, presumably, to the Spanish colony in Florida. No doubt they also reached California from Mexico with the establishment of the Spanish missions in the 18th century (Hume, 1957; Tolkowsky, 1937). By the end of the 18th century they had reached Australia and much of the rest of the world. Since then, whether for whole fruits or its juice, the popularity of the sweet orange has gone from strength to strength. It is certainly outstanding amongst the gifts of the Chinese horticulturists to the rest of the world.

C. ichangensis 宜昌橙 (Yichang Cheng) *Yichang Orange*

ICHANG PAPEDA

C. ichangensis is small tree with sharp spines and very broadly winged petioles, nearly or quite as large as the leaf blade. The lemon-like fruits contain 8–11 segments and have rough skins and very large seeds. It is native to western and south-western China and is the most cold hardy of the *Citrus* species. Wilson (1913) reported that it was grown in the Yichang Gorge but was not common. Meyer collected scions on 10 December 1917 in Hubei, noting that the Chinese used the fruits only to perfume rooms but that foreign residents made lemonade from them. Three weeks later he collected 'an especially fine variety' at Yichang, recording that it exuded a delightful fragrance and made superior lemonade (Cunningham, 1984).

In the early Zhou period fruits called *Ju* and *You* were sent annually by fast runners from south of the Yangtze as tribute to the court at Changan (now Xian). According to Cooper (1989), there seems little doubt that the *Ju* mentioned was a mandarin orange and, according to Swingle and Reece (1967), the *You* was a cultivated variety of a naturally occurring hybrid between *C. reticulata* and *C. ichangensis*, although the name now refers to the pomelo. Swingle and Reece record that whereas *C. ichangensis* is not edible this hybrid is sparingly so. Another unusual citrus variety known as *Shangyuan* is considered to be a hybrid between *C. ichangensis* and *C. maxima* that arose under cultivation. It has been suggested that this is the *Xiangyuan* described in the *Ju Lu* (Needham, 1986) and mentioned by Shen Fu (1983) as differing from the Buddha's Hand (a form of *C. medica*) in requiring no special treatment to stop it spoiling. Presumably he used it as an ornament for scenting rooms.

C. japonica

KUMQUAT

The cultivated kumquats were separated from the genus *Citrus* by Swingle in 1915 and given the name *Fortunella* in honour of Robert Fortune, who brought a plant to England from China in 1846. As mentioned earlier, Mabberley (1998) considers that they should be returned to *Citrus*, in which case they should be regarded as cultivars of *C. japonica*. The cultivated kumquats probably originated in south-eastern China, perhaps having been derived from the wild kumquat, and possibly having been in cultivation for as long as other citrus fruits. The wild kumquat was originally described as *Sclerostylis hindsii*, and later transferred to *Atalantia* before being incorporated in *Fortunella* by Swingle as *F. hindsii*. The relationship of this plant to the well-known cultivated forms is still unclear. It will undoubtedly be incorporated in the genus *Citrus* but under what name is uncertain, so for the time being I have included it here under *C. japonica*.

The *Jin Gan* recorded in the *Ju Lu* is certainlyly a kumquat. Of it Han Yanzhi writes: 'The *jin gan* fruit has a golden colour, a very fine-grained skin, and a spherical form; with its ruddiness it is agreeable and enjoyable to handle. One eats it without peeling off its golden coat. When preserved in honey the flavour is even better.' (Needham, 1986). Han Yanzhi described how the fruit should be slit and squeezed to remove the seeds before immersing it in the honey

Figure 18.4 Meiwa kumquats (*Citrus japonica*)
on sale in Beijing

(Hagerty, 1923). Nowadays the name *Jin Gan* is applied to the wild kumquat.

Not only do the kumquats exhibit a considerable tolerance of temperatures below freezing but also a pronounced winter dormancy. This allows them to pass through periods of warm weather in winter without starting growth or flowering. Hence they could be of importance in breeding new types of acid citrus fruits able to grow in much colder regions than the lemon and lime which exhibit little or no dormancy (Swingle & Reece, 1967).

Kumquats have remained popular in China both as pot plants, particularly for the New Year, and for their fruits which in more recent times are usually eaten preserved with sugar (Meyer, 1911; Wilson, 1913). And outside China a great demand has now developed for kumquats, along with calamondins, as ornamental fruiting plants for growing in containers. It may be that there are kumquat cultivars in China other than those described below, for example the 'Changshou' kumquat mentioned by Swingle and Reece (1967). However, the following types are those usually seen:

1. Meiwa Kumquat (*Fortunella crassifolia*)
金弹 (Jindan) *Gold Marble*

The Meiwa kumquat is a shrub or small tree, with small leaves, very narrowly winged petioles, and broadly ovoid or subglobose fruits, usually a little larger than those of the kinds described below and with about 7 segments. It is possibly a hybrid of the round and oval kumquats and, according to Cooper (1989), is the *Jin Gan* described in the *Ju Lu*. Although it is little known outside eastern Asia, it is generally considered to be of better eating quality than the other kumquats. It is grown widely throughout southern China and in south-western Japan (Saunt, 1990). The fresh kumquat fruits which I saw on sale in Beijing in November 1996 all appeared to be those of this type (fig. 18.4), and certainly were much sweeter and more pleasant than other kinds I have tried. According to the *Ju Lu* (Hagerty, 1923), the *Jin Gan* preserved

in honey was considered better than other kinds of orange similarly treated. The Meiwa kumquat was introduced to the United States probably about 1910 or 1912 (Hume, 1957).

2. Round Kumquat (*Fortunella japonica*)
圓金柑 (Yuan Jingan) *Round Golden Orange*

The round kumquat is similar to the Meiwa kumquat but its fruits are usually globose and have 4–7 segments. Although apparently not mentioned in the *Ju Lu*, it has probably existed in China for a long time, although it is certainly not as popular as the other species. There is also a variegated form.

The kumquat which Wilson (1913) called *C. japonica* and observed was sparingly cultivated around Yichang may have been this cultivar, though since Fortune introduced the oval kumquat to Britain as *Citrus japonica* we cannot be certain (Fortune, 1848; Swingle & Reece, 1967).

3. Oval Kumquat (*Fortunella margarita*)
金桔 (Jin Ju) *Gold Orange*

This kumquat is similar to the round kumquat but has slightly larger leaves and ovoid fruits containing 4–5 segments. It is said to be the kind most appreciated in China (Meyer, 1911). There can be little doubt that this was the kumquat introduced to Britain by Robert Fortune in 1846 as

Figure 18.5 (above)
Oval kumquats
(*Citrus japonica*)
on sale in Macao
for New Year

Figure 18.6 (left)
Preserved oval
kumquats

Citrus japonica (Swingle & Reece, 1967; Synge, 1969) and which later became known as *Fortunella margarita*. Fortune (1848) described the fruits as being about the size of a large oval gooseberry and said that the Chinese usually preserved them in sugar. He noted that this kumquat was propagated by grafting on 'a prickly wild species of Citrus, which seems of a more hardy nature than the Kum-quat itself'. Presumably this was *Poncirus trifoliata*.

Fortune (1847a, 1848) found numerous kumquat groves on the hills of Zhoushan and observed that great quantities were grown in pots in southern China. It seems likely too that the 'dwarf mandarin oranges, about the size of walnuts', seen as pot plants in Chongqing by Archibald Little (1888) in April 1883, were this species. It is still widely cultivated in the south where it is an important ornamental pot plant for the Chinese New Year (fig. 18.5). It is grafted onto lemon rootstocks and the young plants are subjected to a period of drought, which stimulates flowering. The fruits ripen in the autumn and the plants are ready for the market in December (Zee & Hui, 1990). Preserved in syrup and with the seeds removed, the fruits are available in almost every food shop in China (18.6). It seems, however, that the fresh fruits are not as favoured as those of the Meiwa kumquat.

4. Wild Kumquat (*Fortunella hindsii*)
金柑 (Jin Gan) *Gold Orange*

The wild kumquat is known from Hong Kong, Guangdong, and Zhejiang (Swingle & Reece, 1967). It is a spiny shrub bearing subglobose fruits only 1–1.5 cm in diameter and with 3–4 segments. It is included here because it has a cultivar, 'Jindou' ('Gold Bean'), which is used for penjing (Hu, 1982; Swingle & Reece, 1967). 'Jindou' differs from the wild type in having larger, thinner, somewhat narrower leaves, shorter, more slender spines, depressed globose fruits 1.2–1.5 cm in diameter, and only two sets of chromosomes instead of four. As mentioned earlier, it may be that this was the plant which Han Yanzhi called *Jin Ju*, saying that its fruits were inedible and it was only suitable for pots on balcony railings. Photographs of both the wild type and 'Jindou' are reproduced by Swingle and Reece (1967).

C. × *limon* 柠檬 (Ningmeng)

LEMON

The lemon is a small tree with more or less wingless petioles and characteristic fruits containing 7–10 segments. It is believed to be a hybrid derived from *C. medica* and an unknown parent (Mabberley, 1997). It has been suggested that it may have arisen in India or been introduced there very early from further east. According to Needham (1986) it was recorded in Italy in the 1st century AD by Pliny, but its general cultivation in the Mediterranean region is believed to date from Arab introductions between 1000 and 1200 AD. It is first recorded as being cultivated in Guangdong in the 12th century AD, perhaps having come from India via Indochina. Its juice was used in the manner of vinegar in Guangzhou at the time and later for other purposes such as making lemon drinks during hot weather. It was established in the Philippines long before the arrival of the Spanish, was taken to Central America by Columbus in 1593, and completed its progression round the world in the 17th and 18th centuries (Glidden, 1937; Laufer, 1934).

Meyer (1911), writing of northern China, noted that it was apparently not grown as a fruit tree as lemons were rare and expensive, being imported. Even in the south it is rarely grown for its fruits but used mainly as a rootstock for other citrus varieties and kumquats. Except for the preparation of lemon drinks, the fruits are not usually used fresh. The rind is used widely in cookery and confectionery, or as a garnish for fish and meat dishes. Needless to say lemon is the principal flavouring in the famous Chinese dish known as 'lemon chicken'. Incidentally, the Chinese name for *C. × limon* is said to be based on the pronunciation of the word 'lemon'.

C. × limonia (of some authors) 黎檬 (Limeng)

RANGPUR LIME

The Rangpur lime is a small tree with very acid, round or broadly obovate fruits which become deep yellow or orange when ripe. These each contain about 7 segments which can be separated readily from one another and from the rind. How widely this is grown in China I do not know, but it is said to be an appropriate gift for mothers. It seems probable that it is the same as or similar to the cultivar 'Otaheite', which is grown in other countries as an ornamental pot plant and for which the correct name is *C. taitensis* (*Flora of Australia*. Rutaceae. In press). Perhaps a hybrid of the lemon and mandarin, it is more tolerant of frost than the true lime (*C. × aurantiifolia*). It was first grown in the United States from seed obtained by Reasoner Brothers, Oneco, Florida (Hume, 1957).

C. maxima 柚 (You)

POMELO, SHADDOCK

Becoming tree of 8 m or more with very large leaves and broadly-winged petioles, the pomelo is easily recognised by its very large, globose or broadly pear-shaped fruits, which become yellow when ripe and contain 11–14 segments. There are kinds with both pale yellow and pink flesh. The fruit may be eaten fresh or the pulp may be dried and candied. The rind is sometimes used as a veg-etable. It has to be parboiled first, after which it is incorporated in various dishes (Zee & Hui, 1990).

The pomelo is also appreciated for the out-standing fragrance of its flowers. Needham (1974) records that in the *Lingwai Daida* of 1178 AD Zhou Qufei mentions the distilling of these flowers to produce an oil quite able to challenge the sweet-ness of jasmine.

C. maxima is cultivated extensively in southern China, probably having been introduced from

Malaysia. According to Needham (1986) it was flourishing in Andalusia by the 12th century AD, having been brought there by the Arabs. Hume (1957) is of the same view and says that it is reasonable to believe that it was taken to America by the Spaniards but that, being of little economic value, it passed unnoticed in the records.

C. medica 香櫞 (Xiang Yuan) *Fragrant Citrus*

CITRON

The citron is another plant the original home of which is a mystery, though it is thought to be a native of India or to have been brought there from further east. Excavations show it was known in 4000 BC in Mesopotamia and it was the first citrus fruit to be introduced to the Mediterranean, reputedly by the returning armies of Alexander the Great in 300 BC (Huxley et al., 1992). It is believed to have been taken to Haiti by Columbus in 1493 (Hume, 1957; Tolkowsky, 1937).

C. medica is a thorny shrub or small tree with large ser-rate leaves, unwinged petioles and enormous lemon-like fruits, which may be up to 30 cm long and weigh more than 2 kg. The fruit was described by Ji Han in 304 AD as being exceedingly fragrant, and he recorded that as decorations for the banquet table women workers carved designs of flowers and birds on them and stewed them in honey (Li, 1979).

Apparently the celebrated variety known as 'Foshou' ('Buddha's Hand') was not well known in Ji Han's time, although a reference to it from the Tang dynasty says that it can be candied. The fruit of this aberrant form is split into finger-like sections (fig. 18.7) and has little pulp, but is very

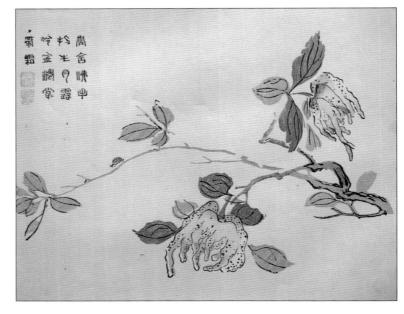

Figure 18.7 *Citrus medica* 'Fingered', depicted in the *Jieziyuan Huazhuan* of 1701

fragrant. These fruits or potted plants bearing them are prized, particularly at New Year, as they are supposed to bring good luck and are a symbol of wealth. And they have come to suggest the wish for a long life, as the character *shou*, meaning hand, is pronounced similarly to the character *shou*, meaning longevity. The fruits are also used for scenting rooms and to connote Buddhism, resembling as they do a classic position of Buddha's hand (Koehn, 1952; Williams, 1975). Dorothy Graham (1938) says that they were usually arranged in pyramids on plates of white porcelain and placed on the scholars table, as the fragrance was believed to be conducive to clear thinking. Shen Fu (1983), writing early last

century, says that there are ways of arranging them and that they should not be smelled by someone who is drunk or they will spoil. The belief that they will spoil if touched by oily hands is a modern version of Shen Fu's warning. They can also be used to flavour sweets and tea, in which case they are considered to be a cure for sore throats (Pratt & Chiang in Shen Fu, 1983). Fortune (1844) noted that this variety of the citron was the most striking plant in the Fa Tee Gardens in autumn and winter, being much admired for its strange form and its perfume. He also saw it 'cultivated in great perfection' in nurseries near Fuzhou, which appeared to him to be its natural locality (Fortune, 1847a). It has at times been given varietal status as *C. medica* var. *digitata* or *C. medica* var. *sarcodactylis*, but is presumably only a mutant which has been vegetatively propagated. In this case it should be known as *C. medica* 'Fingered' (Mabberley, 1997a).

As a result of the esteem in which this cultivar is held, it has often been represented in paintings (e.g. fig. 18.8 and Weidner et al., 1988, cat. no. 48), on porcelain (Krahl, 1987), in carvings (e.g. Batholemew, 1985a, cat. no. 34), and in other ways. Depicted with the peach and pomegranate (e.g. Bartholemew, 1985b, fig. 20) it forms a motif known as The Three Plenties—long life, happiness, and lots of children. Li (1979) says the normal variety is widely planted along the Yangtze valley and southward, and that the fruit is eaten candied or used as an ornament as Ji Han described.

C. × meyeri 香柠檬 (Xiang Ningmeng) *Fragrant Lemon*

MEYER LEMON

Sometimes regarded as a cultivar of *C. × limon*, *C. × meyeri* is able to tolerate lower temperatures than other lemons. It makes a neat bush or small tree and is most handsome when covered with its smooth-skinned, yellow fruits (fig. 18.8). The fruits are less acid than those of other lemons and have a

Figure 18.8 (left) The Meyer lemon (*Citrus × meyeri*)

Figure 18.9 (above left) Calamondins (*Citrus × microcarpa*) on sale in Hong Kong at New Year

Figure 18.10 (below right) **Fruits of *Citrus***
***reticulata*, *Gan* type, on a potted plant, Beijing**
Figure 18.11 (right) ***Citrus reticulata*,**
***Ju* type, on sale in Hong Kong for the Chinese**
New Year

Figure 18.10 (below right) **Fruits of *Citrus***
***reticulata*, *Gan* type, on a potted plant, Beijing**
Figure 18.11 (right) ***Citrus reticulata*,**
***Ju* type, on sale in Hong Kong for the Chinese**
New Year

slightly different flavour. It has been suggested that it may be a hybrid of the lemon and some type of orange. Until more is known about it it may well be sensible to refer to it merely as *Citrus* 'Meyer', as suggested by Mabberley (1997b).

The plants in cultivation outside China have descended from a plant which Meyer bought at Fengtai on the outskirts of Beijing in 1908 and sent to the United States (Cunningham, 1984), from where it has spread around the world. Presumably this is the lemon he recorded as being much appreciated as an ornamental pot plant when dwarfed, those with a dozen or more fruits bringing very high prices in Beijing in winter. He described it as being easily raised from cuttings and having thin-skinned fruit, only slightly sour and with few seeds (Meyer, 1911).

C. × microcarpa (× Citrofortunella microcarpa)
四季桔 (Siji Ju) *Four Seasons Orange*

CALAMONDIN

Thought to be a hybrid between *C. reticulata* and *C. japonica*, this small tree with narrowly winged petioles is often regarded as a kumquat. However, the globose or slightly flattened fruits are a little larger, have loose peel, and contain 7–10 segments. They are used in the same way as those of the kumquats and are on sale in many parts of China in the autumn. Grown in pots, the plants are popular as house plants or movable garden ornaments, especially for the Chinese New Year (fig. 18.9). Presumably the 'tangerine' called 'Four Seasons', which Goody (1993) noted to be the favourite variety marketed at this time, its miniature fruits later being bottled and preserved with salt, was this plant. Forms with variegated or striped leaves exist and, like the kumquats, *C. × microcarpa* will tolerate a few degrees of frost. It is said to have originated in the Philippines and reached China and other parts of the world from there. Although it is not possible to be certain, it is suggested that it may be one of the plants recorded by Han Yanzhi (Needham, 1986). So perhaps it has been in China for 800 years or more.

C. reticulata 桔 (Ju)

MANDARIN, TANGERINE, CLEMENTINE, SATSUMA ORANGE

Included here are two slightly different types, *Gan* and *Ju*, once considered to represent different species. According to traditional usage in China, the *Gan* differs from the *Ju* in the fruit being nearly spherical in shape (fig. 18.10), the rind being thicker and coarser, tight but removable, and in having oval seeds. The *Ju* on the other hand is somewhat flattened at both ends (figs 18.1, 18.11), has a very loose skin, and the seeds are beaked at their tips (Li, 1979). In general appearance and flavour the *Gan* type, which seems to be unknown outside Asia, is intermediate between the sweet orange and the *Ju*, and may perhaps be a hybrid. Both types were described by Ji Han. He recorded under his entry for the *Ju* that since the time of the emperor Wu Di of the Han dynasty there had been a Minister of Oranges responsible for presenting oranges to the royal court, and that during the Three Kingdoms period a specimen bearing 17 fruits on the one stem had been sent and that this had been regarded as an auspicious symbol. It was under his subsequent entry for the *Gan* that Ji Han described the sale and use of the citrus ant, stating that, in the south, if the *Gan* trees do not have this kind of ant, the fruits will all be damaged by many harmful insects and not a single fruit will be perfect. This account is often quoted as the first reference in any literature to the entomological control of plant pests (Li, 1979).

The mandarin is a small tree with narrowly winged petioles and bright orange, depressed-globose fruits, the skin of which is loose and which contain 10–14 easily separated segments. Native to China, it has probably been cultivated there for well over 2000 years, the fruits having formed part of the tribute sent annually to the Zhou kings. There are descriptions from medieval times of mandarins being grown commercially on the island of Dongdongting in Lake Tai near Suzhou, where laden with their brilliant fruit they made a beautiful sight amongst the pines and bamboo (Cooper, 1989). According to Williams (1975), the loose-skinned mandarin is also known in China as the 'cinnabar orange', a name suggestive of alchemy, immortality, and good fortune.

The mandarin seems always to have been the most popular type of orange in China. There are many cultivars and it is the most widely grown of the citrus fruits, both *Ju* and *Gan* types being regularly offered for sale. Also Wilson (1913) records the use of the dried rind and the fibres and pithy substance surrounding the segments as medicines. Meyer (1911) noted that in winter in Beijing one could sometimes buy 'things with the appearance of a warty Satsuma orange' which were used, like quinces, for perfuming rooms. Whether these were a form of *C. reticulata* or something else is not known.

Presumably the oranges sent to Beijing annually from Fuzhou for the imperial sacrifices at the Temple of Heaven at the beginning of the year were the fruits of *C. reticulata* (Williams, 1975). Also the presentation at the Chinese New Year of

Figure 18.12 *Orange Trees* **by Zhao Lingrang (active 1080–1100)**

the fruits or plants bearing them betokens the wish for abundant happiness and prosperity during the ensuing twelve months. In southern China small potted plants covered with fruits are popular ornaments at this time (fig. 18.11). Goody (1993) and Gulland (1928) suggest that this is so because of the similarity of its name to the word for 'blessing' or 'fortune' in Cantonese.

In spite of its popularity as both a fruit and a symbol of good fortune the mandarin seems rarely to have been represented in art. However, Krahl (1987) records its depiction on porcelain and there is a fine Northern Song painting of what is presumably this species in fruit (fig. 18.12).

Curiously enough it is only in modern times that the loose-skinned *C. reticulata* has become known in the West. In 1805 two varieties were introduced to England from Guangzhou by Sir Abraham Hume. It was introduced to continental Europe about 1828 and to North America at some time between 1840 and 1850, when it was brought to Louisiana by the Italian consul at New Orleans (Hume, 1957).

Long before this it was taken to Japan. According to Cooper (1989) it was probably brought there from Zhejiang sometime during the Tang dynasty. Selections suited to the Japanese climate were developed and became known as Satsuma oranges. The first of these to be introduced to the United States from Japan was the cultivar 'Owari' in 1876. It shows greater tolerance of cold than the Chinese cultivars (Hume, 1957).

C. japonica—see *C.* × *aurantium*

FORTUNELLA—see *Citrus japonica*

PONCIRUS Rutaceae

P. trifoliata 枸桔 (Gou Ju) *Gou Orange*

TRIFOLIATE ORANGE

P. trifoliata is a small, much-branched, deciduous tree with trifoliate leaves and large spines. The flower buds form in summer and over-winter on the bare branches, opening in spring before the leaves. The white flowers are 4–5 cm in diameter and profusely borne, so that plants in bloom are most handsome (fig. 18.13). The flowers are followed by small, citrus-like fruits which become yellow when ripe and are deco-

Figure 18.13 *Poncirus trifoliata* blooming in a hedge beside a citrus orchard near Ningbo

rative but, for practical purposes, inedible. However, as mentioned earlier, they are used in medicine, the dried product being known as *zhike* (zhi skin).

Planted closely, *P. trifoliata* forms a thorny, impenetrable hedge and is sometimes grown for this purpose in China, as is mentioned in the *Ju Lu* (Hagerty, 1923). Wilson (1913) described such hedges as giving to the quieter streets of the city he identified as 'Paoning' (now Langzhong, Sichuan) a 'country-lane-like appearance'. I have seen it creating a similar effect in South Korea. In addition, *P. trifoliata* is used as a rootstock for various citrus fruits both in China and in other parts of the world, where it is also occasionally grown as an ornamental. Being deciduous it is more tolerant of cold than its evergreen relatives, and no doubt this is responsible for the belief mentioned earlier that when the orange tree crosses the Huai River it turns into the thorny lime.

It seems likely that this plant was first used as a rootstock to preserve selections of the mandarin as well as to enhance cold hardiness. It is believed that the physiological conditions in the *P. trifoliata* rootstock during cold weather are favourable for the induction of dormancy in the scion, making it more resistant to freezing temperatures (Cooper, 1989). *P. trifoliata* was introduced to Britain by Robert Fortune in 1850 (Coats, 1963).

CHAPTER 19

FURTHER TREES AND SHRUBS

Included in this chapter are many trees and shrubs which are to be found in Chinese gardens but which have not been accounted for in earlier chapters. As it happens, most of the plants dealt with below are as noteworthy as those already described. Amongst the shrubs, for instance, are such favourites as *Chimonanthus praecox*, *Daphne odora*, the forsythias, *Gardenia augusta*, various *Hibiscus* species, *Kerria japonica*, *Nandina domestica*, *Osmanthus fragrans*, spiraeas, viburnums, and *Weigela florida*. In addition there are the palms and numerous broad-leaved trees, including the tree of heaven, the camphor laurel, the persimmon, the loquat, *Gleditsia sinensis*, the Chinese holly, *Koelreuteria paniculata*, the paulownias, *Pistacia chinensis*, the weeping willow, and the Chinese elms. Also here are a few Chinese species, such as *Dimocarpus longan*, *Litchi chinensis*, *Castanea mollissima*, and *Corylus heterophylla*, which, although rarely if ever seen in gardens, have become familiar on account of their fruits or nuts. Likewise, although the white mulberry is usually not encountered in gardens

Figure 19.1 *Catalpa bungei* in the Imperial Garden of the Forbidden City, Beijing

259

of the more sophisticated type, it has such a long history in China in relation to the silk industry and is such a feature of the Chinese countryside that I felt it must be included.

ACER Aceraceae

The maples are much used as garden plants in China. The three species described below are those most frequently encountered, though no doubt others may be planted from time to time. Dorothy Graham (1938) mentioned that in Beijing the Chinese brought into their courtyards maple trees, a few inches in height, and watched the foliage redden as the first frosts arrived, but she gave no indication of what they were.

A. buergerianum 三角枫 (Sanjiao Feng)
Three-lobed Maple

THREE-TOOTHED MAPLE

This maple with its characteristic 3-lobed leaves is a native of eastern China. It is a favourite garden plant, whether planted in the ground or in large pots (fig. 19.2). It is also used for penjing.

A.mono 五角枫 (Wujiao Feng) *Five-lobed Maple,* 色木槭 (Semu Qi) *Painted Maple*

Native to north-eastern China, Korea, and Japan, this species has neat leaves with five broad lobes. It forms a small tree and, like *A. buergerianum*, is sometimes grown in pots as well as in the ground. It is particularly attractive in spring when covered with yellowish flower-clusters and unfolding leaves (fig. 19.3), and again in autumn, when the leaves turn yellow, overlaid to varying degrees with red. In Japan it has given rise to several cultivars.

Figure 19.2 (left) *Acer buergerianum* in the Yu Yuan, Shanghai.

Figure 19.3 (below) *Acer mono* in bloom at the Jietai Si, south-west of Beijing

260

A. palmatum 鸡爪槭 (Jizhua Qi)
Fowl's-claw Maple

JAPANESE MAPLE

Known generally as the Japanese Maple, this small tree is a conspicuous feature of many Chinese gardens in the lower Yangtze region (fig. 19.17). As far as I know it is not native to China but is known to occur in Korea as well as Japan. It seems likely that it was brought from Japan, as many of the cultivars seen in China appear identical to those originating in that country. Those most often seen growing in the ground are green- or purple-leaved forms of the so-called Palmatum Group, which have leaves with 5–7 lobes. Cultivars more widely divergent from the wild types, such as those of the Dissectum Group, are occasionally seen as pot plants, and various forms are trained as penjing.

A. truncatum 元宝枫 (Yuanbao Feng) *Money Maple*, 平基槭 (Pingji Qi) *(Flat-base Maple)*

A. truncatum is a small tree native to northern China and Korea. It is very similar to *A. mono* but the leaves usually have a markedly truncate base and the three anterior lobes often bear two teeth towards their apices. It is commonly seen in the north and one of the main avenues in Beijing is planted with this species (van Gelderen et al., 1994). Like *A. mono* it has given rise to cultivars in Japan. It has been given the name *Yuanbao Feng* because the winged fruits look like the *yuanbao*, a shoe-shaped gold or silver ingot used as money in feudal China. It is occasionally used for penjing (Li, 1956a).

AESCULUS Hippocastanaceae

A. chinensis 七叶树 (Qiyeshu) *Seven-leaved Tree*

CHINESE HORSE CHESTNUT

A. chinensis is native to northern China and, while it is only occasionally planted in gardens, it is a common tree in the grounds of Buddhist and Daoist temples. There is an 11th century record of the planting in the 1st century AD of a horse chestnut in the mountains of Sichuan by a famous Daoist, Zhang Daoling, when he meditated there. As a result it became known as *Tianshi Li* (Heavenly Teacher Chestnut) (Li, 1956c). It seems likely that this was *A. wilsonii*, which occurs widely in western China and, according to Sargent (1917), has been confused with *A. chinensis*. In recent times

A. wilsonii, which has smaller flowers and differs from *A. chinensis* in other respects, was recorded growing in the courtyard of the Daoists' Purple Clouds Temple in the Wudang Shan in western Hubei (Lancaster, 1989), so it seems that it is still a tree of significance for them. It appears that in eastern and northern China they planted *A. chinensis* for the same purpose, fine old specimens of which are still seen in their temples, for example the Baiyunguan in Beijing.

It is suggested that the Buddhists may not have have invested the tree with significance until the 16th century, as in the *Bencao Gangmu* of 1596 Li Shizhen identifies the *Tianshi Li* with plants called *Solo*, a name which he says was unknown in former times (Li, 1956c). This is a transliteration of the Sanskrit *sala*, which originally refers to *Shorea robusta* of India (Burkill, 1946; Li, 1956c, 1963a). Gautama Buddha was born in a grove of *S. robusta* and it was given even greater sanctity when he died, lying on his cloak, between two of these trees. *S. robusta* is not hardy in much of China and it appears that *A. chinensis* has been adopted as a substitute. Specimens are seen in temple grounds from the Yangtze valley north. They are common in and around Beijing, the largest and most handsome examples I have seen being in the principal courtyard of the Tanzhe Si in the Western Hills, and in the interesting burial grounds that lie on the slopes below the temple (fig. 19.4). It was introduced to the United States from this area by F. N. Meyer in 1906 (Cunningham, 1984).

AGLAIA Meliaceae

A. odorata 米籽兰 (Mizilan) *Rice Grain Orchid*

An evergreen shrub or small tree with pinnate leaves, *A. odorata* is similar in appearance to *Murraya paniculata* but has drooping panicles of tiny, cream, scented flowers (fig. 19.5). It is native to southern China and adjoining regions and is common in gardens and as a pot plant in this region on account of its fragrance. Morrison (1822) recorded that it bloomed in the 4th moon (May) in Guangzhou and Fortune (1844) saw it in the Fa Tee Gardens. The flowers are, or have been, used to perfume clothes, for scenting tea, for perfuming tobacco, and by women to ornament their hair (Fortune, 1847a, 1852; Metcalf, 1942; Thrower, 1988; Urban Services Department, 1971). The plant gets its Chinese name because the tiny flowers resemble rice grains and smell like cymbidiums.

Figure 19.4 (opposite page) *Aesculus chinensis* at the Tanzhe Si, south-west of Beijing

Figure 19.5 (above) *Aglaia odorata*

Figure 19.6 (right) *Ailanthus altissima*

AILANTHUS Simaroubaceae

A. altissima 臭椿 (Chouchun) *Stinking Chun*

TREE OF HEAVEN

This tall deciduous tree has pinnate leaves up to 60 cm long. The panicles of inconspicuous flowers give rise to large clusters of reddish, single-seeded, winged fruits (fig. 19.6). The flowers of the male trees are malodorous, hence its Chinese name. It is commonly planted in gardens and temple grounds and grows very rapidly, particularly when young. In China there is a superstitious belief that if the tree of heaven in a courtyard grows too high the family living there will have bad luck (Williams, 1975). In some parts of the country it has been used for raising wild silkworms (Li, 1955). It was one of the first Chinese trees to be introduced to Europe, and it is almost certain that seeds sent by Pierre d'Incarville were received by Philip Miller at the Chelsea Physic Garden in London in 1751 (Spongberg, 1990, 1993)). Soon afterwards it was described from Beijing by Cibot (1777a) as the 'tcheou-tchun' or 'frêne puant' (stinking ash). It has become well known around the world, being tolerant of smog and harsh conditions.

ALBIZIA Leguminosae

The species of *Albizia* described below are deciduous trees with flowers, the conspicuous parts of which are the stamens, borne in mimosa-like clusters, and with leaves which are bipinnate.

A. chinensis 楹树 (Yingshu) *Column Tree*

A. chinensis is a large, fast-growing tree native to India and south-east Asia where it is frequently planted for timber (Macmillan, 1991). Li (1959) says it is the *Albizia* species most frequently planted in southern China. *Ying* is the name given to the principal column in a hall, and in ancient times the timber of *A. chinensis* was used for this purpose.

A. julibrissin 合欢 (Hehuan) *Shut Happy*

SILK TREE

A. julibrissin occurs from Iran across to Japan. It is a small tree with bipinnate leaves, each primary leaflet having 20–30 pairs of secondary leaflets, each up to 1.5 cm long. The leaflets fold together at night and symbolise a couple happy in

bed, a phenomenon which has given rise to its Chinese name. In the form most commonly grown the stamens are pink in their upper third (fig. 19.7). It is a hardy species which can be grown in the open in northern gardens and Dorothy Graham (1938) describes the trees as becoming 'a luminous mass of rose bloom' in summer in Beijing. It has been cultivated for many centuries under many names (Li, 1959). The leaves, according to the *Jiuhuang Bencao*, can be used as a famine food (Bretschneider, 1881).

A. kalkora 山合欢 (Shan Hehuan)
Mountain Albizia

Another species native to the tropics of the Old World, which has white flowers and is frequently planted in southern China (Li, 1959).

A. lebbek 阔荚合欢 (Kuojia Hehuan)
Broad-podded Albizia

This tall deciduous tree has bipinnate leaves, the primary leaflets bearing 4–9 pairs of leaflets, each up to 5 cm long, and scented flower heads which are greenish yellow. Native to the Old World tropics, it is grown for shade and ornament in Hong Kong and other tropical areas of the country.

ALCHORNEA Euphorbiaceae

A. davidii 山麻杆 (Shanmagan) *Mountain Hemp*

A. davidii is a deciduous shrub or small tree producing more or less sessile clusters of small flowers in spring on the wood of the previous season. The heart-shaped leaves are bright red on emergence, turning orange, yellow, and finally green (fig. 19.8). Presumably it is for the colour of the spring foliage that it is frequently

Figure 19.7 (above left) *Albizia julibrissin*
Figure 19.8 (left) *Alchornea davidii* in the Zhuozheng Yuan, Suzhou

264

planted in gardens in the Yangtze delta region. It is also grown in Japan but does not seem to have been introduced to the gardens of the West. Since it is most handsome it would certainly be worth doing so.

ALEURITES Euphorbiaceae

A. fordii 油桐 (Youtong) *Oil Tong*

A small spreading tree, *A. fordii* has deciduous, heart-shaped leaves and is attractive when the white flowers are profusely borne in spring. The fruit is shaped like a small peach and turns brown or black when ripe. *A. fordii* is native to south-western China and is much cultivated in southern regions for the oil extracted from the seeds. This oil has been used by the Chinese since ancient times to waterproof paper and fabrics (Usher, 1974). Other species of *Aleurites* are grown for the same purpose. As well as this, the leaves of *A. fordii* are reputed to have insecticidal properties (Needham, 1986).

ARDISIA Myrsinaceae

The two species described below, both native to eastern Asia, are popular as garden and pot plants on account of the profusion of red berries they produce. The roots are used medicinally (Urban Services Department, 1971).

A. crenata 朱砂根 (Zhushagen) *Cinnabar Root*

CORALBERRY

A small evergreen shrub with leathery leaves to 20 cm long, with 12–18 vein pairs. Its Chinese name refers to its use in medicine.

A. crispa 百两金 (Bailiangjin) *A Hundred Ounces of Gold*

Similar to the above but usually shorter and with leaves to 14 cm long, most with 8 vein pairs (fig. 19.9). The Chinese name refers to the high price pharmacists charge for the roots.

ASPARAGUS Asparagaceae

There are several wild species of *Asparagus* in China, and the tuberous roots of some are used in medicine or as food. Of these, *A. lucidus* is sometimes seen as a pot plant

Figure 19.9 *Ardisia crispa*

(fig. 19.10). The South African species *A. densiflorus* and *A. setaceus* have also been used as ornamentals in recent times. As well as these, the herbaceous *A. officinalis* has been introduced and is grown as a vegetable, as it is elsewhere.

A. lucidus 天门冬 (Tianmendong) *Heaven's Gate Winter [Wheat]*

According to Neeham (1986) this may be the plant called *Mendong* in the ancient literature. It was listed as one of the famine foods in the *Jiuhuang Bencao*, the roots being edible as well medicinal (Bretschneider, 1881). It is found in mainland China, Taiwan, Korea, and Japan. Its flexuous, woody stems may reach up to 3 m in height, but in China it is usually grown as a pot plant and kept low. The character *dong* forming part of its Chinese name refers to *Ophiopogon japonicus*, which is known as *Maidong* (Winter Wheat) and is used for the same purposes in Chinese medicine.

Figure 19.10 (top) Pots of *Asparagus lucidus*, surrounding a display of scarlet salvias, aspidistras, and an oleander at the Baiyunguan, Beijing

Figure 19.11 (above) *Aucuba japonica*, with *Michelia figo* (behind), at the Wangshi Yuan, Suzhou

AUCUBA Cornaceae

A. japonica 桃叶珊瑚 (Taoye Shanhu) *Peach Leaf Coral*
SPOTTED LAUREL

This well-known plant is native to mainland China, Taiwan, and southern Japan. It is now grown throughout the world on account of its tolerance of dry shade. The yellow-variegated form, *Sajin Taoyeshanhu* (Sprinkled With Gold Aucuba), is common in Chinese gardens (fig. 19.11).

BAUHINIA Leguminosae

The species listed below form small trees with spreading branches and characteristic orchid-like flowers. They are known in Chinese as *Yangtijia* (Sheep's Foot), an allusion to the shape of their orbicular to cordate leaves, which are bilobed to one third of their length.

B. × *blakeana* 红花羊蹄甲 (Honghua Yangtijia)
Red-flowered Bauhinia

From late autumn to early spring this species produces large flowers of a rich magenta purple with whitish veins, the uppermost petal darker towards the base (fig. 19.12). It was discovered in 1908 near the ruins of a house on the Hong Kong seashore by the fathers of the French Mission at Pokfulam and named after Sir Henry Blake, Governor of Hong Kong from 1898 to 1903. It rarely sets seed and is thought to be a hybrid of *B. purpurea* and *B. variegata* (Thrower, 1988). It is now commonly planted in Hong Kong and has been taken to other tropical parts of the world.

B. × *blakeana* was approved for use as Hong Kong's emblem in 1965 and has recently received much exposure as such with the handing back of the colony to China. Since this event brought to an end the awarding of British imperial honours, it was announced that Hong Kong would have its own award, the Grand Bauhinia Medal, GBM for short, which joins such coveted Chinese awards as the Order of the Precious Tripod and the Order of the Brilliant Star.

B. purpurea 紫羊蹄甲 (Zi Yangtijia)
Purple Bauhinia

This species is similar to *B.* × *blakeana* but the flowers have narrower petals and are usually paler. It blooms in late autumn and carries fruits in spring. It is native to India and south-east Asia, including southern China.

B. variegata 羊蹄甲 (Yangtijia)
Sheep's Foot

ORCHID TREE

The mauve-pink flowers of this species are a little smaller than those of the two described above. However, they are produced abundantly in spring after the old leaves have been shed and before the new ones develop. The tree is thus particularly showy. Also it is more tolerant of cold, being able to withstand a few degrees of frost. Its natural distribution is similar to that of *B. purpurea*. There is a white variant, *Baihua Yangtijia* (White-flowered Sheep's Foot), known in the West as 'Candida'.

BERBERIS Berberidaceae

B. thunbergii 日本小檗 (Riben Xiaobo) *Japanese Barberry*

This well-known Japanese species has been introduced to China and is frequently seen in public plantings in its purple-leaved form. I have not seen it in traditional gardens.

BROUSSONETIA Moraceae

B. papyrifera 构树 (Goushu)

PAPER MULBERRY

Usually this species is seen as a large spreading deciduous shrub with pithy, pubescent stems and soft, broadly ovate leaves up to 20 cm in length. These coarsely toothed leaves range from being unlobed to those with 3–5 lobes. The latter are most characteristic in appearance when the sinuses

Figure 19.12 *Bauhinia blakeana*

between the lobes are expanded at their bases to become oval or circular. The flowers appear in early spring with the leaves and are unisexual and borne on separate plants. The male flowers are grouped in pendulous catkins and the female in spherical heads. The latter develop into an aggregate about 1.5 cm in diameter of orange-red, one-seeded fruits, ripening in summer. The inner bark has been used for making paper in the East, a procedure noted by Marco Polo (Bretschneider, 1898), and is used for making tapa cloth in Polynesia. The plant and its uses were described by Cibot (1786b), who used the name 'Chou-keou'. Also it is recorded in the *Jiuhuang Bencao* that the leaves and fruit can be eaten in times of food shortage (Bretschneider, 1881).

In China the paper mulberry is a common plant in gardens and temple grounds, appearing to be self-sown in some instances. It was observed by John Barrow (1804) to be conspicuous on the shores of the West Lake at Hangzhou, when he was returning to Guangzhou from Beijing with Lord Macartney in 1793. In the West several cultivars have been selected on account of the shape or variegation of their leaves.

BUDDLEJA Buddlejaceae

B. davidii 大叶醉鱼草 (Daye Zuiyucao)
Large-leaved Buddleja

BUTTERFLY BUSH

Steward (1958) records this native plant as a cultivated ornamental in China. However, it seems likely that its use as such is a recent development, probably as a result of Western influences. I have not seen it in traditional gardens.

B. lindleyana 醉鱼草 (Zuiyucao) *Intoxicating-fish Plant*

B. lindleyana is an upright shrub which produces tubular, curved, purplish-violet flowers in nodding spikes up to 25 cm long in late summer (fig. 19.13). It gets its Chinese name because the crushed flowers, when thrown into water, will stupefy fish (Li, 1959). It was introduced to Britain in 1843 by Robert Fortune who first encountered it growing wild in Zhoushan (Lindley, 1846). He had brought it down to Hong Kong in 1844 and later (1852) reported that '... it is now common in several of the gardens on the island, where it thrives well, and is almost always in bloom, although the flower spikes are not so fine as they are in a colder climate'.

Fortune makes no mention of it in cultivation on the mainland, but Li (1959) says that it is the species most commonly cultivated in China. Also it is listed by Steward (1958) as a cultivated ornamental in the lower Yangtze valley. Whether it was cultivated as an ornamental in China before it was popularised by Fortune thus remains uncertain. If it is planted at all in gardens nowadays this must certainly be a rare occurrence, as I have not seen it anywhere.

While in Western gardens it seems to have been overshadowed by the more flamboyant *B. davidii* and its cultivars, *B. lindleyana* is a charming species worthy of wider cultivation in climates which suit it.

Figure 19.13 *Buddleja lindleyana*

BUXUS Buxaceae

B. microphylla var. sinica 黄杨 (Huangyang)
Yellow Poplar

CHINESE BOX

The Chinese Box can be seen in many old gardens, often as a small tree (fig. 18.2). It is also seen as a rounded shrub in modern plantings. A small-leaved dwarf form is also widely used but does not seem to be encountered in traditional gardens, unless perhaps as a pot plant. However, according to Li (1956a), it is popular for planting in miniature tray gardens, where specimens trained as tiny trees are placed among rocks to portray mountain scenery. *B. microphylla* var. *sinica* does not seem to become brownish in cold weather as does the Common Box (*B. sempervirens*), nor does it have the same unpleasant smell. The name *Huangyang* presumably refers to the colour of its wood.

CAESALPINIA Leguminosae

C. pulcherrima 金凤花 (Jinfenghua)
Golden Phoenix Flower, 蝴蝶花 *(Hudiehua) Butterfly Flower,*
洋金凤 *(Yang Jinfeng) Foreign Golden Phoenix*

C. pulcherrima is a species from tropical America which has become popular on account of its showy racemes of flowers with yellow or orange petals and scarlet stamens. Presumably this was the plant recorded by Morrison (1822) as a poinciana, 'kin fung hwa', blooming in the 4th lunar month in Guangzhou.

CANARIUM Burseraceae

Canarium album 橄榄 (Ganlan)

CHINESE OLIVE

The Chinese Olive is an evergreen tree native to southern China and Vietnam. It has glossy pinnate leaves and produces pale green, oval fruits (fig. 19.14). Like those of the true olive these are fleshy with a central stone. Although it is not grown as an ornamental plant I have included it because the fruits are sold throughout the country, either fresh or salted, being much favoured as a condiment or side dish (Meyer, 1911; Thrower, 1988; Zee & Hui, 1990). According to Thrower they have a rather strong resinous flavour in the raw state. She says that the salting is done by first soaking

Figure 19.14
Fruits of the Chinese Olive (*Canarium album*) bought in Beijing

them in tubs of salt water and then drying them.

Ji Han recorded in 304 AD that the fruits had been sent from the south each year to the court of the Emperor Wu of the Han Dynasty and that the tradition persisted at the time of his writing (Li, 1979). Ji Han's is the first known description of this plant. Princess Der Ling (1911) recorded that at the Chinese New Year the ladies of the court gave the Dowager Empress plates of olives, apples, and lotus seeds, which signified long life, peace, and blessing, respectively. In spite of its use as an auspicious object, the Chinese olive rarely appears in art. However, it can be seen in the painting *New Year's Day* (fig. 2.12), and Bartholemew (1980) illustrates a pair of 18th century porcelain wine cups, each with a leafy branch bearing three fruits painted on the outside.

CARAGANA Leguminosae

C. arborescens 树锦鸡儿 (Shu Jinjier) *Tree Caragana*

A large shrub or small tree with pinnate leaves, each with 4–6 pairs of leaflets and terminating in a spine, *C. arborescens* produces axillary clusters of up to 4 yellow pea flowers. It is native to northern China and Siberia.

C. chamlagu (C. sinica) 锦鸡儿 (Jinjier) *Golden Pheasant*

This plant is similar to the above but the leaves have only two pairs of leaflets, the flowers are borne singly, and it usually remains a small shrub. It is sometimes used for penjing (Hu, 1982). It is native to northern China and is depicted in an 18th century handscroll reproduced by Capon and Pang (1981, cat. no. 65).

CATALPA
Bignoniaceae

The catalpas have been cultivated since ancient times, being considered good timber trees, the wood having similar uses to that of the paulownias. It is recorded in early times that they are easily raised from seed (Needham, 1996).

Figure 19.15 Nuts of *Castanea mollissima*, the Chinese Chestnut, bought from a street vendor in Beijing

CASTANEA Fagaceae

C. mollissima 板栗 (Ban Li) *Board Chestnut*

CHINESE CHESTNUT

C. mollissima occurs both wild and planted in Korea and much of China. According to Wang (1988) it was cultivated in China before the middle of the 3rd century BC. Cibot (1778b, 13) gave an account of this chestnut, noting that it was used medicinally and that the catkins were sometimes made into lamp wicks. He recorded that the nuts of the best kinds were as big as apricots. Large-fruited selections are now grown extensively and the nuts (fig. 19.15), both fresh and dried, are consumed domestically and also exported. In autumn and winter roasted chestnuts are offered for sale in most parts of China and as far afield as Singapore. The nuts are regarded as a symbol of fecundity (Bartholemew, 1985a).

C. bungei 楸树 (Qiushu) *Autumn Tree*

C. bungei forms a large deciduous tree which produces its flowers in small clusters of 3–12 in late spring. These are pink, variously marked inside with brown-purple and yellow, but tend to be somewhat sparsely produced. It is planted widely, especially in the north. The name *Qiushu* refers to the fact that in northern China the leaves of *C. bungei* fall at the end of summer or the beginning of autumn. Such indications of seasonal change were of great importance to an agrarian people in ancient times, and during the Tang dynasty catalpa leaves were worn ceremonially at the time of the autumn equinox. Also it is recorded in the *Jiuhuang Bencao* that the flowers can be eaten in times of famine (Bretschneider, 1881).

Old trees of this species are quite common in Beijing. Specimens may be seen in Behai Park, the Forbidden City (fig. 19.1), and various old gardens. These vary considerably in the colour and markings of the flowers. *C. bungei* was introduced to the United States by F. N. Meyer who sent cuttings collected at the Yellow Temple in Beijing in 1905 (Cunningham, 1984).

C. ovata 梓树 (Zishu) *Native-place Tree*

C. ovata produces its flowers in summer, after *C. bungei* has finished. They are small, dull white, and marked with yellow and purple within. Like *C. bungei*, it has been cultivated since ancient times for its timber, as a shade tree, and in gardens (fig. 19.16). It usually blooms more prolifically than *C. bungei*, though its small whitish flowers are not particularly attractive, and it is conspicuous in autumn when covered with its long slender seed pods. As with *C. bungei*, the falling of the first leaves was once taken to indicate that autumn had commenced (Morris, 1983). Also Needham (1986) records that its bark is regarded as an effective insecticide.

C. ovata appears from time to time in paintings, for example the 15th century *Three Old Catalpas at the Temple of Fan* reproduced by Barnhart (1972). This is an example of the tradition of symbolising people by old trees: the inscriptions on the painting are about Lord Fan (987–1052), who planted them.

This tree and the mulberry, *Sang*, were universally planted in the grounds of Chinese homes. The mulberry was used for raising silkworms and the catalpa wood was valued particularly for making coffins. These two trees thus care for both the living and the dead. According to Li (1959) this accounts for the sentimental term *sangzi*, mulberry and catalpa, for one's home or native place.

CELTIS Ulmaceae

C. sinensis 朴树 (Pushu)
Plain Tree

CHINESE HACKBERRY

This large, deciduous tree is the east-Asian equivalent of the European Nettle Tree, *C. australis*, and the North American Hackberry,

C. occidentalis, and very similar to them in appearance. It gets the name *Pushu* on account of its dense, evenly grained wood, which is favoured for finely detailed carving. It is frequently planted as a roadside tree and in parks and gardens (fig. 19.17), and appears to have been in cultivation for a very long time. It is hardy, long-lived, and tolerant of poor conditions.

The insect galls which form on this tree can be eaten in times of food shortage according to the *Jiuhuang Bencao* (Bretschneider, 1881). It seems likely that this is the tree

Figure 19.17 *Celtis sinensis* at the Huanxiu Shanzhuang, Suzhou, with *Acer palmatum* (foreground)

which Cibot (1786b) called '*Mou-chou-kuo-tsée*', as he described it as bearing insect galls which were eaten. It was apparently rare in Beijing as he knew of it from only a single temple.

Other species of *Celtis* are native to China and may occasionally be planted. For instance, a specimen of *C. julianae* in Hangzhou in the grounds of the tomb of Yue Fei, a hero of the 11th century, is claimed to be 700 years old.

CERCIS Leguminosae

C. chinensis 紫荆 (Zijing) *Purple Thorn*

CHINESE REDBUD

A shrub of upright habit, *C. chinensis* produces a mass of flowers, similar to those of the Judas Tree (*C. siliquastrum*) and the Redbud (*C. canadensis*), on the bare branches in spring. These are usually a bright magenta but vary from mauve to purplish crimson (fig. 19.18). There is also a white form. The bark, flower stalk, and seed have medicinal value and an insecticide can be prepared from the seed (Wang & Ma, 1995).

The Chinese Redbud is planted in almost every Chinese garden and was in cultivation at least by the 6th century AD (Wang, 1988). Fortune (1844) noted it in the Fa Tee Gardens.

C. yunnanensis (*Yunnan Zijing*) is another beautiful Chinese species. I have seen it in the Western Hills outside Kunming where it forms a small tree with pale mauve-pink flowers (fig. 19.19), but whether it was planted or growing naturally I could not tell.

CHIMONANTHUS Calycanthaceae

C. praecox 腊梅 (Lamei) *Twelfth-month Mei*

WINTERSWEET

This sweetly scented shrub has long been a favourite. Producing its waxy, pale yellow flowers in winter, it has featured as a decoration for the Chinese New Year (Metcalf, 1942) and is widely planted in gardens (fig. 19. 20). According to Li (1959) it was not brought into cultivation until the Song. Huang Tingjian composed a poem praising this flower in the 11th century, which made it immediately popular in the capital, Kaifeng. Li says there are several distinct varieties in Chinese gardens with names such as 'Pale Heart', 'Urn-mouthed', 'Lotus Flower' and 'Nine-flowered'. These are propagated by grafting onto the typical form.

As in the case of many other flowers, it was discovered that those of the wintersweet could be eaten in times of famine. This was recorded by Zhu Xiao in the *Jiuhuang Bencao*

Figure 19.18
(right)
Cercis sinensis
in Shanghai
Figure 19.19
(opposite page)
Cercis yunnanensis
in the Western
Hills, Kunming
Figure 19.20
(far right)
Chimonanthus
praecox, the
Wintersweet

and by Bao Shan in his *Yecai Bolu* (Comprehensive Account of Edible Wild Plants) of 1622 (Needham, 1986). The flowers were also worn by women to decorate their hair (Fortune, 1850f), and are used in the preparation of perfumes (Wang & Ma, 1995).

The first European to have noticed the wintersweet seems to have been Alvaro de Semedo, a Jesuit whose *Relatione della Grande Monarchia della Cina* was published in 1643. In it he recorded that the yellow, wax-like flowers appeared in winter (Bretschneider, 1880). It has traditionally been regarded as the floral symbol of the 12th lunar month, *La*, and is frequently depicted as such in paintings (e.g. Barnhart, 1983, cat. no. 21; Weidner et al., 1988, cat. nos 34, 39) and on porcelain (e.g. Li, 1959, pl.1). It has also been used as a substitute for *Prunus mume* in representations of the Three Friends of the Cold Season (Bartholemew, 1980, 1985a, b) and in paintings expressing good wishes for the New Year (e.g. fig. 2.12). According to *Curtis's Botanical Magazine* (1800, t. 466), it was introduced to Britain from China in 1766.

CHIONANTHUS
Oleaceae

C. retusus 流苏 (Liusu)
Tassels

CHINESE FRINGE TREE

When covered with masses of creamy-white, narrow-petalled flowers the Chinese Fringe Tree rivals the best of the crabapples and cherries (fig. 19.21). Although it seems to be rare in Chinese gardens, in some parts of the country the young leaves are valued as a substitute for tea. Apparently this is sweet-smelling like some of the best green teas (Li, 1959). It was amongst the plants introduced to Britain by Robert Fortune, who obtained it from a garden near Fuzhou. He observed that it was much prized in Fujian, that the Chinese propagated it by grafting on *Osmanthus fragrans*, and that, being cold hardy, it probably came from further north (Lindley & Paxton, 1853).

Figure 19.21 *Chionanthus retusus,* **the Chinese Fringe Tree**

Figure 19.22 (right) *Chloranthus spicatus,* from *Curtis's Botanical Magazine* 1820, t. 2190 (as *C. monostachys*)

Figure 19.23 (far right) 700-year-old trees of *Cinnamomum camphora* at Wu Shan, Hangzhou

CHLORANTHUS Chloranthaceae

C. spicatus 金粟兰 (Jinsulan) *Golden Millet Orchid*

C. spicatus (fig. 19.22) is a subtropical shrub which can be grown only in southern gardens. Morrison (1822) lists it as blooming in the 4th moon (May) in Guangzhou. It was introduced to Britain by John Reeves (*Curtis's Botanical Magazine* 1820, t. 2190) as *C. monostachys*, for which the Chinese name 'Chu Lan' was recorded. It is, or was, cultivated in southern China for its inconspicuous blossoms which have been used as hair decorations (Metcalf, 1942) and for perfuming tea (Barrow, 1804; Fortune, 1847a; Li, 1959). Barrow described the flavour of this tea as 'peculiar' but, according to Li (1959), it is the favourite tea of many people. He says, however, that the petals, after they have imparted their aroma, should be carefully removed, since their use is considered to be deleterious if they are left in.

Li (1959) also recorded the name 'Chu Lan' (*Zhu Lan*) for the plant as did Fortune (1857) before him. Fortune suggested that, owing to the similarity of their names, the plant may have been confused with *Aglaia odorata*, which he said was called 'Lan-hwa', and that claims of the use of *C. spicatus* for scenting tea may have no basis. The name *Jinsulan* (Golden Millet Orchid) has been given to *C. spicatus* as the tiny flowers look like grains of millet and smell like cymbidiums. *A. odorata* is called *Mizilan* (Rice Grain Orchid) for similar reasons, so there is certainly room for confusion.

CINNAMOMUM Lauraceae

C. aromaticum 肉桂 (Rougui) *Flesh Cassia*

CASSIA BARK TREE

C. aromaticum is similar to *C. camphora* (see below) but has bigger leaves, flowers in summer, and usually does not grow as large. It was mentioned by Ji Han in 304 AD and is cultivated in southern China for its bark as a substitute for cinnamon (*C. zeylanicum*). It has, however, an inferior flavour. The famous tourist destination, Guilin, was named after this tree, which once must have formed extensive forests in parts of southern China (Li, 1979). It is not, as far as I know, planted in traditional gardens. It is believed to be native to Burma (Huxley et al., 1992).

C. camphora 樟树 (Zhangshu) *Camphor Tree*

CAMPHOR LAUREL

The Camphor Laurel, which is native to eastern Asia, is one of the best known of all Chinese trees. It sheds its leaves in spring and simultaneously produces new ones together with its inconspicuous greenish flowers. It is widely planted in gardens in those areas of the country where the winters are not severe. It can become an enormous tree and specimens claimed to be several hundred years old can be seen in various parts of the country, for example at Wu Shan in Hangzhou (fig. 19.23). It is now cultivated in most tropical and subtropical parts of the world, and has become naturalised in some places.

Commercial camphor is obtained from the wood, roots, and leaves, and the wood is popular for making chests and furniture, not only because of its beauty but because of its insect-repellent properties.

CLAUSENA Rutaceae

C. lancium 黃皮 (Huangpi) *Yellow Skin*

WAMPI, WAMPEE

The wampee is a small evergreen tree with pinnate leaves and terminal panicles of small white flowers. These are followed by round or oval fruits 2–2.5 cm in diameter, often in clusters of 30–50. These are brownish yellow when ripe with sweet white pulp surrounding several jade-green seeds. They may be eaten raw or as a jam or preserve. *C. lancium* is extensively cultivated in southern China, to which it is native. The fruits are not only delicious but, being thought to promote digestion, are also taken medicinally (Thrower, 1988).

CLERODENDRUM Verbenaceae

C. chinense (C. philippinum) 臭茉莉 (Choumoli)
Foul-smelling Jasmine

FRAGRANT GLORYBOWER

A shrub up to 1.5m tall, this species has large, soft, unpleasant-smelling leaves and fragrant, white flowers in terminal clusters about 10 cm across. It is usually the double-flowered form, *Zhongban Choumoli* (Double Foul-smelling Jasmine), which is cultivated in China. *C. chinense* is native to southeast Asia but is now found throughout the tropics. The root is used in the treatment of skin diseases (Urban Services Department, 1971). According to *Edwards' Ornamental Flower Garden* (1854), 'Few plants are more deserving of cultivation than this, which in the gardens of China is one of the handsomest of their flora; as is attested by a beautiful Chinese drawing preserved in the library of the Horticultural Society'.

C. kaempferi 赬桐 (Chengtong) *Red Tong*

This erect shrub with large heart-shaped leaves and small, bright red flowers in terminal panicles up to 35 cm long is apparently native to southern Asia, including India and southern China, where it is also cultivated (Urban Services Department, 1971; Li, 1979). It was first described in 304 AD by Ji Han, who recorded that it was found everywhere in southern China, had leaves resembling the paulownia, and flowers, including their stalks and sepals, of the deepest scarlet (Li, 1979). Almost 1500 years later Loureiro (1790) recorded it from Guangzhou. Li suggests that perhaps it was more common in earlier times than at present and thus was especially noted by Ji Han, who gave it the Chinese name still used. In China many trees with large, broad leaves are referred to by the general name *tong*.

CORDYLINE Dracaenaceae

C. terminalis 朱蕉 (Zhujiao) *Red Banana*

A native of tropical south-east Asia, much cultivated for its variously coloured and patterned leaves, it forms a shrub of up to 4 m, but Livingstone (1822) records instances of its being kept only a few inches tall by the Chinese. Morrison (1822) recorded that it flowered in the 9th moon (October) in Guangzhou, though it is principally grown for its leaves. It was often called *Dracaena ferrea* by early Western visitors.

CORNUS Cornaceae

C. kousa var. chinensis 四照花 (Sizhaohua)
Four-shining Flower

Li (1959) says that this is the common species in Chinese gardens and that it is native to the Yangtze valley. Steward (1958) also records this dogwood as a cultivated ornamental in this region but I have not seen it used as such, nor have I found any evidence of its use as a garden plant in earlier times. According to Morris (1983), sprigs were worn as amulets to ward off evil during the festival held on the 9th day of the 9th month. It was collected by Ernest Wilson in 1907 from places north and south of Yichang and introduced to cultivation, probably first to the United States and then to Britain three years later (Lauener, 1996). Its Chinese name refers to the four white bracts which subtend each flower cluster.

CORYLUS Betulaceae

C. heterophylla 榛 (Zhen)

CHINESE HAZEL

Meyer (1911) recorded that, at the time of his visit, the nuts of this native species were collected and sold but that the plants were never cultivated. Whether they are now I do not know, but the nuts are commonly offered for sale in shops and markets (fig. 19.24).

C. sieboldiana var. *mandschurica*
毛榛 (Mao Zhen) *Hairy Hazel*

In this species the involucre surrounding the nut is prolonged into a tube covered, particularly in the lower part, with bristly hairs. The nuts themselves are very small, usually no more than 1.25 cm wide. Meyer (1911) recorded that they were collected and sold, and they still are (fig. 19.25). I was offered some in Chengde, where I observed the Chinese cracking them open between their teeth. Were I to have adopted this technique it would undoubtedly have led to a dental disaster.

COTINUS Anacardiaceae

C. coggygria 黄栌 (Huanglu) *Yellow Lu*

SMOKE BUSH

The Smoke Bush occurs naturally from southern Europe across to China, and is easily recognised by its rounded leaves and its conspicuous feathery panicles of small flowers (fig. 19.26). Its Chinese name refers to the colour of its wood. It is a common plant in the Western Hills near Beijing and a major contributor to the display of autumn colour in that region. It is also frequently planted in gardens, though the purple-leaved forms so popular in the West do not seem to be used. As seen in the wild in China it is a variable plant, many individuals showing a little purple pigmentation in their leaves and inflorescences. The plants seen in gardens show a similar range of variation and appear to be seedlings. It is recorded in the *Jiuhuang Bencao* that the leaves can be eaten in times of famine (Bretschneider, 1881).

COTONEASTER Rosaceae

Although there are many beautiful species of this genus native to China, they seem to have been ignored by Chinese gardeners until recent times. *C. microphyllus* and *C. pannosus* are used for penjing (Hu, 1982), but the only one I have seen being grown in the ground other than in a botanic garden is *C. multiflorus*.

C. multiflorus 水旬子 (Shui Xunzi) *Water Cotoneaster*

This species from northern China is a deciduous shrub with graceful arching or pendulous branches, which are said to suggest flowing water. When wreathed with clusters of white flowers in spring it is most attractive, as it is with its red berries in autumn. Several plants may be seen in the Huifang Yuan in Zhongshan Park, Beijing.

Figure 19.24 (right) Shelled nuts of *Corylus heterophylla* on sale in Chengde

Figure 19.25 (centre) Nuts of *Corylus sieboldiana* var. *mandschurica* bought from street vendor in Chengde

Figure 19.26 (below) *Cotinus coggygria* in bloom at Badachu, southwest of Beijing

CUDRANIA—see *Maclura*

DAMNACANTHUS Rubiaceae

D. indicus 虎刺 (Huci) *Tiger Spine*

D. indicus is a small east-Asian shrub with slender branches bearing needle-like spines, which are longer than the small leaves. It produces small, white, axillary flowers, solitary or in pairs, which are followed by red berries (Steward, 1958). Li (1956a) records it from Fujian and says that it is used for penjing and in flower arrangements. Its use for penjing is also mentioned by Hu (1982). The Chinese name for *D. indicus* may mean that the spines are too sharp to be touched even by a tiger, or that the spines are as dangerous as a tiger. It is also called *Funiuhua* (Lying-down-cow Flower). Since the character *fu* can mean both to lie down and to admit defeat, it seems possible, in view of the spiny nature of the plant, that its name might better be translated as 'Defeating-cow Flower'.

DAPHNE Thymelaeaceae

D. genkwa 芫花 (Yuanhua) *Yuan Flower*

When it is in bloom the stems of this charming deciduous shrub resemble panicles of lilac (fig. 19.27). It greatly impressed Robert Fortune (1847a) when he first came across it on the hillsides of Zhoushan in 1844. Subsequently Ernest

Wilson (1913) described it as 'by far the finest species of the genus' when he saw it abundant on the hills around Yichang. While it has deservedly become a popular garden plant in the West, it does not seem to have found favour in this regard in its home country, although Wang and Ma (1995) say that it is grown in parks and courtyards. Its main use seems to have been as a fish poison (Li, 1959) and as an insecticide for forcing into the holes made by borers in the trunks of trees. Though its poisonous nature was mentioned in the literature as early as the 1st century BC, the earliest known record of its use as a pesticide is in a book of 1273 AD (Needham, 1986). The flowers, bark, and root are also used in Chinese medicine (Wang & Ma, 1995).

It was taken to Japan and cultivated there, but does not seem to have been grown in the West until Fortune sent it to England in 1844. The specific epithet *genkwa* is a transliteration of the Japanese pronunciation of its Chinese name.

D. odora 瑞香 (Ruixiang) *Lucky Scent*

DAPHNE

Unlike *D. genkwa*, this species has been grown and valued since the Song. Being evergreen, flowering in winter and early spring, and having a strong fragrance, it is not surprising that it satisfies the canons of Chinese horticultural taste (fig. 19.28). Loureiro (1790) noted it as a cultivated plant in China, and Morrison (1822) listed the typical form as flowering in Guangzhou in the 10th moon (November), and the form with silver-edged leaves as doing so in the 12th (January).

Figure 19.27 *Daphne gengkwa*

Figure 19.28 *Daphne odora*

Li (1959) says that it is often grown in pots in northern China and there are varieties with white, red, purple or spotted flowers. He also records that it is used to make certain kinds of paper. Fortune (1847a) saw it cultivated on Putuo Shan, offshore from Ningbo.

According to the *Bencao Gangmu*, the famous pharmacopoeia written by Li Shizhen, *D. odora* was first brought into cultivation during the Song from Lu Shan in Jiangxi. There is a legend which tells of a monk who fell asleep below a cliff on Lu Shan and dreamt of an intense fragrance. On awakening he searched about and found the plant which he named *Shuixiang* (Sleeping Scent). This name was later changed to another of similar pronunciation, meaning Lucky Scent (Li, 1959).

As is the case with many plants cultivated in China, it is now hard to trace plants which approximate to the wild form and its exact distribution is uncertain. Hemsley recorded collections from Zhejiang, Hubei, and Sichuan but, according to Rehder, the typical form does not occur wild (Brickell & Mathew, 1981). While this may be so, I have seen plants of a similar type in the Western Hills near Kunming. Clearly further research is needed if the position is to be clarified.

Presuming that it did not exist there as a wild plant, *D. odora* must have been taken to Japan relatively early as it is depicted in a Japanese scroll of 1309, along with other plants of Chinese origin (Ogisu, 1996). It was sent to Kew from China in 1770, arriving in 1771 (Brickell & Mathew, 1981).

DEUTZIA Hydrangeaceae

Many handsome species of this genus are native to China but do not seem to have been brought into gardens. Visitors to China are likely to notice them when they are in bloom on roadsides and hills. For instance in spring *D. grandiflora* is conspicuous in many areas of northern China, standing out like patches of snow on the hillsides. It is the earliest flowering species and has the largest individual flowers but, striking though it is, the Chinese have left it to its own devices. The only species recorded in cultivation in China until recent times seems to be the Japanese *D. scabra*.

D. scabra 溲疏 (Soushu) *Urine Sparse*

Steward (1958) lists this species as a cultivated ornamental in the lower Yangtze valley. Li (1959) says it has long been

Figure 19.29 *Dimocarpus longan*

cultivated but that in ancient times it was used only for medicinal purposes. It gets its name because the leaves, flowers, and fruits are taken as a diuretic. I have not seen it in traditional gardens, though this probably means little. Although it is native to China, it was from Japan that it was introduced to the West in 1922 (Lauener, 1996).

DIMOCARPUS Sapindaceae

D. longan 龙眼 (Longyan) *Dragon's Eye*

LONGAN

D. longan is an evergreen tree native to southern China and adjacent areas. A close relative of the lychee, it is distinguished by its larger leaves, rougher, more corky bark, and smaller fruits, which are brown not red (fig. 19.29). It is also slightly hardier and less exacting than the lychee, tolerating occasional light frosts. The fruits are eaten fresh or dried, and nowadays are canned as well. Although grown principally for its fruits, in tropical and subtropical climates it also makes an attractive ornamental tree, with its spreading habit and

Figure 19.30 **Persimmon trees at the Wuta Si, Beijing**

DIOSPYROS Ebenaceae

D. kaki 柿树 (Shishu) *Shi Tree*
PERSIMMON

One of the most important tree fruits in the north of the country, the persimmon, which is not known wild, has been cultivated for hundreds of years in China and Japan. It forms a tree up to 10 m or more in height, and is particularly handsome when bearing its ripening fruits and when its leaves turn orange before falling. The varieties grown in the West are usually Japanese and form small, spreading trees, whereas those most commonly seen in China are more upright and become much larger (figs 19.16, 19.30, 19.120). Meyer (1916) saw one over 24 m tall. As well as being grown in orchards, *D. kaki* is frequently planted in temple grounds and gardens. It is said that there are about 240 varieties in China (Zee & Hui, 1990). Meyer (1911) noted that to keep them

through the winter in northern China the fruits were piled in heaps and allowed to freeze, being kept frozen until needed. Some of the less juicy varieties are dried. Persimmons have also been used in medicine and for making wine and vinegar. The juice of green persimmons has been used for waterproofing paper (Bartholemew, 1980).

Persimmons were eaten as far back as the Han. They were mentioned in the literature of the time and seeds have been excavated from a tomb dated 168 BC. During the Tang, Duan Chengshi in his *Youyang Zazu* (Miscellany of Youyang) wrote that the tree had seven virtues—it lives for a long time, provides shade, birds do not nest in it, it is resistant to worms and grubs, the autumn leaves are enjoyable, and one can practise calligraphy on the fallen leaves (Bartholemew, 1980).

glossy pinnate leaves. The longan can live up to 400 years and old trees can still yield well (Zee & Hui, 1990). It gets its Chinese name because the shiny, black seed is suggestive of a dragon's eye.

Presumably the longan has been cultivated for a very long time. According to Wang (1988) both it and the lychee were brought to Changan from the south in the 2nd century BC and cultivated in the heated conservatories of Emperor Wu. The first description of the plant in Chinese is that of Ji Han in 304 AD, who recorded that, as with the lychee, the fruits were sent annually as tribute to Emperor Wu (Li, 1979). It does not seem to have attracted the attention of artists in the way the lychee has, but Krahl (1987) records its appearance as a decorative motif on porcelain.

In 1335 the man who was later to found the Ming dynasty is said to have wrapped a red robe of investiture around the trunk of a persimmon tree, giving it the title 'Marquis of Ice and Frost'. His tribute was in recognition of the fact that, during his military campaigns some years earlier, he had been without food for two days and, happening on this tree in a ruined garden near what is now Dangtu in Anhui, had eaten 10 of its ripe fruits (Tun, 1965).

Meyer (1911) reported that the persimmon is propagated in China by grafting or budding onto *D. lotus*. He noted that of the many different varieties one stood out above all others in excellence, being 7.5–12.5 cm in diameter, bright orange-red, seedless, and not astringent. He described it as having a peculiar shape, perfectly round but with an equatorial constriction, which makes it look 'as if two fruits had been joined, or, to use a more terse expression, as if somebody had sat upon it'. No doubt this was what Francis Xavier d'Entrecolles described in 1736, in a letter from Beijing to Du Halde, as being a 'fruit of two stories' (Bretschneider, 1880). This variety is called 'Da Shizi' ('Big Persimmon') or 'Damoban Shizi' ('Big Millstone Persimmon'), which Williams (1975) says sometimes reaches 'a pound in weight' (fig. 19.31). Meyer sent cuttings of this variety to the United States in 1905, collected in the valley north of Beijing where the Ming Tombs are situated (Cunningham, 1984), but it is puzzling that, in spite of its outstanding qualities, it appears not to have taken the Western world by storm. Persimmons of this type which I ate in Beijing in November 1996 were delicious and much superior to any I had previously encountered elsewhere. A fruit exhibiting this 'double' characteristic appears in a 13th century painting of six persimmons (Froncek, 1969) and others can be seen in later works (e.g. She & Yuhas, 1973, no. 48), so this cultivar, or something similar, has existed for at least 700 years.

Persimmons were popular as offerings for the Autumn Festival, which fell on the 15th day of the 8th month, since they are both round and orange-red. Red is an auspicious colour and the character for round is *yuan*, which forms part of a phrase meaning 'reunion'. As well as this, the name *Shi* is pronounced in the same way as the character meaning 'affairs' or 'business'. In paintings the depiction of two fruits suggests 'all affairs'. They are often combined with other symbols to suggest wishes such as 'May there be no limits to your good fortune in business' (Bartholemew, 1980).

Although the persimmon is portrayed in early paintings,

Figure 19.31 'Big Millstone' persimmons bought in Beijing

it does not seem to have become a popular subject until the Qing dynasty, when it frequently appears on porcelain and elsewhere. In this period it is often included amongst the auspicious symbols depicted on the scrolls hung at New Year (fig. 2.12).

D. lotus 黑枣 (Heizao) *Black Date*

CHINESE DATE

D. lotus is a tree similar in appearance to *D. kaki*, growing to 25 m or more in height and producing small round fruits, only about 2 cm in diameter, which turn black when fully ripe. These fruits are listed among the famine foods recorded in the *Jiuhuang Bencao* (Bretschneider, 1881) and are sometimes collected and eaten. The seeds are planted to raise the stocks on which the cultivars of *D. kaki* are grafted or budded. It is native to temperate Asia from Asia Minor to China (Brickell, 1996).

It may be that other species of *Diospyros* are cultivated in China from time to time. For instance, Meyer said that *D. sinensis*, which he introduced to the United States as a potted plant acquired at Hankou in 1907, is rarely seen as a garden shrub but rather grown dwarfed in pots (Cunningham, 1984). This species, which Steward (1958) renamed *D. cathayensis*, is an evergreen shrub or small tree with greenish-white, bell-shaped flowers.

EDGEWORTHIA Thymelaeaceae

E. papyrifera (*E. chrysantha*) 结香 (Jiexiang) *Fragrant Knot*

This deciduous shrub is a member of the daphne family and is common in Chinese gardens. It is grown for its round heads of slightly scented, yellow flowers, which are produced in late winter and early spring. The stems are so pliable that they can be tied in knots without apparently causing the plant any harm. Occasionally one sees plants treated in this manner, although their appearance may not be to everyone's taste (fig. 19.32). The stem and bark can be made into paper, the whole plant is used in Chinese medicine, and an insecticide can be prepared from the stem and leaves (Wang & Ma, 1995). *E. papyrifera* was introduced to England from Zhoushan in 1845 by Fortune (Lindley, 1846), who also saw potted plants in nursery gardens in Shanghai (Fortune, 1852). There is a form with reddish-orange flowers known in the West as 'Red Dragon'.

ENKIANTHUS Ericaceae

E. quinqueflorus 吊钟花 (Diaozhonghua)
Hanging Bell Flower

E. quinqueflorus is amongst the most beautiful of the *Enkianthus* species but, being native to south-eastern China, it can tolerate only the lightest of frosts. The flowers, large for the genus, appear in clusters at the ends of the bare branches in late January and early February. They are waxy and glistening, the corollas, calyces, pedicels, and bracts all being some shade of pink (fig. 19.33).

E. quinqueflorus is the type species of the genus and was named by the 18th century Portuguese botanist Ioannes de Loueiro. *Enkianthus* means 'pregnant flower', a name he gave to the plant because he said its flowers are pregnant with others. He was mistaken, as what he took to be a single terminal flower was actually the assemblage of coloured bracts with five true flowers within (Coats, 1963).

Figure 19.32 (left) *Edgeworthia papyrifera* at the Wannian Si, Emei Shan, Sichuan. Note the branches tied in knots.

Figure 19.33 (below) *Enkianthus quinqueflorus* in the Orchid Garden, Guangzhou

Curtis's Botanical Magazine (1814, t. 1649) reports that Loureiro said that flowering branches were preserved many days in porcelain vessels filled with water in the houses of the mandarins, and that Mr Kerr said it was native in the province of Guangzhou 'where it is held in a kind of veneration, and the flowers deemed an acceptable offering to the gods; and, accordingly, at the Chinese new-year, which happens at the season of their blossoming, large branches with flowers are hung up in all the temples'. Likewise Clarke Abel (1819), who saw it in the Fa Tee Gardens and identified it as *Vaccinium formosa*, said

'The *Vaccinium formosa* is a sacred plant; its flowers are gathered at the commencement of the Chinese New Year, and placed in all the temples as an acceptable offering to the gods'. While it is occasionally planted in gardens, it is mainly as a decoration for the Chinese New Year that it is used, a practice which became established in Guanzhou early in the Qing dynasty (Goody, 1993). For this purpose, as noted by Fortune (1847a, 1852), cut branches are brought from the hillsides, just as the buds are opening, and sold in markets and shops in southern China. Fortune (1852) wrote that if the branches were cut and placed in a jar of water before the flowers were fully expanded, they remained in perfection for two or three weeks. Thus it is ideally suited to be a symbol of good fortune, opening on the 1st day of the 1st lunar month and lasting until the 15th day, the day on which the New Year festivities end.

As with the other New Year favourites, its popularity was enhanced by the favourable meanings extracted from its various characteristics. The bell-shaped flowers suggested the Cantonese expression *chong yuen ko chun*, meaning 'came first in the imperial examination', and the great number of these flowers and the seeds which follow stood for *dor tze dor suen*, 'having numerous descendants' (Goody, 1993).

Louriero (1790) said that it was cultivated in Guangzhou, but it does not seem likely that this was a widespread practice. However, according to Herklots (1932c), Professor McClure of Lingnan University, Guangzhou, said that it was cultivated in a crude way in certain places, small plants being brought in from the wild and set out on the slopes overlooking the villages where they could be protected from unauthorised collectors. Before the plants leafed out in spring the branches destined for cutting at the end of a given year were 'ringed', a strip of bark about 2.5 cm wide being removed from around the stem at approximately the point at which it was to be severed. This ensured the abundant production of flower buds.

Its use as a decoration for the New Year was widespread in Hong Kong in the late 19th century but, to prevent its extinction in that territory, an ordinance prohibiting its possession and sale was introduced in 1913. However, cultivated and imported plants were exempted, so an industry sprang up in the mountains about 100 km north of Guangzhou producing the branches by the method described above. Nevertheless the Japanese occupation, followed by changes in land use and the period during which the growing and use of

Figure 19.34 *Eriobotrya japonica* in the Yi Pu, Suzhou

flowers was banned on the mainland, led to a decline in its use as a New Year flower (Goody, 1993). Even so cut branches were still conspicuous in Macao and Guangzhou when I was there at this time of year in 1980. And apparently in relatively recent times it was still suffering from illegal cutting in Hong Kong (Urban Services Department, 1971).

ERIOBOTRYA Rosaceae

E. japonica 枇杷 (Pipa)

LOQUAT

The large, dull-green leaves of this small tree (fig. 19.34), hairless above but rusty tomentose beneath, strongly resemble the leaves of many of the large-leaved *Rhododendron* species. The white flowers are usually produced in autumn in terminal clusters and are followed by the yellow fruits which ripen in spring. Loquats are often grown in southern and central China for their fruits. Meyer (1911) reported that the finest kind is considered to be the 'White Loquat', which is said to have fruit of a pale yellowish colour and of very fine

Figure 19.35 *Euonymus japonicus* in the garden of the Lingyin Si, Hangzhou

Figure 19.36 *Fatsia japonica* at the Tuisi, Tongli

flavour. While this may be so, any loquats I have tasted have proved rather insipid. As well as being grown for its fruit, the loquat is frequently grown as an ornamental plant in Chinese gardens, and is depicted in paintings from the Song dynasty on (e.g. Barnhart, 1983, cat. no. 8) and on porcelain (e.g. Fourcade, 1965, no. 21). It is also grown in Japan, from where

it was originally described by Thunberg, and its cultivation has spread to many parts of the world.

ERYTHRINA Leguminosae

The species of *Erythrina* are known as Coral Trees and several species are grown as ornamentals in southern China, their popularity no doubt resulting from the bright red colour of their flowers. These species include *E. caffra*, *E. corallodendrum*, *E. crista-galli*, *E. speciosa*, and *E. variegata* (Thrower, 1988). All of these except the last have been introduced, presumably no earlier than the 16th century, as *E. caffra* is from southern Africa and *E. corallodendrum*, *E. crista-galli*, and *E. speciosa* are from Central or South America. *E. variegata* has a wide distribution in southern Asia and adjoining regions and is believed to be native to southern China. It is easily distinguished from the other species mentioned by its calyx, which is completely split in the opened flower, whereas in the others it is not or is only partly split.

E. variegata 刺桐 (Citong) *Thorny Tong*

CORAL TREE

The first description of this tree in the Chinese literature was recorded in 304 AD by Ji Han, who called it *Citong* (Li, 1979). Li says that it is a familiar plant in gardens and along roadsides in many parts of southern Asia, and that it was extensively planted along the city wall of Quanzhou in Fujian in the southern Song and Yuan dynasties. This city was the most important port in China at this time and, as the Chinese sometimes called it Citongcheng (Coral Tree City), it came to be known to the Arabian merchants as Zayton. Marco Polo described Zayton as the greatest seaport in the world.

EUONYMUS Celastraceae

E. japonicus 冬青卫矛 (Dongqing Weimao) *Holly Euonymus*

Dongqing (Winter Green) and *Weimao* (Defending Spear) are the Chinese names for holly and the winged euonymus, respectively. *E. japonicus* is a well-known evergreen shrub which is native to China, Japan, and Korea. Both the plain green and variegated varieties are commonly planted in Chinese gardens (fig. 19.35).

284

FATSIA Araliaceae

F. japonica 八角金盘 (Bajiao Jinpan) *Eight-lobed Gold Plate*

This Japanese shrub with large, leathery, palmate leaves has long been admired for its bold architectural appearance. It is seen from time to time in Chinese gardens, very often planted against rocks (fig. 19.36).

FICUS Moraceae

In China numerous native and introduced species of *Ficus* occur, principally in the south. Several of these are planted as roadside trees and in temple grounds. However, they are rarely used in gardens and hence I have included only those species most likely to be encountered.

F. carica 无花果 (Wuhuaguo) *Flowerless Fruit*

FIG

According to Needham (1954), the fig is thought to have been brought to China somewhere between the 3rd and 7th centuries AD, but Laufer (1967) says that it was introduced from Persia and India not earlier than the Tang. Whatever its date of introduction it does not seem to have achieved the popularity that it has elsewhere. Meyer (1911) noticed black and white varieties but reported that the fruits were not great favourites. Perhaps the fact that it was listed among the famine foods recorded in the *Jiuhuang Bencao* (Bretschneider, 1881) is an indication of its failure to appeal to Chinese taste. Laufer (1967) says that it is cultivated in the Yangtze valley as a shrub bearing fruits smaller and inferior in quality to those found in Persia. It is also planted as an ornamental in some gardens, in the ground or in large pots. At his residence in Beijing in the 1930s, George Kates (1967) 'never ceased to marvel how quickly their leaves grew each spring, when the dumpy tubs were rolled out from their winter storage in a small "ear room" attached to the main house'.

F. erecta var. *beecheyana* 天仙果 (Tianxianguo) *Heavenly-immortal Fruit*

This many-stemmed, straggling shrub or small tree bears leaves up to 20 × 10 cm and globose to pear-shaped figs up to 2.5 cm in diameter. It is native to eastern Asia and its fruits were described as extremely sweet by Song Qi in his

Yibu Fangwu Lueji (Notes on Things from Yibu) of 1057 (Needham, 1986). This book is an account of the natural curiosities of Sichuan and the south-west at the time. Presumably this plant is hardy only in subtropical areas and it is not known whether it is or was cultivated in gardens in China.

F. microcarpa 榕树 (Rongshu) *Rong Tree*

CHINESE BANYAN

A large evergreen tree with smooth bark and small leaves and fruits, *F. microcarpa* is widely distributed in south-east Asia and adjoining regions as far south as Australia. Numerous aerial roots grow down from the branches and, on reaching the soil, develop into new supporting trunks, as happens in many species known as banyans. Ji Han in his *Nanfang Caomu Zhuang* of 304 AD said that Chinese Banyans were frequently planted in the south and that, because the wood is useless for fuel or carpentry, they lasted a long time without being cut down and became very large (Li, 1979; Needham, 1986). Ji Han's is the first record of this tree (Li, 1979). Many famous places are associated with these banyans, for example the Liurong Si (Six Banyan Temple) in the centre of Guangzhou. In spite of the huge size of mature trees it can be dwarfed (Li, 1956a).

F. religiosa 菩提树 (Putishu) *Buddha Tree*

BODHI TREE, BO TREE, PEEPUL

F. religiosa is an evergreen or briefly deciduous tree native to southern Asia, with poplar-like leaves, each with a long point (fig. 19.37). It was under a tree of this species that Gautama

Figure 19.37 *Ficus religiosa*, the Bodhi Tree

Buddha, seated cross-legged, is said to have received enlightenment about 528 BC, so it has great significance for Buddhists (Burkill, 1946). Schafer (1963, 1967) says that it had been sent to the Chinese emperors from the Mahabodi Temple in the Magadha Kingdom, India, before Tang times, and was often planted in temple grounds. It can be seen in such places throughout southern Asia, where it is revered as a symbol of the Buddha and the enlightenment he offered to all men. The most famous of these trees in China was one planted at the Guangxiao Si in Guangzhou in 502 AD. This tree subsequently became venerated for its association with the Tang monk, Hui Neng. Schafer (1967) recorded that the tree at the Guangxiao Si at the time of his writing was a late

18th century replacement of the original. However, whether this still exists I do not know. Sampson (1869c) had earlier commented on this tree and mentioned that skeleton leaf pictures of flowers, butterflies, and suchlike made from the leaves of this species were exposed for sale to foreigners in almost every curiosity shop in Guangzhou. In those parts of China which are too cold for *F. religiosa*, *Tilia* species, which have similarly shaped leaves, are sometimes planted in the grounds of Buddhist temples as a substitute.

F. rumphii 假菩提树 (Jia Putishu) *False Buddha Tree*
FALSE BODHI TREE

The leaves of this large deciduous tree, native to south-east Asia, are heart-shaped with long points and closely resemble those of *F. religiosa*. For this reason *F. rumphii* has also been planted in the grounds of Buddhist temples and monasteries in southern China, no doubt because it has been mistaken for the true Bodhi Tree (Thrower, 1988).

F. virens 笔管榕 (Biguan Rong)
Pen-cap Banyan

A briefly deciduous tree with large leaves, smooth bark, and a spreading crown, *F. virens* sometimes produces aerial roots which develop into secondary trunks after reaching the ground. Wilson (1913) referred to it as the Chinese Banyan and observed large specimens in Sichuan and elsewhere. It is often planted as a shade or roadside tree in southern China.

FIRMIANA Sterculiaceae

F. simplex 梧桐 (Wutong)
PHOENIX TREE, CHINESE PARASOL TREE

A native of eastern Asia, this tree has a very characteristic appearance. It is deciduous, somewhat sparsely branched, and has smooth bark which usually stays green for long periods,

Figure 19.38 *Firmiana simplex* in autumn at Gongwangfu, Beijing

eventually ageing to light grey. The leaves have 3–7 lobes, are 30 cm or more in diameter, and turn a soft yellow before falling (fig. 19.38). The small cream flowers are borne in large terminal and axillary panicles.

F. simplex is widely planted in China as a roadside and garden tree from the south to at least as far north as Beijing. In ancient times its wood was used for making a special seven-stringed musical instrument (Wang, 1988). Meyer (1911) noted that the seeds were sold sparingly as delicacies in central China but that 'They are far from being of good flavour, however, and as a food will never become of any importance to western races'. This prognostication seems to have been borne out. However, Bartholemew (1980) says that in Hong Kong in autumn a popular dish is chicken steamed with these seeds, which taste rather like chestnuts.

The phoenix tree is much represented in art (e.g. fig. 19.74; Capon & Pang, 1981, cat. no. 39; Kao, 1988, no. 96; Krahl, 1987), and it has considerable mystique since it is the only tree upon which the fabled Chinese phoenix is believed to alight (Couling, 1917; Koehn, 1952). This fabulous bird is supposed to have arisen from the sun or from fire, hence is often pictured gazing at a ball of fire. The sun being yang, an active force, the phoenix has a great influence on the begetting of children. It has a fowl's head, a human eye, a serpent's neck, a locust's viscera, a swallow's brow, and a tortoise's back. Its tail has 12 feathers except in years with an intercalary month, when there are 13. The male is called *feng* and the female *huang*.

The first recorded appearance of this extraordinary creature is said to have occurred in the reign of Huang Di, some 2600 years BC, and subsequent appearances are commonplace in Chinese history. These are believed to occur only in times of peace and prosperity and have been interpreted as glorifying a peaceful reign or flattering a successful ruler. In spite of the abundance in China of suitable trees upon which it might rest, it does not seem to have favoured the country with a visit in recent times.

F. simplex was often planted near the study of a scholar, as the phoenix was regarded as the bringer of inspiration. The tree also exerted a salutary influence on the scholar, urging him to hasten the accomplishment of his task by foretelling, like the catalpas, the approach of autumn by starting to drop a leaf or two in summer. Morris (1983) says that it 'sheds just one leaf a day until the land is locked in winter's embrace'. As well as all this, since the character for 'together'

is also pronounced *tong*, the parasol tree brings to mind a variety of sayings involving this word (Koehn, 1952).

FONTANESIA Oleaceae

F. philliraeoides subsp. *fortunei* 雪柳 (Xueliu)
Snow Willow

A privet-like, deciduous shrub with glossy leaves up to 10 cm long, it bears small white flowers in panicles up to 5 cm long in late spring. Steward (1958) recorded that it was cultivated as an ornamental plant in the lower Yangtze valley.

FORSYTHIA Oleaceae

Forsythias have been cultivated in Chinese gardens for a very long time (Li, 1959) and are conspicuous wherever one goes in spring. Schafer (1965) says that they were sent to Changan during the Tang, so presumably they were cultivated at least as early as this. However, they are little mentioned in the literature and rarely appear in paintings before the Qing, although Li (1959, pl. 7A) reproduces a work attributed to the 15th century in which a forsythia is shown. Another can be seen in an 18th century *Hundred Flowers* reproduced by Capon and Pang (1981, cat. no. 65).

While in Tianjin, Fortune (1863) noted that forsythias were amongst the few plants stocked by nurseries which did not require protection in winter. No doubt it was because they are unsuited to the climate of Guangzhou that they were not introduced to Europe until the middle of the 19th century, when it became possible to investigate the regions to the north. Since then these species, together with the hybrids subsequently raised from them, have achieved immense popularity in the gardens of temperate regions throughout the world.

F. suspensa 连翘 (Lianqiao)

F. suspensa is a vigorous shrub flowering in early spring and has ovate leaves which, on vigorous shoots, are sometimes 3-lobed. A selection with pendant branches was originally described from Japan by Thunberg in 1784 as *Syringa suspensa*, but was transferred to the new genus *Forsythia* in 1804 (Lauener, 1996). It was introduced to Holland from Japan in 1833 but did not reach England until much later (*Curtis's Botanical Magazine* 1857, t. 4995). A vigorous upright form of this species is the forsythia seen most frequently in the

FRAXINUS Oleaceae

F. chinensis 白蜡树 (Bailashu) *White Wax Tree*
CHINESE ASH

Although *F. chinensis* is a typical ash its leaves have only 5 leaflets. In China it is often planted as a street or shade tree. In earlier times it was widely cultivated for the production of insect white wax, as was *Ligustrum lucidum*, particularly in the west of the country. According to Mrs Archibald Little (1901), who visited western China late last century, the young insects were taken from privets near Tachienlu (now known as Kangding) to Kiating (now Leshan) and put on pollarded *F. chinensis*. She said that the undersides of the boughs and twigs became coated with white wax a quarter of an inch thick after three months. The boughs were then lopped off and the wax removed, chiefly by hand, placed in an iron pot of boiling water, skimmed off, and placed in a rough mould. Wilson (1913) gave a similar account, stating that insect white wax became known to the Chinese about the middle of the 13th century AD. Lancaster (1989) records that it was mentioned in the notes of the Jesuit missionary Matteo Ricci, published by Christiana de Trigault in 1615. During succeeding centuries several accounts appeared, including one in Magaillan's *History of China* of 1688, where the wax is described as 'the most beautiful, the clearest, and the whitest that ever was seen', and another in Salmon's *Modern History* of 1725 (H.F.W.H., 1868).

As this wax has a high melting point, it was principally used to mix with the fats and oils employed in the manufacture of candles, a thin coating being applied to the outside as well. It was also used to impart gloss to the higher grades of paper, for coating pills, and as a medicine itself. F. P. Smith (1869) said that, at the time of his writing, the wax industry had become an imperial monopoly, the price being five or six times what it was before the Taiping Rebellion. The industry remained an important one until relatively recent times.

gardens of northern China (fig. 19.39). No doubt this is what Fortune (1864) saw on his visit to Beijing in 1861.

F. viridissima 金钟花 (Jinzhonghua) *Golden Bell Flower*

F. viridissima is similar to the above but the leaves are lanceolate, the stems square in cross-section, the flowers, which are usually smaller, appear later, and the branches remain green for several seasons. It was sent to England in 1844 by Robert Fortune, who first encountered it in a garden in Zhoushan and wrote: 'I afterwards found it wild amongst the mountains of the interior in the province of Chekiang, where I thought it even more ornamental in its natural state among the hedges than when cultivated in the fairy gardens of the Mandarins' (Lindley, 1846). Steward (1958) recorded it as a cultivated ornamental in the Yangtze delta region, but it does not seem to be common there nowadays.

F. mandschurica 水曲柳 (Shuiquliu) *Water-ripple Willow*

MANCHURIAN ASH

This species is planted occasionally in northern China. It differs from *F. chinensis* in having leaves with up to 11 leaflets. The clusters of winged fruits are said to resemble rippling water.

GARDENIA Rubiaceae

G. augusta 栀子 (Zhizi) *Wine Cup*

GARDENIA

Being a southern plant, the gardenia (fig. 19.40) seems to have been hardly mentioned in the classical literature. However, Wang (1988) says that, according to the history of the Han dynasty, the yearly income of about 66 hectares of gardenia was equal to the taxes collected from 1000 peasant families, though what evidence there is that the plant mentioned in this history was really a gardenia I do not know. However, it appears in paintings at least as early as the Song, both as the wild type (e.g. Keswick, 1978, figs 46, 203) and as forms with double flowers (e.g. Hulton & Smith, 1979, fig. 5; Sirén, 1949, facing p.100), so has clearly been cultivated since early times. Both kinds are shown in the *Song Hundred Flowers* in the collection of the Palace Museum, Beijing, and it is still a favourite subject (fig 19.41). It is also seen as a motif on lacquerware, beginning in the Yuan, and on Ming and Qing blue-and-white porcelain (e.g. Bartholemew, 1985a, cat. nos. 23, 24).

John Barrow (1804) noted gardenias in the nurseries at Guangzhou in 1794, and a double with unusually large flowers, which became known as 'Fortuniana', was sent from China by Robert Fortune to the Horticultural Society of London in 1844 (*Edwards' Ornamental Flower Garden*, 1854). Single forms are rare in Chinese gardens nowadays but the double ones are commonly grown.

According to Li (1959) gardenia flowers were much used by women as a hair ornament, for making cosmetics, and for scenting

Figure 19.40 (above right) *Gardenia augusta*, wild type

Figure 19.41 (right) A Qing painting showing a double gardenia and a double hollyhock

tea, a use also noted by Fortune (1852, 1857) and Williams (1975), and Needham (1974) records their use as a constituent of incense. The roots, leaves, and fruits are used in Chinese medicine (Wang & Ma, 1995), and in former times a yellow dye was made from the fruits (Herklots, 1932a; Li, 1959).

GLEDITSIA Leguminosae

The genus *Gleditsia* is best known outside China on account of *G. triacanthos*, the Honey Locust, a North American tree which is now grown in many parts of the world and which has given rise to numerous cultivars. However, several species of this genus occur in eastern Asia and are now in general cultivation. In China the most important cultivated species is *G. sinensis*, which is planted in various situations and occasionally as a garden tree.

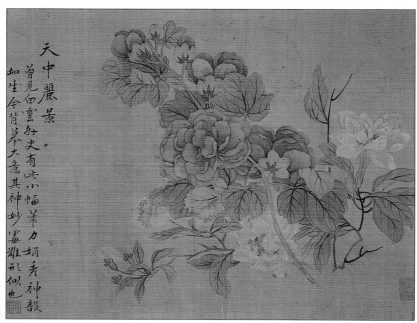

G. sinensis 皂荚 (Zhaojia) *Soap Pod*

CHINESE SOAP-POD TREE

G. sinensis is the famous Chinese Soap-pod Tree, mentioned in the literature as early as the 1st century BC (Needham, 1986). It is a deciduous tree with large, branched thorns and pinnate leaves. In general appearance it resembles the American Black Locust, *Robinia pseudoacacia*. The pods have been used since ancient times as a soap by boiling them in water to produce a lather. Its use for washing clothes is recorded in the *Qimin Yaoshu* (Important Arts for the People's Welfare), which appeared in the middle of the 6th century AD (Needham, 1986). Cibot (1786d) gave an account of the tree and recorded that the seeds and pods were used in the preparation of skins, particularly sheepskins and those destined as furs. According to Li (1963a), it is also useful for cleaning furniture as it does not harm wood. The leaves and pods can be eaten in times of famine according to the *Jiuhuang Bencao* (Bretschneider, 1881).

HIBISCUS Malvaceae

Several *Hibiscus* species are of Chinese origin and seem to have evoked interest in ancient times in various parts of the

Figure 19.42 A fading bloom of the double form of *Hibiscus mutabilis*

country, no doubt because of their numerous showy flowers, which usually last for just one day. As a result they have a long history of cultivation, enjoying in the past even greater popularity than they do now. They were frequently depicted in paintings and were praised in the poetry of the Song and Ming dynasties (Li, 1959). They enjoyed special favour because of their flowering in late summer and autumn. Li says that not only are they treated in general horticultural works but also in special manuals on autumn flowers, flowers to which much attention has been paid in the Chinese literature of the past.

Cibot (1786d), writing from Beijing on *Gleditsia sinensis*, included a few remarks about *Hibiscus*. He said that there were two kinds, '*Mou-kin*' and '*Fou-tsang*', the former of which he had already seen in Bordeaux. These names accord with the present Chinese usage of *Mujin* for *H. syriacus* and with an old name, *Fusang*, for *H. rosa-sinensis*. He said that they were multiplied by seed except for the very beautiful doubles, which were regarded as rarities and were propagated by air-layering.

H. mutabilis 木芙蓉 (Mufurong) *Tree Lotus,* 拒霜花 (Jushuanghua) *Resisting Frost Flower*

COTTON ROSE

In China this is the most celebrated species, much renowned for its flowers which open white in the morning and change to pink before fading in the evening. Bartholemew (1985a) says that, on account of this phenomenon, another name for it is *Zuijiufurong* (Drunken Hibiscus). It is easily recognised by the characteristic appearance of its flowers and by its broad downy leaves, which usually have 3–7 shallow, triangular lobes. There are single, semi-double, and double forms (fig. 19.42). The bark has been used for textile and paper making and the flowers, leaves, and root-bark are ingredients of anti-inflammatory preparations (Wang & Ma, 1995).

H. mutabilis is one of the most frequently depicted subjects in Chinese art. The 12th century catalogue, *Xuanhe Huapu*, lists several examples from the Five Dynasties period (Bartholemew, 1985a) and it appears regularly from then on, usually as the double form, not only in China but also in Japan, to which it was introduced. Burkill (1946) says that Buddhists sometimes decorate altars with its flowers when the lotuses are not in bloom.

Since *H. mutabilis* is the floral emblem of Chengdu, where formerly the city wall was covered with it, Chengdu is

Figure 19.43 A cultivar of *Hibiscus rosa-sinensis*

sometimes called Rongcheng (Hibiscus City). Li (1959) says that when this hibiscus was in bloom the whole city appeared like a piece of pink silk when viewed from a distance and hence another name for the city is Jincheng (Brocade City). Also it was one of the plants which John Barrow noted as conspicuous on the shores of the West Lake at Hangzhou and in nurseries on the outskirts of Guangzhou, when he was returning from Beijing with Lord Macartney's embassy in 1793 (Barrow, 1804). It was one of the earliest Chinese plants to reach Europe, being grown in Rome before 1632 (*Edwards's Botanical Register* 1821, t. 589).

H. rosa-sinensis 朱槿 (Zhu Jin) *Red Hibiscus*
HIBISCUS

Thanks to Paul Gauguin and others, this hibiscus is associated in many people's minds with the South Seas. Nevertheless it seems probable that it is native to south-east Asia and that, along with the dog, fowl, banana, breadfruit, yam, and taro, it spread from there to the Pacific Islands at some time before they were first visited by Europeans (Merrill, 1954). Nowadays it is immensely popular as an ornamental plant in almost all the warmer parts of the world and has been the subject of much further development, particularly in Hawaii. As well as this, it is recorded that its bark has been used for making paper (Laufer, 1967).

According to Li (1959) it is mentioned in the *Li Sao* (Elegy on Encountering Sorrow), which Needham (1986) dates at about 295 BC, and in other ancient Chinese works. It was given its alternative name, *Fusang*, after a fabulous island in the eastern sea, behind which the sun rose and on which a tree 6000 m tall bore fruit which made men immortal. *Zhujin* is the name recorded for it by Ji Han in his *Nanfang Caomu Zhuang* of 304 AD. Needham (1986) gives us a translation, using transliterations according to the Wade Giles system, of Ji Han's description of it as follows:

> The China rose (*chu chin*) has stems and branches resembling the mulberry (*sang*), the leaves on which are glossy and thick. The bushes are only four to five feet tall, but densely branched and leaved. They put forth blossoms from the second month onwards, not ceasing to do so until nearly mid-winter. The flowers are deep red in colour, pentamerous (*wu chhu*), and as large as those of the hollyhock (*shu khuei*). There is a single style (*jui*) exserted beyong the petals (*hua yeh*), and with golden shreds or specks (*chin hsieh*) attached to it. When seen in bright sunlight one would think that the flowers were blazing flames. On a single bush several hundreds of blooms appear each day, opening in the morning and fading towards nightfall. The plant is easily propagated by grafting (*chha chih chi huo*). It grows (especially) in the commandery of Kao-liang. Other names for it are *chhih chin* (the scarlet *chin*) and *ji chi* (lasting but a day).

A comparison of this description with figure 19.43 confirms the accuracy of Ji Han's observations.

The original plant appears to have had red flowers, but subsequently white, pink, cream, apricot, yellow, and orange varieties arose in one place or another, as did doubles and semi-doubles. Because it is frost tender, it can be grown in the ground out of doors only in the south. However, such is its popularity that it is grown as a pot plant in central and northern China and brought indoors in winter.

Figure 19.44 A cultivar of *Hibiscus syriacus*

H. sinosyriacus 华木槿 (Hua Mujin) *Chinese Hibiscus,* 大花木槿 (Dahua Mujin) *Large-flowered Hibiscus*

This species is similar to *H. syriacus* but has larger leaves and differs in other minor details. The flowers are usually white marked with red or purple towards the bases of the petals.

H. syriacus 木槿 (Mu Jin) *Woody Hibiscus*

ROSE OF SHARON

A hardy deciduous shrub or small tree, *H. syriacus* has leaves which are usually 3-lobed and coarsely toothed. The flowers are white, reddish purple, or lavender, usually marked with crimson at the base, and may be single, semi-double, or double (fig. 19.44). In some parts of China the leaves are substituted for tea and the petals of white flowers are sometimes eaten as a vegetable (Li, 1959). Apparently the leaves can also be eaten when there is nothing better on offer, as is recorded in the *Jiuhuang Bencao* (Bretschneider, 1881).

It is mentioned under various names in the earliest Chinese literature, for example in the *Er Ya*, a dictionary of ancient material dating from the Zhou dynasty (Li, 1959). It has also been depicted in Chinese paintings. A single form appears in the *Song Hundred Flowers* in the collection of the Palace Museum, Beijing, and doubles appear in later works.

Not only has it been cultivated in China for a long time but also in neighbouring countries such as Vietnam and Korea. Li (1959) says that an old Chinese name for Korea is 'Chin Yueh' (*Jinyue*), which means the 'Land of Hibiscus', on account of the extensive cultivation of this species there. In fact *H. syriacus* has been adopted by South Korea as its national flower. In the West it is commonly known as Rose of Sharon, a name presumably appropriated from the lines 'I am a rose of Sharon, and a lily of the valleys' in the Song of Solomon. It was given the specific epithet *syriacus* in the 18th century by Linnaeus, as it had reached Europe in the 16th century via the Middle East. It was later realised that it was native to eastern Asia and that it had probably been brought west along the Silk Road. It is now widely grown throughout the world in climates which suit it and many cultivars exist.

Like *H. mutabilis*, *H. syriacus* was noted in 1793 on the shores of the West Lake at Hangzhou by John Barrow (Barrow, 1804). Later, Robert Fortune saw it forming a shrub, with light 'blue' flowers, in the hedges and on hillsides on Putuo Shan and other islands near Ningbo.

H. tiliaceus 黄槿 (Huang Jin) *Yellow Hibiscus*

This evergreen shrub or small tree has heart-shaped leaves and yellow flowers marked with purple at their bases. It occurs widely on seashores in the Old World tropics and can be found in such places in southern China. Li (1959) says it is sometimes planted there as an ornamental or windbreak.

HOVENIA Rhamnaceae

H. dulcis 拐枣 (Guaizao) *Twisted Chinese Date*

RAISIN TREE

H. dulcis is a handsome deciduous tree with glossy leaves, native to north-eastern Asia. It is notable for the way its small, spherical fruits are borne on pedicels which become swollen, fleshy and sweet at maturity (fig. 19.45). They are brown and have a texture and flavour similar to that of raisins. These 'fruits' are listed as a famine food in the *Jiuhuang Bencao* (Bretschneider, 1881) but otherwise appear not to be highly valued in China. Whether it has ever been planted in gardens I do not know.

HYDRANGEA Hydrangeaceae

H. macrophylla 绣球 (Xiuqiu) *Embroidered Ball*

HYDRANGEA

Although this species is a Japanese plant, some of its garden varieties have been grown in China for centuries. In fact the first living hydrangea to reach England appears to be the one Sir Joseph Banks introduced from China in 1789 (Haworth-Booth, 1984). This introduction has round heads of sterile flowers, as apparently do most of the forms grown in Chinese gardens. The Chinese appear to have noted the changes in flower colour which hydrangeas undergo, as Li (1959) quotes an author who states that the flowers, lasting often for several months, are greenish at first, becoming pink and often changing to blue and finally to bluish-green. Whether this means that he understood that the flowers of a single variety may be pink or blue according to soil conditions, or merely according to their age, is not, however, clear. Fortune (1847a) noted that the hydrangeas in Fuzhou invariably produced flowers of the deepest blue, much deeper than he had ever seen them in England.

I have not seen hydrangeas in bloom in Chinese gardens but the plants are quite common (fig. 19.46). A variegated cultivar in Hangzhou looked similar to the cultivar known as 'Tricolor' in the West, and was probably a lace-cap type.

Presumably the existence of lace-cap forms in China, often with eight sterile flowers in each inflorescence, has led to such types being called *Baxianhua* (Eight Immortals Flower), a name also applied to the superficially similar *Viburnum macrocephalum* f. *keteleeri*. In fact there is much confusion in China between the hydrangeas and *Viburnum macrocephalum*, the sterile form of which is also called *Xiuqiu*. This frequently appears in paintings and Western writers have often further compounded the situation by identifying it as a hydrangea. Hydrangeas, however, are rarely seen in Chinese paintings, the only one I have come across being a hortensia type depicted in an 18th century album leaf (Weidner, 1988, cat. no. 40.9). According to Wang and Ma (1995), dried hydrangea flowers are used to treat heatstroke and malaria.

The smaller more slender forms known as *H. macrophylla* subsp. *serrata* or *H. serrata* are also said to be cultivated in Chinese gardens (Li, 1959).

HYPERICUM Guttiferae

H. monogynum 金丝桃 (Jinsitao) *Gold-silk Peach*

Although there are many species of *Hypericum* native to China, this is the only one I have seen in traditional gardens.

Figure 19.45 (right) *Hovenia dulcis* in fruit

Figure 19.46 (far right) *Hydrangea macrophylla* in the Liu Yuan, Suzhou

It forms a low shrub with arching branches bearing sessile leaves. The flowers have relatively narrow, spreading petals, and the stamens, which are almost equal in length to the petals, are grouped in 5 bundles (fig. 19.47). Although it is frequently planted it does not seem to have been paid any particular attention by the poets and painters. However, it does turn up in various Qing dynasty paintings (e.g. Barnhart, 1983, cat. no. 33; Fourcade, 1965, no. 49; Weidner et al., 1988, cat. no. 34). Wang and Ma (1995) record that the root is used in Chinese medicine.

ILEX Aquifoliaceae

I. cornuta 构骨 (Gougu) *Dog Bone*

CHINESE HOLLY

In Chinese gardens this holly is usually seen as a rounded shrub, though in some places it has been allowed to grow into a small tree. The leaves often have 1–3 spines on each side, and those bearing 2 near the apex have the characteristic appearance described by the specific epithet *cornuta*, which means 'horned'. On some plants, however, most of the leaves are spineless. The berries are red and about the same size as those of the common holly *I. aquifolium*, but are slightly flattened and not as glossy (fig. 19.48).

It is native to China and is often cultivated there, except in the north which is too cold for it. However it will grow satisfactorily and fruit more in warmer climates than will *I. aquifolium*, so it has proved useful in such circumstances in various parts of the world. It was introduced to England from the Shanghai region by Fortune in 1846 (Nelmes,

1944). Its Chinese name refers to its white wood, which is compared to a dog's bone.

I. rotunda var. *microcarpa* 微果冬青

(Weiguo Dongqing) *Small-fruited Holly*

I. rotunda is an evergreen tree which has small red berries and spineless leaves. It occurs widely in eastern Asia where it sometimes reaches 20 m or more. The variety *microcarpa*, which has berries only 5 mm in diameter, is included here on account of a large tree which I saw in the garden of the Huanglong Si at Hangzhou (fig. 19.49). A label affixed to this tree asserts that it is over 700 years old. It certainly has the appearance of great age. Also, it is recorded by Lindley and Paxton (1853) that it was found by Fortune at 'Tein-tung' (the temple now known as Tiantong Si near Ningbo), so it seems that it has found favour as a tree for planting in temple grounds.

INDIGOFERA Leguminosae

I. decora 庭藤 (Tingteng) *Courtyard Vine*

I. decora is a small, spreading, deciduous shrub with pinnate leaves and erect axillary racemes of pale pink, pea-like flowers. Fortune (1850f, 1852) saw it growing in pots in a nursery near Shanghai, and sent it to England from there in 1845 (Lindley, 1846). The Chinese name alludes to its use as a decoration for courtyards and to its slender, vine-like branches. It is apparently native to central China.

Figure 19.47 *Hypericum monogynum*

Figure 19.48 *Ilex cornuta* in the Flower Nursery, Hangzhou

Figure 19.49 (right) The almost leafless tree in the centre of the background of this scene at the Huanglong Si, Hangzhou, is a 715-year-old specimen of *Ilex rotunda* var. *microcarpa*.

Figure 19.50 (far right) *Ixora chinensis* in Hong Kong

Figure 19.51 (below right) *Ixora coccinea*

IXORA Rubiaceae

The ixoras are tropical plants grown for their brightly coloured flowers borne in large terminal clusters above their handsome, glossy leaves. *I. chinensis* is commonly seen in southern China and *I. coccinea* is occasionally encountered. Fortune (1844, 1847a) remarked that the ixoras were very ornamental at Fuzhou and in the Fa Tee Gardens.

I. chinensis 龙船花 (Longchuanhua)
Dragon-boat Flower

IXORA

A dense evergreen shrub with large glossy leaves, *I. chinensis* bears small orange red-flowers in compact heads up to 12 cm or more across (fig. 19.50). It is commonly planted in southern China but is suitable only for tropical or subtropical areas. It is native to tropical eastern Asia, including Taiwan and southern mainland China (Herklots, 1937). Colour variants are known, including white, cream, and pale pink.

I. coccinea 红仙丹草 (Hongxiandancao)
Red Immortal Powder Herb, Red Elixir of Life

RED IXORA

Said to be native to India and Sri Lanka, this species requires fully tropical conditions to be successful. It is similar in appearance to the above but looser in habit and with larger individual flowers, which are usually bright red (fig. 19.51). However, it is very variable, pink, yellow, and orange forms being known. It is much cultivated in warm regions and is occasionally seen in the tropical parts of China. Fortune (1847a) remarked on it in Hong Kong.

Figure 19.52 Foliage of a walnut, *Juglans regia*, frames this view at the Beitang, Beijing

of the temperate area of the country, usually in orchards on hillsides and occasionally in gardens and temple grounds (fig. 19.52). Meyer (1911) observed that the Chinese had not learned the art of grafting or budding the walnut so that, at the time of his visit, all the trees were seedlings. It must be said, however, that while grafted or budded plants are produced in the West, the walnut has proved to be difficult to propagate vegetatively. It gets its Chinese name because the fruit looks like a peach but only the kernel is edible.

KERRIA Rosaceae

K. japonica 棣棠花

(Ditanghua) *Ditang Flower*

KERRIA

The double form, known in the West as 'Pleniflora', is seen in almost all Chinese gardens (figs 19.53, 19.120) and has obviously been in cultivation for a long time. The wild single form is also occasionally cultivated there but is probably a modern introduction. As its botanical name indicates, it first became known to the West from Japan, where it is also common in gardens and from where it was described by Kaempfer (1712). However, it was not introduced to Europe until 1805, when the double form was sent to Kew from China by William Kerr (Coats, 1963). There are also forms with variegated leaves. The double form can be seen in the 18th century *Hundred Flowers* reproduced by Capon and Pang (1981, cat. no. 65), and no doubt a search would reveal its presence in other paintings.

JUGLANS Juglandaceae

J. regia 核桃 (Hetao) *Kernel Peach*

WALNUT

According to Laufer (1967) the walnut was probably not generally known in China earlier than the 4th century AD. Presumably it was introduced by seed from the Iranian and Tibetan regions of central Asia. It is now cultivated in much

KOELREUTERIA Sapindaceae

K. paniculata 栾树 (Luanshu) *Luan Tree*

GOLDEN RAIN TREE

Known since the most ancient times as *Luan*, *K. paniculata* is native to China and Korea. It is one of the earliest planted trees in China, being designated some 3000 years ago as one of the five official memorial trees, and the one that was once chosen to plant on the graves of scholars (Li, 1963a).

It forms an impressive tree with large, usually bipinnate leaves with shallowly lobed leaflets. The small yellow flowers are produced in terminal panicles and are followed by characteristic bladder-like, papery fruits. In the past the flowers have been used to make a yellow dye and the seeds have been made into necklaces (Usher, 1974). Also the *Jiuhuang Bencao* records that the leaves can be eaten in times of food shortage (Bretschneider, 1881). It is native to northern China and has

long been cultivated in Japan. It was sent to Europe with a Russian caravan by Pierre d'Incarville sometime before 1847 (Spongberg, 1990, 1993), being introduced to St Petersburg about 1750 and to England in 1763 (Lauener, 1996).

K. paniculata is still a common tree in Chinese gardens and temple grounds, where many fine old specimens are seen (fig. 19.54). One sometimes hears it said that it is the tree represented on the willow pattern plate. Why this is so I find hard to imagine, as the stylised tree depicted there looks more like a weeping willow than anything else. If it had been otherwise surely the pattern would have been given a different name.

Figure 19.53 (below) *Kerria japonica* in the Sheng Yuan, Shaoxing

Figure 19.54 (right) **Old trees of** *Koelreuteria paniculata* **at the Wofo Si, Beijing**

LAGERSTROEMIA Lythraceae

L. indica 紫薇 (Ziwei) *Purple Rose,* 百日红 (Bairihong) *Hundred Days Red*

CREPE MYRTLE

L. indica is the most cold-hardy member of its genus. It is native to eastern Asia and has been cultivated by the Chinese since early times. It was a favourite ornamental tree during the Tang and was extensively planted in the grounds of the Emperor's palace and around official buildings in the capital (Li, 1959). It is much admired for its flowers, the crimped petals of which are each borne on a slender claw, and may be white, pink, mauve, or purple (fig. 19.55). The smooth, mottled trunks are often very handsome and the leaves turn yellow orange, and red before falling in autumn.

Cibot (1778b, 10) recorded it in Beijing, saying that it had been brought there from the south with the Ming court and had been conserved ever since in the palace gardens. He described how the plants were pruned back hard in the autumn so that they would produce vigorous flowering shoots the following season, a practice which many people in other parts of the world carry out today. He mentioned only a variety with carmine-red flowers, which began to bloom in Beijing at the end of July and continued to do so until the end of September, provided the plants were protected from intense sunlight. As a result of its long flowering season it was known as 'Hundred Days Red'. Li (1959) says that it is also called 'Monkey Tree', because monkeys cannot climb up its smooth, slippery bark. He adds that it is also known as 'Itching Flower' because when the trunk is scratched the smaller branchlets shiver and move. Liu (1982) says that in the past its planting was favoured as it represented a high official position. As well as being grown in gardens it is also used for penjing (Li, 1956a). A large plant trained to represent a deer is a feature near the entrance to the Flower Nursery in Hangzhou (fig. 19.56).

It remains a common plant in China and has become popular in those other parts of the world where the climate suits it, forming a spectacular, summer-flowering tree. It is one of the plants which arrived in Britain comparatively early, being brought from China in 1759 (*Curtis's Botanical Magazine* 1798, t. 405). Robert Fortune (1847a) saw it in various places in China, said it was very ornamental in the Fa Tee Gardens in autumn, and noted red, white, and purple kinds in Hong Kong. It appears in Chinese paintings, at least from the 16th century on (e.g. Kao, 1988, no. 35.7; Weidner et al., 1988, cat. nos 34, 44).

Figure 19.55 (above left) *Lagerstroemia indica*
Figure 19.56 (left) *Lagerstroemia indica* trained to represent a deer at the Flower Nursery, Hangzhou

Figure 19.57 *Birds on a Branch in Snow* by Wang Dingguo, Song dynasty. The fruits upon which the birds are feeding appear to be those of *Ligustrum lucidum*.

LIGUSTRUM Oleaceae

L. lucidum 女贞 (Nuzhen)
Women's Chastity

LARGE-LEAVED PRIVET

A small evergreen tree to 10 m with leaves to 10 cm long, *L. lucidum* produces white flowers in terminal panicles up to 20 cm long in summer. These are followed by large clusters of small blue-black fruits (fig. 19.57). This must be the plant which Du Halde (1736) called '*Tong tsin*' in his *Description Géographique, Historique, Chronologique, Politique, et Physique de l'Empire de la Chine et de la Tartarie Chinoise*. He noted that the tree was evergreen, produced white flowers in the 5th moon, and had black fruits in bunches. He wrote that the white wax produced in the southern part of the country was made by insects found on it, giving details of its cultivation and recording that eggs wrapped in leaves were hung in the trees to infest them. And, as recorded in the *Jiuhuang Bencao* (Bretschneider, 1881), in times of food shortage the leaves can be eaten.

The Large-leaved Privet is frequently seen in Chinese gardens, as well as having been cultivated as a host for the white wax scale, as has *Fraxinus chinensis*, under which heading a summary is given of Mrs Archibald Little's (1901) description of white wax production.

L. lucidum was introduced to Britain from China in 1794 by Sir Joseph Banks (*Curtis's Botanical Magazine* 1825, t. 2565) and has found its way around the world. Variegated forms and other variants have arisen in cultivation. In some countries, for example Australia and New Zealand, it has become naturalised and is regarded as a pest.

L. sinense 小蜡树 (Xiao Lashu) *Small Wax Tree*

SMALL-LEAVED PRIVET

This deciduous shrub or small tree has thin, oval leaves up to 7 cm long and produces white flowers in small terminal and axillary panicles in late spring. It is planted occasionally in Chinese gardens. Like *L. lucidum* it has given rise to a number of cultivars and has become a weed in some places, Australia for example.

LIQUIDAMBAR Hamamelidaceae

L. formosana 枫香树 (Fengxiangshu) *Fragrant Maple Tree*

CHINESE LIQUIDAMBAR

L. formosana is a handsome, straight-trunked tree, very similar in general appearance to the better known *L. styraciflua* of North America. However the fruit clusters are smaller, about 2.5 cm in diameter, and the leaves usually have only 3 lobes, though Sampson (1869a) says that they may be 5-lobed on young trees. Like its American counterpart, many plants produce striking autumn colour (fig. 19.58). It can grow up to 40 m tall and many magnificent specimens exist, particularly in temple grounds in the Yangtze valley region (fig. 4.7). It is possible that, according to Needham (1986), it came into prominence in Chinese botany during the Tang. The trunk yields a fragrant gum, Chinese storax, which is used in medicine and perfumery (Schafer, 1967; Usher, 1974). It is occasionally planted in gardens and, along with the maples, features in literature and art (Li, 1963a). It is

Figure 19.58 *Liquidambar formosana* in autumn

LITCHI Sapindaceae

L. chinensis 荔枝 (Lizhi)

LYCHEE

The lychee is the best known of all Chinese fruit trees and is widely cultivated in the south of the country, where it is usually a short-trunked tree with a low spreading crown (fig. 19.59). The evergreen leaves are pinnate with 2–4 pairs of leathery leaflets, and the tiny flowers are produced in terminal panicles in spring. These are followed by the round fruits, 3–4cm in diameter, which have a red, brittle skin covered with rough tubercles and which ripen in summer (fig. 19.60). Each contains a single seed surrounded by white pulp. As well as being eaten fresh, the fruits are dried or canned in large quantities, particularly for export.

Although lychees have been cultivated in the south from earliest times they cannot, of course, be grown in the north. Nevertheless they have been known there since the Han emperor Wu Di conquered the south (Li, 1979; Schafer, 1963; Wang, 1988) and built in his capital, Changan, the Fuli, a palace named after the lychee, as a sort of botanical garden for southern plants. Writing in 304 AD Ji Han said that over 100 lychee trees

possibly the tree with 3-lobed and 5-lobed leaves depicted in the 18th century painting *Eagle on a Maple Tree* reproduced by Fourcade (1965, no. 34). Chinese botanists distinguish a variant, also with 3-lobed leaves, as *L. acalycina*, and perhaps some of the plants seen in gardens and temple grounds are this species.

There is an ancient superstition that tumours on the trunks suddenly enlarge during violent thunderstorms and assume the form of a man. Ji Han in 304 AD noted that these tumours were used in witchcraft and later writers record that they featured in rain making ceremonies (Li, 1979; Sampson, 1869a). Whether these objects were woody tumours, accumulations of gum, or perhaps even a mistletoe is unclear.

were planted there but none survived until, after a few years, a single tree became established, although it never flowered or fruited. One day this tree suddenly died and as a result dozens of gardeners were put to death. Not only this but many of the couriers who brought the fruits sent from the south as tribute to the court every year died of exhaustion along the way (Li, 1979). So what with this and ignorance about differences in the cold-hardiness of plants, it is clear that early enthusiasm in the north for the lychee cost a considerable number of lives.

Even by the Tang dynasty, lychees were still treated in poetry and art as exotics, colourful and romantically charming but, unlike the old familiars, only feebly expressive of ordinary dreams and passions (Schafer, 1963). Nevertheless they enjoyed a prestige which has never been challenged by their cousins the longans (*Dimocarpus longan*), which were first brought to the north about the same time (Wang, 1988). An attempt to boost the reputation of the lychee has been made by Lai (1977), who has published a selection of poems and musings about it.

Li (1979) records that several monographs have been written about the lychee, the earliest and most famous being Cai Xiang's *Lizhi Pu* of 1059, the first monograph on any fruit tree by a Chinese writer, in which over 30 varieties are mentioned. The lychee appears in Chinese paintings from at least as early as the Song (e.g. Keswick, 1978, fig. 46; Lai, 1997, pp. 215, 216, 220) and also appears as a motif on porcelain (Krahl, 1987). When it is depicted together with the *Ling* (*Trapa bicornis*), this suggests *lingli*, an expression meaning 'cleverness'. Also a dried lychee is an important item for the

marriage bed as, like the chestnut, it is a symbol of the birth of a son (Bartholemew, 1985a). And it is said that, because the character for 'leave' is also pronounced *li*, and *zhi* means 'branch', this where the Chinese name for the plant came from, as the fruits are harvested with their stalks attached.

LIVISTONA Palmae

L. chinensis 蒲葵 (Pukui)

CHINESE FAN PALM

Occasionally seen in gardens or as a tub specimen, this Chinese palm used to be cultivated quite extensively in southern China for its young leaves, which were made into fans and hats. The leaves are fan-shaped and may be up to 2 m across in mature specimens (fig. 19.61). *L. chinensis* was mentioned in 304 AD by Ji Han, who noted that it grew in the Guangzhou area and resembled *Trachycarpus fortunei*, but

Figure 19.59 (right) *Litchi chinensis*

Figure 19.60 (above) Fruits of *Litchi chinensis*

Figure 19.61 (far right) *Livistona chinensis*, the Chinese Fan Palm

had soft leaves which could be made into rain-hats (Li, 1979). It was also used for roofing and for making raincoats and mats as well as the fans which were sold all over the country (Sampson, 1869d).

Although in general the palms, so characteristic of the tropics, escaped the attention of the poets, as early as Tang times the Chinese Fan Palm was familiar even in northern poetry about elegant women and expensive courtesans (Schafer, 1967). In representations in paintings and on porcelain a fan made from a palm leaf is generally carried by retired scholars and by alchemists (Williams, 1975).

LONICERA Caprifoliaceae

L. maackii 金银木 (Jinyinmu) *Gold and Silver Tree*

This species from Japan, Korea, and northern China is an upright shrub, growing up to 4 m tall. It is a popular garden plant, particularly in northern China, where it blooms in late spring. The fragrant white flowers are borne in pairs in the axils of the leaves along the upper sides of the branches and, like those of *L. japonica*, they become yellow as they age

(fig. 19.62). They are followed by small, red, translucent fruits which are conspicuous on the bare branches immediately after the leaves fall in autumn.

LOROPETALUM Hamamelidaceae

L. chinense 檵木 (Jimu) *Silk-ribbon Tree*

L. chinense is a very common wild plant in southern China. It is a dense, evergreen shrub which produces a mass of cream, narrow-petalled flowers in spring. It is usually grown in gardens as a pot plant, being popular for penjing (fig. 19.63). Clarke Abel (1819), who noticed it in various parts of China, gives a figure of the plant. It gets its Chinese name because the character *ji* appears to be made up of four silk ribbons.

There is a cultivar with purple leaves and carmine flowers which I first saw as a pot plant in Guangzhou in 1980, since when it has become popular for municipal plantings. It is now conspicuous, for instance, along the streets of Hangzhou (fig. 19.64). It was described as *L. chinense* f. *rubrum* by H. T. Chang in 1973 but is probably worthy of no

Figure 19.62 *Lonicera maackii* in the Daguan Yuan, Beijing

Figure 19.63 *Loropetalum chinense* in the penjing garden, Suzhou

Figure 19.64 (right) A pink
cultivar of *Loropetalum
chinense* in Hangzhou

Figure 19.65 (below right)
Lycium chinense trained over a
rock in the Liu Yuan, Suzhou

more than cultivar status. In
Australia it has recently been
offered for sale under the
names 'Roseum', 'Rubrum',
and 'China Pink'. No doubt
it also has a Chinese name
and this all goes to show
how little influenced many
distributors are by the International Code of Nomen-
clature for Cultivated Plants.

LYCIUM Solanaceae

L. chinense 枸杞 (Gouqi) *Holly Willow*

CHINESE MATRIMONY VINE

As it occurs in Chinese gardens this is a lax, spiny, deciduous shrub with arching branches, ovate leaves, and small
purple flowers. These flowers are followed by small, ovoid,
scarlet fruits which are pendulous. Although from a Western
point of view it is an inconspicuous and unexciting plant, it
is commonly planted in Chinese gardens to grow over rocks,
no doubt having some symbolic significance (fig. 19.65). It
is also used for penjing as it is easy to train (Li, 1956a). As
well as this it is cultivated for its leaves and fruits. The young
leaves are often used in stews and soups or fried with eggs.
Herklots (1972) gives a recipe for cooking them with pork,
shrimps, and egg. The fruits, often called wolfberries, are
sweet and taste like liquorice. They are used in soup with
chicken, duck, pork, beef, or turtle meat. They are also an
important Chinese medicine, believed to be capable of re-
storing energy (Zee & Hui, 1990). Herklots (1972) says that
when this shrub was first introduced to England some
thought it was the true tea plant and it was sometimes called
Lord Macartney's Tea. *Gouqi* (Holly Willow) is a combination of parts of the name of the Chinese Holly and that of
Salix integra, a willow with soft flexible branches.

Figure 19.66 (left) *Maclura tricuspidata* at the Tanzhe Si, south-west of Beijing
Figure 19.67 (below) *Mahonia bealei* in the Zhuozheng Yuan, Suzhou

Figure 19.66 (left) *Maclura tricuspidata* at the Tanzhe Si, south-west of Beijing

Figure 19.67 (below) *Mahonia bealei* in the Zhuozheng Yuan, Suzhou

MACLURA Moraceae

Maclura tricuspidata 柘树 (Zheshu) *Zhe Tree*

Often known as *Cudrania tricuspidata*, this relative of the mulberry has been used at times for feeding silkworms. Wilson (1913) recorded it in various parts of western China and noted that, around Kiating Fu (now Leshan) in Sichuan, silkworms were fed for the first 22 days of their lives on the finely chopped leaves of *M. tricuspidata* before being transferred to mulberry leaves. This procedure was believed to produce silk of a more durable quality.

From an ornamental point of view *M. tricuspidata* is nothing to write home about. It is a small deciduous tree, the leaves of which are sometimes 3-lobed towards the apex. The orange-yellow fruits are said to be edible (Huxley et al., 1992) and the bark is believed to cure women of sterility (Buckley et al., 1994).

The fine old Tanzhe Si 45 km south-west of Beijing was once famous for these trees. 'Tanzhe' means 'Pool and Zhe' in reference to the pool and the *Maclura* trees. By 1900 both

the pool and the trees had gone (Tun, 1965), although Juliet Bredon (1931) said that a few stumps remained. In a footnote to his translation of Tun's (1965) *Annual Customs and Festivals in Peking*, Derk Bodde also recorded that part of a trunk still stood in a courtyard there in the 1930s. In recent times a group of young trees has appeared, along with an explanatory notice, outside the main entrance (fig. 19.66).

MAHONIA Berberidaceae

M. bealei 阔叶十大功劳 (Kuoye Shidagonglao)
Broad-leaved Mahonia

No doubt this species has had a long history in Chinese gardens, where it can often be seen today, usually placed with other plants against rocks (fig. 19.67). The pinnate leaves consist of 9–15 broad spiny leaflets, each with 3–4 teeth on either side. The yellow flowers are borne in short, more or less upright racemes at the ends of the previous seasons growths. These are followed by the grey-green fruits.

M. bealei was first found in a Chinese garden by Robert Fortune during his visit in 1848–49. He had some difficulty in obtaining a plant as the owner said it had medicinal virtues, but eventually the local people brought him several small ones (Fortune, 1850c). These were cared for in the garden of T. C. Beale at Shanghai, after whom the species was subsequently named, before being dispatched to Standish and Noble's nursery at Sunningdale, England. In due course their descendants were put into commerce as both *M. bealei* and *M. japonica*, a procedure that has caused much confusion. *M. bealei* has since been found wild in western China (Bean, 1973), but the status and distribution of the true *M. japonica* remains uncertain.

M. fortunei 十大功劳 (Shidagonglao) *Ten Great Virtues*

It seems that hardly any garden or public planting in China omits this unassuming plant, except in the north where it is not hardy out of doors. It grows as a mass of suckering, upright stems, usually only about 1 m tall. The dark green, upwardly pointing leaves have, in most cases, 7 very narrow leaflets and the flowers, borne in short racemes are hidden amongst the foliage (fig. 19.68). Presumably its popularity rests with its association with its Ten Great Virtues rather than with its appearance, though I have not so far ascertained what these virtues are. Nevertheless it can be quite effective when used boldly, as it has been in some recent plantings such as in the garden of Lotus in the Breeze at Crooked Courtyard, one of the ten famous sites on the edge of the West Lake at Hangzhou. *M. fortunei* was originally named *Berberis fortunei* by Lindley, in honour of Robert Fortune who introduced it to England from a nursery in Shanghai in 1845 (Lindley, 1846). According to Lindley it was a favourite plant with the Chinese, who called it the Blue and Yellow Bamboo on account of the bluish green of the leaves and the yellow of the flowers.

M. japonica 华南十大功劳 (Huanan Shidagonglao)
South China Mahonia

The confusion over this species has already been mentioned under *M. bealei*, from which it differs chiefly in having the flowers in longer, pendulous racemes, and leaflets which are narrower and have up to 6 teeth on each side. It was originally described from Japan by Thunberg and later introduced to Britain from cultivated plants there. However, it is probably not native to that country, to which it is said to have been introduced during the Tang. Li (1963b) says that it is probably indigenous to the south-east of the mainland and to Taiwan. The fact that it is now known as *Huanan Shidagonglao* tends to support this contention, but whether it has ever been seen as a wild plant in this region I do not know. Perhaps *M. japonica* and *M. bealii* will turn out to be states of one variable species.

MELIA Meliaceae

M. azedarach 楝树 (Lianshu) *Lian Tree,*
苦楝 (Kulian) *Bitter Lian*

CHINABERRY, WHITE CEDAR

M. azedarach occurs widely in the Asian region, extending as far south as Australia. Long cultivated in southern China, it does not seem to have claimed the attention of either the naturalists or the poets of the Tang dynasty, though it was noted by the pharmacologists (Schafer, 1967). The bark, leaves, and fruit are bitter and used in Chinese medicine. It is a fast-growing, deciduous tree, now found in all the warmer parts of the world. The loose panicles of small, mauve flowers appear soon after the large, bipinnate leaves begin to expand in spring (fig. 19.69, see over), and the clusters of round, yellow fruits become conspicuous as the season progresses. The bark, leaves, and fruits are used as insecticides (Usher, 1974 ; Needham, 1986) and the light, strong wood has been used for making wooden pattens (Li, 1979).

Figure 19.68 *Mahonia fortunei* in Hangzhou

MORUS Moraceae

M. alba 桑 (Sang)

WHITE MULBERRY

Because of its involvement with the silk industry *M. alba* is among the most influential of all Chinese plants. The development of this industry was a technical triumph and one of the greatest achievements of the ancient Chinese people, who were the first to discover that the 600 or so metres of thread composing the cocoon of the silkworm could be wound off, spun, and woven. The domestication of silkworms and the cultivation of mulberries to produce the leaves to feed them took place at some time in antiquity, at least as early as the 2nd millennium BC. By the Han dynasty silk production and weaving had become major industries and silk textiles one of China's chief exports, remaining so until recent times. Silk, along with tea and porcelain, became one of the three Chinese products which Westerners came to want more than anything else—and would go to any lengths to get (Needham, 1988).

In China mulberries are very often grown on banks or on raised ground which is unsuitable for irrigation. They are invariably kept pollarded (fig. 19.70), the leaves being harvested by cutting off the shoots of the previous year about 1 m above ground. This procedure, which was described by Fortune (1847a), gives the mulberry plantations a very characteristic appearance. Curiously the fruits, which may be white, pink, or purple, seem to be little valued. In the neighbourhood of Beijing Meyer (1911) noted a variety which produced large, white fruits that were very sweet to the taste. He recorded that these ripened in June but were not often offered for sale. Perhaps this was because the fruits, along with the leaves and bark, were regarded as famine foods, as recorded in the *Jiuhuang Bencao* (Bretschneider, 1881).

As well as being grown for its leaves and fruit, the wood has apparently sometimes been used for making pattens (Li, 1979) and the bark for making paper in the same manner as that of *Broussonetia papyrifera*. Laufer (1967) says that mulberry bark was used for making paper money by the Mongols and again during and after the Ming.

The mulberry is emblematic of the comforts of home and also of industry. In addition it is a symbol of filial piety and, in earlier times, people carried staffs made of mulberry wood when mourning for a mother. In spite of all this it appears that the mulberry should not be planted in front of a house because the character *sang*, for mulberry, sounds the same as *sang*, meaning funeral. Also the 'transformation of the sea into mulberry orchards' is a figure of speech applied to the changing of the times (Williams, 1975). In spite of its great importance and significance in China, it is rarely represented in art, though it occasionally appears as a decorative motif on porcelain (Gulland, 1928).

Figure 19.69 (left) *Melia azedarach*
Figure 19.70 (opposite page) A mulberry plantation in Zhejiang in early spring

Figure 19.71 (above) *Murraya paniculata*

Figure 19.72 (left) **Bananas at Gongwangfu, Beijing**

MURRAYA Rutaceae

M. paniculata 九里香 (Jiulixiang) *Nine Li Scent*

This relative of *Citrus* has a long history of cultivation in southern China. It forms an evergreen shrub with pinnate leaves and can ultimately grow to about 5 m. It is chiefly valued for its scented, orange-blossom-like flowers (fig. 19.71), which are followed by small red fruits. European visitors to Guangzhou last century, including Robert Fortune (1847a), commented on its presence in gardens and nurseries there. Fortune also noted that it was grown in Fujian for its flowers, for mixing with tea. Its Chinese name indicates that the scent can be detected from a considerable distance

(1 li = approx. 0.5 km). It is widely distributed in the south-east Asian region and was introduced to Britain in 1771 (*Edwards's Botanical Register* 1818, t. 434).

MUSA Musaceae

Although the various bananas are not trees at all but giant herbs, I have included them here as I assume this is where most readers would expect to find them. The so-called 'trunk' is made up of of tightly packed and concentrically arranged leaf bases, through which the inflorescence emerges.

Mentions of the banana in Chinese literature have been surveyed by Reynolds and Fang (1940), who point out that, although the word *jiao* is now a generic term for plants of the Musaceae, it did not originally mean a fruit but one of the many plants grown for fibres used for weaving. Later it was applied to the banana plant and to the fruit. Early writers have made clear that it was grown for both fibre and fruits, and that it was planted in central China not later than the early part of the 3rd century AD, although it had been cultivated in the south in much earlier times.

M. basjoo 芭蕉 (Ba Jiao)

JAPANESE BANANA

This Japanese species is recorded as a planted ornamental by Steward (1958), but it is certainly not nowadays widely grown as such. Occasionally, however, it is planted in areas too cold for the ordinary banana, *M. × paradisiaca*, which otherwise seems to be an essential component of the Chinese garden. In general appearance it is like a smaller version of *M. × paradisiaca*, except that the bracts subtending the flowers are yellowish rather than dark red, and the fruits are usually no more than 6 cm long. It seems probable that the bananas in the ground in a courtyard of Gongwangfu in Beijing (fig. 19.72) are this species, though, since it does not flower and fruit in this climate, it is difficult to be certain. The Chinese name for *M. basjoo* is made up of *ba*, the term for a fragrant plant in the ancient literature, and *jiao*, the generic name mentioned above.

M. coccinea (M. uranoscopus) 树头芭蕉 (Shutou Bajiao)
Tree-trunk Bajiao, 红蕉 (Hong Jiao) *Red Jiao*

This banana from Vietnam is relatively short-stemmed and has smaller leaves than the other species described here. The 'stem' and the midribs of the leaves often exhibit varying degrees of red pigmentation and the inflorescences are upright, rather than drooping, with magenta or scarlet bracts. The fruits are often orange-yellow and usually no more than 5 cm long. No doubt its small stature makes it popular as a pot plant and it has been used for penjing (Li, 1956a).

It is unclear whether this species was included in the writings of the Tang poets. However, it may well have been the red banana much admired in medieval times (Schafer, 1967). It has certainly been cultivated as an ornamental plant at least as far north as the Yangtze valley (Steward, 1958), but is rarely seen today, except perhaps in the south. Morrison (1822) listed it as flowering in the 6th moon (July) in Guangzhou. Although it is not depicted in art as frequently as the ordinary banana, it has appeared in paintings from time to time (e.g. fig. 19.73; Kao, 1988, no. 59).

M. × paradisiaca 香蕉 (Xiang Jiao) *Fragrant Jiao*

BANANA

There is every reason to believe the banana has been a staple food in south-east Asia since prehistoric times (Reynolds, 1951). The first literary evidence of it was recorded in India

several centuries BC, and the Greeks, under Alexander, made mention of it after they had crossed the Indus in 327 BC. Migrations from south-east Asia carried it far into the Pacific before European explorers arrived; it reached Africa many centuries ago; and it was taken to the New World by the Portuguese in 1516.

Although it is probably indigenous to southern China there is no early mention of it in Chinese literature, no doubt

Figure 19.73 *Lohan* by Liu Songnian, Song dynasty. *Musa coccinea* can be seen in the background.

because Chinese civilisation was centred further north. The first written record of it seems to be that of Yang Fu in his *Jiaozhou Yiwu Zhi* (Record of Unusual Things in Jiaozhou [modern Annam]) (Reynolds, 1951), which Needham (1986) dates at about 90 AD. Yang Fu mentions not only the fruits but the preparation and use of the fibres for making cloth. Ku Hui, a writer of the 3rd century, makes it clear that by then its culture had spread north to Guangdong (Needham, 1986), and Ji Han in 304 AD described it in his *Nanfang Caomu Zhuang*. He said that there were three kinds, 'sheep's horn banana', 'cow's udder banana', and a kind with larger fruits which are not sweet and are inferior but with a stem which could be separated into fibres and woven into cloth (Li, 1979), as noted two centuries earlier by Yang Fu. This cloth remained a popular item on the market in southern China until at least the 18th century, but eventually the importation of foreign machine-made textiles caused its disappearance. The plant concerned, presumably, was the kind now considered to be a separate species, *M. textilis*, Manila Hemp, much cultivated in the Philippines but most probably originating on the mainland (Li, 1979). The literature of the years following Ji Han's publication refers to several different cultivars of fruiting bananas (Reynolds & Fang, 1940).

In due course the banana came to be extravagantly praised in medieval Chinese poetry and was grown in gardens for the beauty of its tall, waving leaves and drooping inflorescences. From Tang poetry we learn that it was the sounds made by the huge leaves even more than their shape and colour which captured the Chinese imagination. The wind rustling them, the raindrops falling on them, and the water dripping from them enhanced the slightly melancholy or introspective mood associated with darkness, storms, decay, or autumn (Schafer, 1967). Also the legend of a student who, for want of better material, wrote on the leaves has made the plant an emblem of self-education (Williams, 1975).

The banana, then, became an essential feature of the garden as far north as it would grow, even though those in the north produced no fruit. It can be seen in innumerable depictions of gardens from the past (fig. 19.74), whether in paintings, on porcelain, or on screens, and, where it can be

Figure 19.74 *Enjoying Antiquities* by Du Jin, Ming dynasty, showing bananas (left), *Firmiana simplex* (rear), and *Alcea rosea, Dianthus chinensis*, and *Abelmoschus manihot* (foreground).

grown, it is almost invariably present in the gardens which have survived to the present day. It also has a long history in Chinese medicine, which has used the fruit, root, sap, leaf, and flower for one purpose or another (Reynolds & Fang, 1940).

MYRICA Myricaceae

M. rubra 杨梅 (Yangmei) *Poplar Mei*

A small evergreen tree, *M. rubra* is native to eastern Asia, has leaves up to 12 cm long, and produces spherical fruits up to 2.5 cm in diameter. Fruits preserved in liquid were found in a tomb from the 2nd century BC in Hunan, and it was mentioned during the Han dynasty along with grapes, cherries, and Chinese dates, so presumably was cultivated at that time (Bartholemew, 1980, 1985a). According to Meyer (1911) there are several varieties ranging in colour from dull white to carmine and blackish red, and in flavour from acid to refreshingly sweet. Meyer (1911) also noted that the best varieties were grafted on wild or seedling stocks, a procedure earlier noted by Fortune (1853a). The plant also has various uses in Chinese medicine.

While *M. rubra* is occasionally grown as an ornamental plant in the West this is rarely the case in China, where it is principally valued for its fruit. It was included in Ji Han's *Nanfang Caomu Zhuang* of 304 AD, which notes that the fruits are red, ripen in the 5th month, and have a sweet-sour taste (Li, 1979). I have included it here in case anyone should come across it. The incorporation of the character *mei* in its Chinese name alludes to its flowering in late winter, at the same time as *Prunus mume*. Representations of it are rare in Chinese art, but Bartholemew (1980, fig. 9; 1985a, cat. no. 27) figures an 18th century *famille-rose* bowl on which the fruit is depicted.

NANDINA Berberidaceae

N. domestica 南天竹 (Nan Tianzhu)
Southern Heavenly Bamboo

HEAVENLY BAMBOO

The Heavenly Bamboo has been cultivated for centuries in Chinese and Japanese gardens. It resembles bamboo when viewed from a distance and was used for decorating altars, hence the name. It was sent from Zhejiang to the garden of

Figure 19.75 *Nandina domestica* in fruit

Li Deyu near Luoyang during the Tang dynasty (Schafer, 1965), so may well have been cultivated in the south earlier than this. The typical form has green bi- to tripinnate leaves and terminal panicles of small white flowers, which are followed by bright red berries (fig. 19.75).

According to Li (1959) there are in cultivation forms with variously shaped leaves and with flowers and fruits of different colours. Apparently in Japan there are some 60 named cultivars (Huxley et al., 1992), but in China one usually encounters the typical form, though occasionally a relatively low-growing variety with purplish leaves is seen. I have not come across there the very dwarf form with leaves which turn red in winter, so widely planted in other countries, but perhaps it is tucked away somewhere.

Wilson (1913) found *N. domestica* growing wild in western China during his visit of 1910–11. Coats (1963) says it was introduced to Japan before the 16th century and it was seen there by Engelbert Kaempfer at the end of the 17th. It reached England in 1804 in a consignment of plants sent from Guangzhou by William Kerr, and since then has become popular in many parts of the world.

Observing branches for sale in winter in Shanghai, Fortune (1849c) said it 'takes the place of our holly'. It seems it was in demand for the Chinese New Year in the same way that *Enkianthus quinqueflorus* was in Guangzhou, but whether this is still the case I do not know. It has been a

Figure 19.76 *Nerium oleander*

seen as an emblem of beauty and grace, and I am told they were worn as a hair ornament in the old days. Williams (1975) also records that the bark of the root is used in medicine. In view of the extremely poisonous nature of the plant one assumes that the doses are carefully calculated.

The oleander was introduced to China, presumably from the Middle East, and it soon became absorbed into the group of plants considered appropriate for gardens. One might assume that it found its way overland along one of the trade routes, but there seems to be no evidence of this. Li (1959) records that Li Kan's *Zhu Pu* of 1299 AD mentions a bamboo-like plant with a foreign name which was grown in pots. Li (1959) goes on to say that in a later work about Fujian Province this is identified as the oleander, a plant said to have originated to the west of China and which could not be grown in the north. It is suggested that it probably arrived in Fujian by the sea route and spread from there to the warmer parts of the country and, in due course, to the colder parts as a pot plant. John Barrow (1804) saw it in 1793 in gardens along the Grand Canal near the Yellow River and in nurseries at Guangzhou. Both single and double forms have been represented in paintings (e.g. Barnhart, 1983, cat, no. 26; Capon & Pang, 1981, cat. no. 65; Suzuki, 1982, vol.3, JM1–075).

popular motif in Chinese art since the Song, and is often included in compositions which form rebuses expressing good wishes at New Year (fig. 2.12) and for birthdays (Bartholemew, 1980, 1985a,b). And, as mentioned earlier, it is sometimes grouped with podocarpus and wintersweet as a variation of the 'Three Friends of the Cold Season'.

NERIUM Apocynaceae

N. oleander 夹竹桃 (Jiazhutao) *Mixed Bamboo and Peach*

OLEANDER

The oleander's Chinese name alludes to the bamboo-like appearance of its stems and leaves and its peach-like flowers (fig. 19.76). It is grown in most parts of China, even in the north where it is treated as a pot plant and sheltered in winter. For instance Tun (1965), writing in Beijing about 1900, said that oleanders and pomegranates flowered there in the 5th month, and that householders arranged them together so as to have something fresh to enjoy. And John Blofield (1961), describing Beijing in the 1930s, mentions the removal from a courtyard in autumn of a massed arrangement of pink and white oleanders in pots to make way for the first batch of chrysanthemums. Potted oleanders remain popular in Beijing today. According to Williams (1975) the flowers are

OSMANTHUS Oleaceae

O. fragrans 木犀 (Muxi) *Tree Rhinoceros,* 桂花 (Guihua) *Gui Flower*

O. fragrans is one of the most popular garden plants in China on account of the delicious perfume of its tiny flowers, which pervades the air for a considerable distance. It remains in bloom from autumn until spring and is one of the many plants appreciated by the Chinese for making its presence felt inconspicuously. It is also one of the numerous species with fragrant flowers which have been used for scenting tea (Barrow, 1804; Fortune, 1852, 1857; Williams, 1975). Fortune (1852) said that they were also used to make hair ornaments or dried and kept in ornamental jars in the manner of pot-pourri. As well as this the flowers are used to flavour wine and to make perfume and medicines, the bark is used in dye-making, and the timber is good for carving (Wang & Ma, 1995). Meyer (1916) noted that in the nursery gardens of Tianjin *O. fragrans* was grafted onto privet.

To the Chinese mind the name *Gui* suggests a similarly pronounced character meaning 'noble', so a branch or a depiction of it can suggest this quality (Koehn, 1952). Thus a boy holding an osmanthus branch represents 'noble son'. *O. fragrans* is also a symbol of literary merit and an emblem of studious pursuits (Gulland, 1928). 'To pluck the Guihua from the Moon Palace' is an expression which meant 'To pass the civil service examinations' (Bartholemew, 1985a).

In many lists or depictions of flowers of the months *O. fragrans* represents the 8th, and in pictorial representations of the four seasons it figures as an emblem of autumn (Williams, 1975). I have noted its presence, along with camellia, narcissus, and plum, as a motif in a decorative border found in a 13th century tomb at Fuzhou and in a *Hundred Flowers* handscroll of around the same date, both reproduced in Bickford (1996). It also appears frequently in later paintings. And because it blooms at the time of the mid-autumn or moon festival it has long been associated in China with lunar legends (Li, 1959). To the Chinese the object visible in the moon is an osmanthus bush. It seems that an immortal, Wu Gang, was banished to the moon for infidelity and spends his time there continuously trying to chop it down. However, it posseses the power to heal its wounds immediately, so is unaffected by his efforts. Another ancient legend describes how osmanthus seeds fall from the moon on bright moonlit nights.

Most plants in Chinese gardens are of the form with single greenish-white flowers (fig. 19.77) but Li (1959) says there are also double-flowered ones. He also records that there is a yellow form called 'Yin Gui' ('Silver Osmanthus') and that the rare 'Hong Gui' ('Red Osmanthus') originated during the Ming dynasty in the garden of a scholar living on the coast of Zhejiang. And he notes that some early authors

have stated that red flowers could be obtained by grafting osmanthus onto the pomegranate. As mentioned earlier, this type of advice seems characteristic of the older Chinese horticultural literature, even though attempts to follow it must surely have met with disappointment. I have not been able to find any evidence of a red osmanthus in China, though an orange cultivar is now quite well known in the West. *O. fragrans* is depicted in an 18th century handscroll reproduced by Capon and Pang (1981, cat. no. 65) and possibly in other paintings. Although other Chinese species of *Osmanthus*, such as *O. delavayi*, are far more showy than *O. fragrans*, they have not as yet become popular.

PAULOWNIA Scrophulariaceae

The paulownias are amongst the most impressive of the Chinese trees and have been cultivated in China for more than 3000 years, both for their useful wood and their handsome flowers. It was early recognised that the wood was both strong and light, ideal for constructing such things as chests and musical instruments. Hu (1959) tells of the custom of planting a paulownia at the birth of a daughter, as by the time the girl was old enough to marry the tree would be large enough for making her wardrobes, which were an important part of the dowry. Dorothy Graham (1938) says that a variety of paulownia known as the 'official hat catalpa' was thought to be a sign of good luck and was planted wherever

Figure 19.77 *Osmanthus fragrans*

a son was working for a literary degree. The fruit resembled the official hat of the mandarinate to which the young man aspired. What species this was is unknown, but they all have fruits of the same general shape.

Hu records that reference to the genus occurs in the *Er Ya* (Literary Expositor), based on material from the 1st millennium BC, in which it is included in a list of useful woody plants with the name *Yongtongmu* (Glorious Paulownia Tree). The name *Tong* continued to be used for the paulownias and in 1049 AD a treatise called *Tong Pu* was published by Chen Zhu, in which he described two distinct kinds and some other trees with similar wood. It seems that paulownias were considered difficult to raise from seed but they could be propagated from suckers (Needham, 1996). However, judging by the variability of the paulownias seen in gardens and urban situations at present it appears that most of them are seedlings. The general name given to these trees nowadays is *Paotong*, which means 'Spongy Tong', an allusion to their light porous wood.

Not only are paulownias valued in China as rapidly growing timber trees, but the bark, wood, leaves, flowers, and fruits are all employed for various medicinal purposes, including promoting the growth of healthy hair, turning grey hair black, and treating various external and internal ailments (Hu, 1959). A study of the chemical composition of these remedies may well lead to the identification of compounds which, as Hu puts it, 'will bring profit to some and well-being to many people'.

All this aside, visitors to China at the time of their flowering cannot be other than greatly impressed by the beauty of the paulownias and by an astonishing range of variation for which no study of the literature could prepare them. Confronted with panicles ranging in length from 20 cm to well over 1 m, with individual flowers from 5 cm up to 10 cm long, varying in colour from white, through pale greyish mauve to a clear strong lilac and marked internally in various ways, how can one decide what is what? Besides all this, some trees bloom when completely leafless and others with the developing leaves, which themselves may be broad or narrow and entire or shallowly lobed. Since interspecific hybrids are known in China, it seems likely that many of these plants are of hybrid origin and that to attempt to place them in any particular species would be unwise. Remember, too, that paulownias have been in cultivation for thousands of years and that it is impossible to tell whether trees observed

in the countryside are growing there naturally or have become naturalised. What is clear, however, is that in China there are forms which produce a display of flowers quite the equal of that exhibited by *Jacaranda mimosifolia*, something which I have not seen outside that country. Nor have I seen paulownia flowers of so rich a colour as those borne by one or two of the trees I observed in China.

Forms of uncertain identity are being marketed outside Asia as 'Princess Trees', presumably because the genus was named for Princess Anna Paulowna, or 'Powtan Trees', a name which seems to be a version of *Paotong*. These have an astonishing growth rate and when young produce leaves of enormous size. They are grown for ornament and for timber-producing ventures and are said to be forms of *P. × taiwaniana*, believed to be a hybrid population derived from *P. kawakamii* and *P. fortunei* (Hu & Chang, 1975). Some of them are most handsome and produce panicles of enormous size, though the flowers are usually somewhat wishy-washy in colour. As far as I know, no-one has yet selected and introduced vegetatively propagated clones of an ornamental appearance equal to that of some of the variants I have seen in China. However, trees of a much-vaunted cultivar called 'Sapphire Dragon' may prove to be their equal as they mature.

Regardless of, or perhaps because of, the complexities alluded to above, botanists have recognised within the genus varying numbers of species. I suspect that there is much more work to be done before the situation is fully clarified. The most comprehensive study carried out so far appears to be that of Hu (1959), whose monograph deals with the history, geographical distribution, taxonomy, cultivation, and economic uses of these plants. Hu recognises six species. These are *P. elongata*, *P. fargesii*, *P. fortunei*, *P. glabrata*, *P. kawakamii* and *P. tomentosa*, and for the purposes of this book these are the names I shall use. All these species are recorded from mainland China; *P. fortunei* extends south into Laos and North Vietnam, *P. kawakamii* and *P. fortunei* also occur in Taiwan, and the range of *P. tomentosa* includes Korea and Japan. Those travelling in southern China and the Yangtze delta region are likely to see in cultivation only *P. fortunei* and forms close to it, whereas in northern China *P. elongata*, or something like it, is the predominant form, and in the west, where paulownias seem to be cultivated less than in the east, it is *P. fargesii* which occurs. However, this statement may well be an over-simplification, as a careful

Figure 19.78 *Paulownia elongata* in Changan Avenue, Beijing

survey of the whole country would be needed to clarify the position. Since *P. glabrata* and *P. kawakamii* are not likely to be seen by visitors to Chinese cities, towns, and traditional gardens, I shall say no more about them here, except that *P. glabrata* is known as *Guang Paotong* (Smooth Paulownia), and *P. kawakamii* as *Taiwan Paotong* (Taiwan Paulownia).

In spite of their ubiquity and their long history, the paulownias have rarely been represented in Chinese art. So far the only examples I have found are the mention by Krahl (1987) of their depiction on porcelain and the reproduction by Fourcade (1965, no. 48) of an impressionistic 19th century painting.

P. elongata 兰考泡桐 (Lankao Paotong) *Lankao Paulownia*

This species was established by Hu (1959) to include specimens collected in Hubei, Henan, Hebei, and Shandong. Nevertheless he points out that it has features indicating affinity with both *P. fortunei* and *P. tomentosa* and that it may be of hybrid origin. He says that it is a native of Henan where it is much cultivated and that it has been introduced to Beijing. It produces pyramidal panicles of pale, greyish-

mauve flowers and has leaves which may be up to twice as long as broad. According to Hu, J. Hers, who collected the type specimen in Henan early this century, recorded the vernacular names *Tong* and *Paotong* for this plant. Hers also reported that it is considered in northern China to be the best paulownia, its wood costing 20 per cent more than that of the round-leaved form, whatever that may be or have been. Hu suggests that *P. elongata* is probably the *Tong* of the classics.

The paulownias at present planted in Beijing seem to be close to *P. elongata* as described by Hu. All those I have seen have the general appearance of *P. tomentosa* but usually have narrower, unlobed leaves, and flowers of a pale colour (fig. 19.78). As a result the trees are unimpressive when in bloom, particularly when viewed from a distance. However *P. elongata* is a valued timber tree in China and hybrids with *P. tomentosa* are recorded.

P. fargesii 川泡桐 (Chuan Paotong) *Sichuan Paulownia*

This species from western China has been reported by various collectors to be common at altitudes between 1200 m and 2290 m, apparently growing wild. It was first collected

by the French missionary Paul Farges, who was stationed in Sichuan in the late 19th century, and it was subsequently described and named for him by Adrien Franchet. According to Hu (1959), the flower colours reported by various collectors range from white through mauve to violet and blue-violet. Presumably it is also cultivated in western China, as Hu says that it seldom produces any fruit in Chengdu but flowers abundantly there. As seen in cultivation outside China it has lilac flowers, 5–7cm long and marked with yellow within.

P. fortunei 白花泡桐 (Baihua Paotong)
White-flowered Paulownia

Most paulownias seen in the Yangtze delta region and further south can be safely referred to this species. Although it has been grown in China since ancient times and its wood is used in the same manner as that of the other species, it was unknown in the West until it was described from a specimen collected by Robert Fortune. Fortune gave no locality for his specimen but it is known from as far north as Yantai in Shandong (Herklots, 1964).

P. fortunei forms a tall, straight-trunked tree with a rounded crown and elongate-ovate leaves. The flowers commence to open before the leaves appear and are borne in terminal panicles which are usually somewhat rounded. These flowers are the largest of all the species, being 8–11 cm long and up to 7 cm broad at the mouth, and range in colour from white or cream, through mauve to soft violet (figs 19.79, 19.80). The white forms, which are usually marked with yellow and spotted with purple within, tend to present a greyish and unattractive appearance, whereas the clear mauves and soft violets can be most handsome.

Since Fortune's specimen and subsequent introductions to the West were white-flowered, *P. fortunei* in cultivation outside China usually has white flowers and is described in Western literature without mention of any variation, although Hu (1959) says that the flowers may also be light purple. It was introduced to Hong Kong from nearby Guangdong in 1905 and is now planted there as a roadside tree (Herklots, 1964). This form has white or cream flowers heavily marked with deep purple inside and flushed with lilac above on the outside. However in the Yangtze delta region the coloured forms seem to predominate, though whether this is natural or the result of hybridisation perhaps requires investigation, especially since hybrids with *P. tomentosa* and *P. kawakamii* are known. It may be that variability within *P. fortunei* has been recognised for a very long time as, according to Needham (1986), varieties of it are mentioned in the *Yu Gong* (Tribute of Yu), thought to date from not later than the 5th century BC. However, it is impossible now to be certain that this mention referred to one species only.

Variants blooming at the same time, which presumably are hybrids and which have smaller individual flowers borne in much larger panicles, are quite common. On occasional plants the panicles are enormous, 1 m or more long (fig. 19.81). None of those which I encountered had white flowers, though many were pale. However, some trees of this type had flowers which transformed their leafless crowns into dense clouds of a soft clear violet, quite the most impressive paulownias I have seen (fig. 19.82, see over). These are the types mentioned above which would give the jacarandas a run for their money and which would be well worth introducing elsewhere.

Figure 19.79 (above) **A white form of** *Paulownia fortunei* **in Jiaxing**

Figure 19.80 (opposite page) *Paulownia fortunei* **overhanging a canal in Suzhou**

Figure 19.81 (right) **A paulownia with huge panicles in Suzhou**

Figure 19.82 **A paulownia in the Xi Yuan, Suzhou.**

Both *P. coreana* and *P. lilacina* are regarded by Hu as cultivars of *P. tomentosa*, with the names 'Coreana' and 'Lilacina'. This nomenclature has been followed in *The New Royal Horticultural Society Dictionary of Gardening* (Huxley et al., 1992) and presumably is acceptable if vegetatively propagated clones with such names exist. However, the name *P. coreana* was given to plants cultivated in Korean gardens and presumably was not intended to represent any particular clone. Trees I have seen in South Korea appear to be seedlings and show some slight variation but, in their general appearance at least, there is not enough variation from *P. tomentosa* to warrant any distinction. Likewise, as recorded by Hu (1959), the name *P. lilacina* was given to a pale-flowered plant growing at Kew which had been raised from seed originally sent to France. Whether there are any descendants of this plant and whether they are sufficiently close to it to deserve the same cultivar name I do not know. Not only this but, with its narrow, unlobed leaves, 'Lilacina' appears to have much in common with *P. elongata* as described by Hu (1959).

P. tomentosa 毛泡桐 (Mao Paotong) *Hairy Paulownia*

This is the first species to become known in the West, having been introduced to France from Japan in 1834 (Bean, 1976). Although several collections identified as *P. tomentosa* have been made in China (Hu, 1959) and there are records of its being grown in various parts of the country, its use as an ornamental seems to be rare. I have not seen it, but a thorough survey may show it to be more common in the far north, as it is the only species known from Korea and Japan. It is easily distinguished from *P. fortunei* by its smaller flowers, no more than 6 cm long, which are usually a clear lilac colour and marked within with cream. Its downy leaves are broad and often exhibit 3–5 shallow lobes.

PHILADELPHUS Hydrangeaceae

P. pekinensis 太平花 (Taipinghua)
Peace and Tranquility Flower

This species is cultivated in and around Beijing. Li (1959) records that it is said to have originated in Qingcheng, Sichuan, and that it was propagated and taken to the north in the 11th century by the Song emperor Ren Zong who bestowed upon it the name 'Flower of Peace and Royal Fortune'. Dorothy Graham (1938) says it was a favourite of the Dowager Empress Cixi and was grown in the Imperial Garden of the Forbidden City surrounded by a marble balustrade 'chiselled with the clouds of immortality'.

In spite of all of this, it is a plant which would fail to impress those acquainted with the mock oranges of the West, as its flowers are small and of an unexciting yellowish white. However, they are pleasantly fragrant and one should perhaps remember that plants which are modest yet make their presence felt by their scent have always been highly regarded in China. Plants raised in my family's garden from seeds collected at the railway station near the Ming tombs north of Beijing have produced yellowish flowers, but others I have seen in the Daguan Yuan, Beijing, and at the Dajue Si in the Western Hills, which appear to be *P. pekinensis*, have flowers which are almost white (fig. 19.83).

P. sericanthus 绢毛山梅花 (Juanmao Shanmeihua)
Silky-hairy Mountain Mei

This philadelphus is very similar to the above, but the flowers are pure white and scentless. Also the calyx and flower stalk bear adpressed hairs whereas those of *P. pekinensis* are glabrous. Like that species it is native to western China but has been recorded in gardens in the east (Steward, 1958).

PHOTINIA Rosaceae

P. serratifolia 石楠 (Shinan) *Rock Nanmu*

P. serratifolia is usually seen as a large evergreen shrub with leathery oblong leaves up to 20 cm long and small white flowers in flat panicles up to 25 cm in diameter (fig. 19.84). The young leaves are reddish in some variants, yellowish green in others. Under favourable conditions it ultimately becomes a spreading tree 10 m or more in height. A specimen planted in my family's garden about 1880 has reached these proportions and is most handsome, though in spring its habit of shedding daily a wheelbarrow-load of its scarlet dying leaves is not one of its more endearing characteristics. However, because of this and its red berries it always exhibits some red colour, and in China is sometimes called 'Red for a Thousand Years'. It is native to the country and is extensively planted there in areas which are not too cold for it, principally from the Yangtze valley south. Its Chinese name suggests that it grows in rocky places and that it exhibits some similarity to the *Nanmu* (*Persea nanmu*), a Chinese tree renowned for its wood. However, Wilson (1913) noted that several species of what he called *Machilus* (now *Persea*) were called *Nanmu*, so the reference may not have been to any particular species. *Photinia serratifolia* was introduced to Britain in 1804 by Captain Kirkpatrick of the East India Company (Bean, 1976).

Figure 19.83 (below) A philadelphus in the Daguan Yuan, Beijing
Figure 19.84 (right) *Photinia serratifolia*

PISTACIA Anacardiaceae

P. chinensis 黄连木 (Huanglianmu) *Yellow Link Tree*

A large deciduous tree with pinnate leaves, *P. chinensis* produces inconspicuous flowers in panicles near the ends of the shoots. The female flowers are followed by reddish berries about the size of peppercorns (fig. 19.85). While it is a handsome tree when in leaf, its chief beauty is its autumn colour

in shades of red, orange, and yellow. In this regard it ranks as one of the best deciduous trees and is recommended for planting as an alternative to *Rhus typhina*, which provokes in many people an allergic reaction. While it is native to China it does not often seem to be planted there, though Bean (1976) records that the young shoots and leaves are cooked and eaten as a vegetable. As well as this, the leaves, bark, stems, and roots are used in Chinese medicine for the same purposes as the rhizome of a *Coptis* species known as 'Gold Thread', and this is said to account for its Chinese name.

PITTOSPORUM Pittosporaceae

P. tobira 海桐 (Haitong) *Sea Tong*

This compact evergreen shrub has glossy leaves and clusters of highly scented, cream flowers which turn yellow as they age (fig. 19.86). It is native to eastern Asia, growing near the sea as its Chinese name suggests, and is much cultivated both in the ground and in containers. No doubt it is greatly appreciated for its scent, and there are several distinctive cultivars including a variegated form. It is now widely dispersed in the milder parts of the world, and plants up to 8 m tall are known in areas favourable to it.

PLATANUS Platanaceae

P. × hispanica (*P. × acerifolia*) 悬铃木 (Xuanlingmu)
Hanging Bell Tree

LONDON PLANE

Presumably this tree was introduced to China by those living in the European concessions, as is suggested by the common names 'English Firmiana' and 'French Sycamore', which one sometimes hears in China. It was extensively planted as an avenue tree in the French Concession in Shanghai and is now similarly used in many cities and towns in the Yangtze delta region. It is particularly conspicuous and handsome in

Figure 19.85 (above left) *Pistacia chinensis*
Figure 19.86 (left) *Pittosporum tobira*

Figure 19.87 **An avenue of London planes in Nanjing**

Nanjing where, according to Morris (1983), an effort was made during the 1920s to mitigate the summer heat by planting planes from France.

When grown as a street tree in China it is usually topped at a height of 4–6m to induce the production of spreading branches which arch over the roads. This also allows overhead cables to pass inconspicuously through the crowns. Trees trained in this manner are to be seen along the principal thoroughfares of Nanjing (fig. 19.87).

PLUMBAGO Plumbaginaceae

P. indica 紫雪花 (Zixuehua) *Purple Snow Flower*

This south-east Asian species, long known as *P. rosea*, is sometimes seen in the south of China. It forms a small, lax shrub with pale red or deep rose-pink flowers in terminal clusters. Morrison (1822) listed it as blooming in the 9th lunar month (October) in Guangzhou.

PLUMERIA Apocynaceae

P. rubra f. *acutifolia* 鸡蛋花 (Jidanhua) *Egg Flower*

FRANGIPANI

The frangipani is so widespread in India, south-east Asia, and Indonesia that people tend to be forget that it is a tropical American plant. It seems probable that it was brought by the Spanish to Manila, from where it spread throughout the tropical regions of eastern Asia. It was taken up on such a scale by the Buddhists that it early became known as a temple tree (Merrill, 1954). In Indonesia it is particularly conspicuous as a shade tree in Moslem cemeteries. It was noted in the nurseries on the outskirts of Guangzhou by John Barrow late in 1793 (Barrow, 1804). Its Chinese name no doubt relates to the appearance of the flowers (fig. 1.4).

Presumably the frangipani reached southern China quite early and it is not surprising that, in a region where the jasmines had proved so appealing, it should have been taken up with enthusiasm. Such is its appeal that it is grown in the north in containers which are removed to shelter in the winter. Such plants, for instance, are put out each summer at Gongwangfu in Beijing. Also it seems likely that various different forms of *P. rubra* are present in the south, as no doubt are other species of *Plumeria*.

POPULUS Salicaceae

Poplars are common in gardens and as roadside trees, their popularity no doubt being due to their speed of growth.

Figure 19.88 An avenue of *Populus* × *tomentosa* in Behai Park, Beijing

is considerable variation in the degree of downiness of the young leaves, but as the leaves mature they usually become glabrous. *P. adenopoda* becomes a tree, 20 m or more in height, with smooth, whitish bark.

P. × tomentosa 毛白杨 (Maobai Yang) *Hairy White Poplar*

Believed to be a hybrid of *P. alba* and *P. adenopoda*, this poplar is very commonly planted in China, often as avenues in the north (fig. 19.88). It forms a vigorous tree up to 30 m tall with large, coarsely serrate leaves, downy below when young but becoming glabrous. It is probably this poplar which is mentioned in the *Jiuhuang Bencao* as having leaves which can be eaten in times of food shortage (Bretschneider, 1881).

PTEROCARYA Juglandaceae

P. stenoptera 枫杨 (Fengyang) *Maple Poplar,* 元宝枫 (Yuanbaofeng) *Money Maple,* 大叶柳 (Dayeliu) *Large-leaved Willow*

Frequently planted as a roadside and garden tree, *P. stenoptera* resembles its relatives the hickories. Its large, pinnate leaves are deciduous and the male and female flowers are borne in separate pendulous catkins in spring. As might be expected from the generic name *Pterocarya* (from the Greek *pteron*, wing, and *karyon*, nut), the fruits are winged nutlets. Amongst many notable examples are the 400-year-old tree at Jiaoshan, Zhenjiang (fig. 19.89), old trees in the Qiuxia Pu at Jiading, and the avenue leading to the main entrance of the Tianyi Ge in Ningbo, the oldest surviving private library building in China. It is widely planted elsewhere, particularly as a street tree and along the banks of canals and ponds.

P. stenoptera gets the first of the Chinese names mentioned above because its leaves make a sound like those of a

They have been cultivated since ancient times and it was discovered early that they could be grown from cuttings (Needham, 1996). Several species are native to China and nowadays there are numerous introduced species and hybrids present as well, so that the situation is confusing for the visitor. In view of this complex situation, together with my own lack of expertise, I have chosen to mention only the two species of Chinese origin most likely to be seen in cultivation.

P. adenopoda 响叶杨 (Xiangye Yang) *Noisy-leaved Poplar*

This is the Chinese equivalent of the aspen, *P. tremula*, differing in the long drawn-out apex of the leaves of mature trees and in the shallower undulations of the margin. There

poplar in the wind; the second because its winged fruits look like the ancient shoe-shaped gold ingots, *yuanbao*; and the third because its leaflets resemble enlarged willow leaves.

PUNICA Punicaceae

P. granatum 石榴 (Shiliu)

POMEGRANATE

Although there is hardly a garden in China without a pomegranate, either growing in the ground or in a container, it is not native to the country. It has been present for so long, however, that it has become incorporated into the traditional garden flora. Tun (1965), who wrote of the Beijing of the late 19th century, said that, when the oleanders and pomegranates began to bloom in the 5th lunar month, almost every household placed plants of each in a symmetrical arrangement with a large bowl of goldfish in the middle. There was an old Beijing saying which spoke of 'a mat covering overhead, an earthenware fish-jar and a pomegranate tree', which drew attention to this clichéd convention. It is also a favourite plant for penjing, the stem naturally assuming a crooked and aged appearance (fig. 13.19).

According to Laufer (1967), its first introduction appears to have taken place in the latter part of the 3rd century AD, as it is not on record until then. He quotes an author of the 5th or 6th century as saying 'The pomegranate, particularly as regards its blossoms, is charming, hence the people plant the tree in large numbers'. No doubt its bright red flowers appealed to Chinese taste, suggesting happiness and good fortune, and the fruits, with their abundant seeds, became a symbol of fecundity (Williams, 1975). In many floral calendars the pomegranate is the emblem of the 5th month, and on the 5th day of that month, considered the most inauspicious day of the year, women wore the flowers in their hair, believing that their fiery red colour would ward off evil (Bartholemew, 1985a).

In the light of all of this it is not surprising to find that both flowering plants and the fruits have frequently been represented in Chinese art (e.g. Capon & Pang, 1981, cat. no. 9; Kao, 1988, no. 50.6; Li, 1959, pl. 1; Sirén, 1973, vol. 3, pl. 251; Weidner et al., 1988, cat. nos 34, 61). The flowers shown are usually those of a double red form, while the fruit is often pictured with a peach and a Buddha's Hand (Finger Citron), 'The Three Plenties', a combination signifying good fortune, long life, and many sons (Bartholemew, 1985b; Koehn, 1952). Laufer (1967) says the pomegranate fruit is still a favourite marriage gift or plays a role in the marriage feast. An open fruit is also likened to the smile of a beautiful girl (Koehn, 1952).

Figure 19.89 A 400-year-old *Pterocarya stenoptera* at Jiaoshan, Zhenjiang

Figure 19.90 (far left)
A pomegranate at the
Baiyunguan, Beijing
Figure 19.91 (below left)
A pyracantha in a penjing
arrangement at the plant
and flower market near the
Temple of Heaven, Beijing
Figure 19.92 (left)
Rhaphiolepis indica, the
Indian Hawthorn, wild
in Guangdong

In spite of all of this, Meyer (1911) was of the opinion that pomegranates were still considered as exotics and were mainly grown as ornamentals, the fruit being considered to be of more or less medicinal value. This is borne out by the fact that the leaves and fruits are recorded as famine foods in the *Jiuhuang Bencao* (Bretschneider, 1881). Meyer recorded that the great part of all pomegranates in China were double flowered, that there were both dwarf and tall forms, and that in colour they ranged from white, striped, and pale red to very dark red, in every degree of variation. Nowadays one usually sees only the typical single form (fig. 19.90), many of the cultivars perhaps having disappeared during the vicissitudes of the present century. In season the fruits occasionally appear in the shops.

PYRACANTHA Rosaceae

P. crenulata 细园齿火棘 (Xiyuanchi Huoji)
Small Round Serrate Firethorn

FIRETHORN

This species is the only firethorn I have found recorded as being in cultivation in China (Steward, 1958), and Wilson (1913) recorded that its leaves were in common use as a source of tea in western Hubei. While not as showy as the species and hybrids favoured by Western gardeners, presumably it is appreciated for its small, bright red berries. I have not seen it planted in the ground in Chinese gardens, but Hu (1982) has published illustrations of penjing specimens which he identifies as *P. crenulata*. I have seen similar specimens (fig. 19.91), but whether these can reliably be placed in this species I am not qualified to say.

RHAPHIOLEPIS Rosaceae

R. indica 车轮梅 (Chelunmei) *Wheel Mei*

INDIAN HAWTHORN

Although commonly known as Indian Hawthorn, this evergreen shrub is native to southern China, where it is abundant on exposed hillsides. Although quite handsome when in bloom (fig. 19.92), it seems to be rarely cultivated in China, though it is now widely planted in other parts of the world.

RHAPIS Palmae

The species of *Rhapis* described below have long been used as garden plants. As well as being grown in China, they were taken to Japan at least 200 years ago and became enthusiastically cultivated. Following World War II its culture there has become something of a craze on a par with the tulipomania of 17th century Holland, and there have been a series of booms and crashes in the prices of unusual variants, particularly variegated forms. Plants have even been purchased as a hedge against inflation, extraordinary prices of up to \$US 80,000 being paid for rare cultivars (Yamaguchi & Barry, 1974). There appears to have been no parallel in China to this state of affairs.

R. excelsa 棕竹 (Zongzhu) *Palm Bamboo*

LADY PALM

The Chinese name of this elegant palm is particularly apt. It forms a cluster of bamboo-like stems, which bear crowns of palmate leaves with 5–12 truncate lobes. The stems are clothed with persistent fibrous sheaths and may reach 4 m or more on plants grown in open ground. It is cultivated in southern China where, according to Sampson (1869d), it is often seen planted in 'the small recesses of temples'. It was introduced to cultivation in the West from Japan, but is believed to be native to southern China. Li (1956a) says that it can be easily dwarfed by restricting the roots and prematurely removing the sheaths at the bases of the leaves, and that a 10-year-old specimen treated thus may be no more than 25 cm tall. Also the typical form has become popular as an indoor plant in the West.

R. humilis 矮棕竹 (Ai Zongzhu) *Short Rhapis*

SLENDER LADY PALM

R. humilis is similar in size and appearance to *R. excelsa*, but the sheath fibres are finer, and the leaves have 8–22 pointed lobes (fig. 19.93). It too is native to southern China and is cultivated there. Like *R. excelsa*, it is amenable to dwarfing (Li, 1956a) and is widely grown as an indoor plant in the West.

RHODOTYPOS Rosaceae

R. scandens 鸡麻 (Jima) *Chicken Hemp*

With its slender, arching stems and deeply veined leaves, this deciduous shrub native to northern China shows a marked resemblance to *Kerria japonica*. In fact specimens from Japan were given the name *R. kerrioides* by Siebold & Zuccarini (1835–41). Its flowers, however, are very different, being pure white and 4-petalled. These are followed by shiny, spherical, black berries about 8 mm in diameter. It is occasionally seen planted in gardens in northern China, for example at the Summer Palace and at the Wofo Si on the outskirts of Beijing (fig. 19.94). The Chinese name refers to the rough surface the leaf, as *ma* also means 'pock-marked' or 'rough'. The name thus alludes to the texture of chicken skin.

Figure 19.93 (below)
Rhapis humilis
Figure 19.94 (right)
Rhodotypos scandens
at the Summer
Palace, Beijing

R. scandens was introduced to the West from Japan by Siebold but may originally have been taken there from China. It became popular during the Victorian era but is now rare in Western gardens (Spongberg, 1990).

Figure 19.95 *Rubus rosifolius* in the Hangzhou Botanical Garden

Figure 19.96 *Rubus rosifolius* 'Coronarius' from *Curtis's Botanical Magazine* 1815, t. 1783

ROBINIA Leguminosae

R. pseudoacacia 洋槐 (Yanghuai) *Foreign Scholar Tree*

BLACK LOCUST

It is not clear when this well-known North American tree reached China. As it has done in many parts of the world to which it has been introduced, it has become naturalised in places. It can be said in its favour, however, that where it is planted in Chinese gardens and temple grounds it looks as if it belongs there, developing into a picturesque tree showing a general resemblance to *Sophora japonica*, with which it frequently finds itself a companion.

RUBUS Rosaceae

R. rosifolius 空心泡 (Kongxinpao) *Hollow Bubble*

This east-Asian species with few thorns, white flowers, and red fruits was seen in the Fa Tee Gardens by Fortune (1844), has later been recorded by Li (1959) as a garden plant in China, and it is still grown here and there (fig. 19.95). The double-flowered form is known in the West as 'Coronarius' and in China as *Zhongban Kongxinpao* (Double Hollow Bubble) (fig. 19.96). I assume the Chinese name refers to the compound fruit, the core of which is left behind when it is picked, as happens with blackberries and raspberries. As mentioned in Chapter 6, it is sometimes called *Tumi*.

SALIX Salicaceae

Both weeping and upright willows have been grown in China since ancient times, having proved useful not only for stabilising banks and dykes but also for their flexible branches and their timber. It has long been known, too, that they can easily be raised from cuttings (Needham, 1996). Willows were symbolic of feminine grace and of Guanyin, who is often depicted holding a willow branch (Morris, 1983). Translations of a selection of poems about willows are given by Lai (1977) along with reproductions of paintings in which they feature.

S. babylonica 垂柳 (Chui Liu) *Pendulous Willow*

WEEPING WILLOW

The Weeping Willow is one of the best known and most widely planted of all trees. It was introduced to the West from

Figure 19.97 *Salix babylonica* **in the Zhuozheng Yuan, Suzhou**

the Near East in the 18th century, which is why Linnaeus gave it the specific epithet *babylonica*. However, it was subsequently realised that it had probably been brought across Asia from China, where it has been grown since ancient times. It remains one of the most characteristic features of the Chinese scene, planted beside ponds and lakes, along causeways (Contents page and fig. 19.97), on the banks of streams and man-made waterways, and in gardens. The slender branches have been used for making baskets and ropes, the leaves have been employed by the poorer classes as a passable substitute for tea, and the bark is used in medicine.

It may perhaps be that *S. babylonica*, at least in its typical state, occurs only as a cultivated or naturalised tree. The weeping forms by which it is known may be mutants of some species of normal habit, although no such species is found wild at the present time (Bean, 1980). In the Near East and in most parts of the West the trees are all female, whereas in north-west India and bordering Tibet male trees are by far the commoner and may all belong to a single clone. However, in Tadjikistan, through which the ancient trade route passed, there are, according to Bean, at least three cultivated clones, one of them male. According to Li (1963a), a male form of

the weeping willow was introduced from China into Japan in early times, and from there it was eventually taken to California and widely planted around San Francisco. No doubt these distributions of the sexes are merely the result of the vegetative propagation of the original introductions.

The weeping willow has played an important part in Chinese garden art, folklore, painting, and literature. It appears in paintings from the earliest times and is well known as a decoration on porcelain (Froncek, 1969, p.134; Krahl, 1987). Due to the early emergence of its fresh green leaves it is regarded as a symbol of spring, light, and vitality, the enemy of darkness. It is seen, too, as a symbol of meekness, and. with its beauty, suppleness, and impression of frailty it provided the poets of the Tang and the painters of the Song with a symbol for a beautiful woman (Bredon & Mitrophanov, 1927; Koehn, 1952; Williams, 1975). A slender branch swaying in the wind has likewise been compared in the traditional Chinese literature with a dancing figure (Li, 1963a) and, on account of its suppleness and pliability, it is also a symbol of prostitution (Graham, 1938). One reads, for instance, of people frequenting the 'willow lanes'.

The interweaving branches also symbolise affection and

Figure 19.98 (left) Upright and weeping forms of *Salix matsudana* on the causeway at the Summer Palace, Beijing

Figure 19.99 (below left) *Salix matsudana* 'Tortuosa' in the Imperial Garden of the Forbidden City, Beijing

However, it is maintained that it should not be planted at the back of the house because its connotations of frailty and lust may exercise an unhealthy influence on the women, who generally occupy the rear apartments (Williams, 1975). And I have already mentioned in Chapters 3 and 7 the remarks of the 17th century writer Wen Zhenheng concerning the vulgarity of planting peaches and willows together, though his opinion seems little heeded today.

S. matsudana 旱柳 (Han Liu) *Dryland Willow*

Both upright and weeping willows occurring in and around Beijing and other parts of northern China are referred to this species (fig. 19.98), which differs only in very minor characters from *S. babylonica*. In fact some botanists prefer to consider it to be a variety of that species. The upright form grows well in dry situations and the pendulous form replaces the typical *S. babylonica* in waterside locations in northern China and shares the same symbolism and folklore. For instance, Princess Der Ling (1911) describes how, during a drought in 1903, the Dowager Empress, the Emperor, the Empress, and their attendants all placed willow leaves on their heads and, in front of a table bearing two porcelain vases of willow branches, prayed for rain, repeating the ceremony daily until the rain came. Presumably it was *S. matsudana* which they used.

Perhaps it may be shown that all the weeping willows in northern China are the one clone, as every one I examined in Beijing was female. The ovaries split open in late spring releasing vast quantities of unfertilised ovules bearing tufts of fine hairs which carry them aloft. In relation to this phenomenon Tun (1965) quotes an old poem which says:

attachment, and in former times a sprig of willow was given as a token to a departing friend. The tree is believed, moreover, to possess power over demons and is thus used in various ways to ward them off. And being a water-loving plant it is a rain charm, the branches being used in various ways in rain-making ceremonies. For example, in parts of the country during a drought men and boys go about, or at least used to, wearing wreaths made of fresh willow branches (Bredon & Mitropohanov, !927). Also the 24th of the 28 constellations of the Chinese zodiac is named *Liu*, after the willow.

The weeping willow is an indispensable component of the Chinese garden, always planted by the side of the pond, as depicted on the willow-pattern porcelain of the West.

At the time when the rush shoots appear, the willow catkins fly about.
When the red cherries become ripe, the wind is cool through the wheat.

Translations of several other poems dealing with the flying of the willow 'catkins' are given by Lai (1977), and it all sounds very romantic. But, alas, one is driven to think otherwise when they blow in one's face, along with the dust and detritus stirred up by the tiresome spring winds characteristic of the region. On the credit side, though, it is recorded by the French Jesuit missionaries in earlier times that the willow 'wool' was used like cotton (Bretschneider, 1898).

In addition to the upright and weeping forms of *S. matsudana*, there is the well-known cultivar 'Tortuosa', known in China as *Longzhao Liu* (Dragon's-claw Willow). This upright clone has curiously twisted branches, twigs, and leaves. It is particularly striking when leafless in winter (fig. 19.99). Its appearance undoubtedly appeals to Chinese taste and it is commonly planted in gardens. There is, for instance, a fine specimen near the western boundary of the Imperial Garden in the Forbidden City. It was introduced to Europe in the 1920s (Bean, 1980) and is now planted in many parts of the world.

SAPIUM Euphorbiaceae

S. sebiferum 乌桕 (Wujiu) *Black Mortar*

CHINESE TALLOW TREE

This interesting tree produces fruits, each of which splits open to reveal three elliptical seeds which are covered with a layer of pure white wax (fig. 19.100). The earliest first-hand observation by a European of the tree and use of the wax was that of Martini in his *Novus Atlas Sinensis* of 1655 (Bretschneider, 1880). The wax was removed by throwing the seeds into boiling water and was chiefly used for making candles (L., 1870). The cultivation of the trees, the harvesting of the seeds, and the preparation of the wax was described in detail by Knowlton (1868). After removal of the tallow the seeds were pressed to extract an oil which, according to Li Shizhen, the author of the *Bencao Gangmu* of 1596, could be used for lamps, as a purgative, to dilute lacquer, and to make oil-paper. The oil and wax were also exported to Europe for making special types of soap. As well as this, the leaves have been used to make a black dye and for preparing a spray for

controlling a variety of agricultural and garden pests, and the root bark is used in medicine (Needham, 1986).

G. T. Lay (1846), writing in Fuzhou in November 1844, said: ' The leaves of the tallow tree turn red about this time, then the tree exceeds in beauty both the Chinese Plane Tree and the Chinese Maple, which in autumn are tinged with carmine and are much talked of by poets under the common name "Fung Shu"'. While it is not altogether clear what plants he was comparing it with, presumably he meant *Liquidambar formosana* and a species of *Acer*, both of which are called *Feng*. The striking autumn tints of the poplar-like leaves of this Chinese tree (fig. 19.101) were also much admired by

Figure 19.100 (right) Fruits of *Sapium sebiferum* opening to reveal the seeds coated with white wax

Figure 19.101 (below) *Sapium sebiferum* in autumn

Ernest Wilson (1913), another to describe how wax and oil were obtained from the seeds. At the time of his visit to Hubei this was a substantial industry, the fruits being harvested and spread in the sun to open. Although the fruits are perhaps still harvested and processed sometimes, it is now chiefly grown as an ornamental tree.

S. sebiferum was first raised in Britain from seeds sent from China by James Cunningham somewhere about 1700 (Cox, 1944). Unlike many deciduous trees it will produce its brilliant autumn colour in regions where the winters are mild, and for this reason it is often planted in warmer parts of the world. It is usually grown from seed but, since seedlings vary considerably in their autumn colour, the vegetative propagation of superior clones is desirable. It has become naturalised in south-eastern USA (Bean, 1980).

The derivation of the Chinese name for *S. sebiferum* is unclear. *Wu* means 'crow' or 'black', and Needham (1986) says that Li Shizhen records that birds like to eat the fruit, which might explain the name. Perhaps this was the source

of Du Halde's (1736) statement that rooks like very much the fruit of ' L'Ou Kieou Mou', the tree which produces tallow. It appears occasionally in Chinese paintings, for example the *Bird on a Wutong Tree in Autumn* dated 1654 and reproduced by Kao (1988, no. 96), in which it is shown in autumn colour with its fruits opening to reveal the white seeds. It is also included in the *Jieziyuan Huazhuan* (1975).

SERISSA Rubiaceae

S. japonica (S. foetida) 六月雪 (Liuyuexue) *Sixth Moon Snow*

S. japonica is a small shrub native to eastern Asia and cultivated in the warmer parts of China for penjing and for hedging. It has small, opposite leaves which are a dark shining green and produces tiny, white, funnel-shaped flowers in summer and autumn. There are several cultivars, including a double form, and the plant is said to be of medicinal value (Urban Services Department, 1971). Loureiro (1790) recorded it as a cultivated plant in China, where Morrison (1822) listed it as flowering in the 8th moon (September) in Guangzhou. It was found both wild and cultivated in Japan in the late 17th century by Engelbert Kaempfer, and *Curtis's Botanical Magazine* (1797, t.361, as *Lycium japonicum*) states: 'if you squeeze a flower-bud, or the top of a young shoot, betwixt your thumb and finger, you will perceive a smell highly disgusting, which Kaempfer likens to that of human ordure.' It is this state of affairs which is responsible for the synonym *S. foetida*.

SOPHORA Leguminosae

S. japonica 槐树 (Huaishu) *Huai Tree*

CHINESE SCHOLAR TREE, JAPANESE PAGODA TREE

One of the most famous of all Chinese trees, *S. japonica* has been cultivated since very early times (fig. 19.102). Many wonderful old specimens are seen in China, particularly in the grounds of temples.

Figure 19.102 An old *Sophora japonica* behind a wall at the Round City, Beijing, with a pear tree (right foreground)

Figure 19.103 (above) *Sophora japonica* **in bloom**
Figure 19.104 (right) *Sophora japonica* 'Pendula'
at the entrance to the Shizilin, Suzhou, with
Camellia japonica **and** *Cycas revoluta*

In the Zhou dynasty it was included amongst the five official memorial trees. It was recognised, too, that it could be easily raised from seed (Needham, 1996). Since early times it has been widely planted in gardens and temple grounds and as a road-side tree. According to Dorothy Graham (1938) it is the symbol of the 8th moon (September), when the official examinations for political preferment took place, and so was cultivated by aspiring scholars. How reliable this piece of information is I do not know, as according to Li (1959) the historical records of the Zhou dynasty describe it as the tree officially designated for planting on the graves of scholars.

S. japonica has also long been cultivated in Japan and was sent to Europe from China by d'Incarville in the middle of the 18th century (Bretschneider, 1880; Spongberg, 1990, 1993). It has deciduous, pinnate leaves and in summer produces large terminal panicles of small, greenish-white flowers (fig. 19.103). It seems that these flowers are edible, as they are listed as a famine food in the *Jiuhuang Bencao* (Bretschneider, 1881). They are followed by pendulous pods which exhibit marked constrictions between each seed, so that they appear like strings of beads.

After his defeat by the peasant army, the last Ming emperor is said to have hung himself in 1644 with his own girdle from a sophora in what is now Jingshan Park, Beijing, rather than fall into the hands of the invaders (Arlington & Lewisohn, 1967; Hookham, 1972). Toeing the party line, *Peking: A Tourist Guide* (1960) says: 'Thus, the Ming court, which had cruelly exploited the peasants, was overthrown by the great revolutionary force of the exploited. In 1957, when Comrade Voroshilov, President of the Presidium of the Supreme Soviet of the U.S.S.R., visited this place, he humor-ously remarked that this old tree ought to be given a deco-ration, for it had given the emperors a good lesson'. He apparently failed to mention that the first Qing emperor seized power almost immediately afterwards.

As well as the typical form there is a clone with stiff, pendulous branches known in China as *Longzhao Huai*

Figure 19.105 (above) *Sorbaria sorbifolia* at the Summer Palace, Beijing

Figure 19.106 (left) *Spiraea cantoniensis* in the Zhuozheng Yuan, Suzhou

(Dragon's-claw Scholar Tree), and in the West as 'Pendula'. It is frequently planted, both in the ground and in large containers. There are numerous fine specimens to be seen, including those at the Summer Palace in Beijing. It is propagated by being grafted high on seedlings of the normal type. It rarely flowers but is much appreciated for its striking and characteristic form (fig. 19.104).

SORBARIA Rosaceae

S. sorbifolia 珍珠梅 (Zhenzhumei) *Pearl Mei*

A suckering shrub with stiff upright stems and pinnate leaves, *S. sorbifolia* produces small, white flowers in upright panicles in summer. The leaves are very like those of species of *Sorbus*, hence the specific epithet. It is very common in gardens in and around Beijing (fig. 19.105). Its Chinese name is derived from the fact that the 5-petalled flowers resemble those of *Mei* (*Prunus mume*) but are much smaller, like pearls. It is native to a considerable area of north-east Asia.

SPIRAEA Rosaceae

There are many spiraeas native to China but, while they are frequently seen growing wild, nowadays they do not seem to be common in gardens. The only one I have noticed in traditional gardens is *S. cantoniensis*, but *S. japonica* and *S. prunifolia*, and perhaps others, have been recorded in cultivation. These are all small deciduous shrubs and are generally known as *Xiuxianju* (Embroidered Chrysanthemum) (Li, 1959). They are chiefly admired for their abundant clusters of small flowers typical of the rose family to which they belong, and have become popular garden plants throughout the temperate parts of the world. Wilson (1913) recorded that in China the leaves of several species are sometimes used as tea.

S. cantoniensis 麻叶绣线菊 (Maye Xiuxianju)
Coarse-leaved Spiraea

S. cantoniensis is a spreading graceful shrub with coarsely toothed leaves. In spring the arching branches produce abundant hemispherical clusters of pure white flowers (fig. 19.106). This appears to be the first of the Chinese spiraeas to be introduced to Europe. It was sent from Canton to London in 1824 by John Reeves and, consequently, became known as *S. reevesiana* or Reeves Spiraea. It is commonly planted in Chinese gardens. Fortune (1857) encountered it in the company of many other flowering shrubs in a garden in Ningbo. There is a double form with lanceolate leaves, known in the West as 'Lanceata', which Fortune (1854) also saw in China and considered more beautiful than the original. This has become a popular garden plant in many parts of the world (fig. 19.107).

S. japonica 粉花绣线菊 (Fenhua Xiuxianju)
Pink-flowered Spiraea

This species is usually seen in Western gardens as cultivars introduced from Japan or developed from them by European nurserymen—the ubiquitous, pink-flowered 'Anthony Waterer', for instance. However, it is also native to China and recent wild collections now growing in Australia are also pink-flowered and appear almost identical to the garden varieties.

S. *japonica* first became known in the West when sent by Robert Fortune to Standish and Noble during his second visit to China (Bean, 1980). This cultivar with deep pink flowers in large compound inflorescences became known as 'Fortunei'.

S. prunifolia 李叶绣线菊 (Liye Xiuxianju)
Plum-leaved Spiraea

This plant is usually seen in its floriferous double-flowered form. With its many-petalled, pure white flowers it has become extremely popular in the West. Although it was introduced to Europe from Japan by Siebold about 1845, it is native to China, from which it was sent to the Horticultural Society of London by Fortune at about the same time (Fortune, 1846). Wilson (1913) recorded that the double-flowered form was commonly planted on graves in western Hubei and Li (1959) says that this is the spiraea most commonly planted in China.

SYMPLOCOS Symplocaceae

S. paniculata 白檀 (Baitan) *White Sandalwood*

S. *paniculata* is a deciduous shrub or small tree, native to eastern Asia. It has small, finely toothed leaves and in late spring produces small panicles of fragrant, white flowers, which are followed by round, blue berries. The Chinese name for this plant refers to its white wood, but the pronunciation of the character *tan* is similar to that of another character meaning 'open and honest', and for this reason it is often planted in temple grounds.

TAMARIX Tamaricaceae

T. chinensis 柽柳 (Chengliu) *Red Willow*

This is a typical tamarisk with drooping juniper-like branchlets and large feathery panicles of tiny pink flowers. It is native to temperate eastern Asia. Steward (1958) records it as a cultivated ornamental in the lower Yangtze valley, and Hu (1982) says that it is used for penjing.

TILIA Tiliaceae

While there are many species of *Tilia* in China, few are cultivated. They are, however, felled for timber and, in the past at least, the bark has been used as a source of fibre for making rope, mats, and sandals (Li, 1958, 1963a). Also, it is recorded in the *Jiuhuang Bencao* that the leaves can be eaten in times of food shortage (Bretschneider, 1881).

Although the lindens are rarely seen in Chinese gardens nowadays, they have been used as courtyard trees in Beijing. George Kates (1967), for instance, in the 1930s went to live

Figure 19.107 *Spiraea cantoniensis* 'Lanceata'

in a house in the old Imperial City where in the first court-yard 'grew a gigantic linden tree so ramified that it spread its benison over a large part of the property'. Nevertheless it is chiefly for reasons other than their value as shade and ornamental trees that the lindens are cultivated in China. Since the bodhi tree (*Ficus religiosa*), as mentioned earlier in this chapter, could not be cultivated in the colder parts of the country, Buddhist monks planted lindens in their temples as a substitute. Burkill (1946) felt that no similarity justified this transfer of the name and that it was done 'in a kind of desperation'. However, as Li (1958) points out, this is not true as the leaves are similarly shaped (see figs 19.37, 19.108). Not only this, but the fruits of the lindens, while very different from those of *F. religiosa*, resemble somewhat those of another Chinese tree, *Sapindus mukorossi*, which bears fruits believed by the Daoists to possess exorcising powers. These were called *Putizi* (Bodhi Fruits) by the Buddhists in later times.

The two species of *Tilia* cultivated as bodhi trees are *T. mandschurica* in northern China and *T. miqueliana* in the east-central parts (Li, 1958, 1963a). However they are rarely seen in the precincts of temples nowadays, presumably having been removed for timber or firewood. Even so this seems hard to understand when *Aesculus chinensis*, grown as a substitute for *Shorea robusta*, another Buddhist holy tree, remains such a common feature of temple grounds.

Figure 19.108 *Tilia mandschurica* **in the Fragrant Hills Park, Beijing**

T. mandschurica 辽椴 (Liao Duan) *Liaoning Tilia*

This linden has large leaves which are coarsely serrate, sparsely pubescent above, and pale grey-tomentose beneath. The downy inflorescences are 7-flowered (fig. 19.108). As mentioned above it is chiefly seen in northern China, where it is native, extending into adjoining Siberia and North Korea.

T. miqueliana 南京椴 (Nanjing Duan) *Nanjing Tilia*

The leaves of *T. miqueliana* are usually smaller than those of *T. mandschurica*, coarsely serrate, shiny dark green above and grey-tomentose beneath. The pendulous inflorescences are tomentose and about 20-flowered. This species is native to eastern China but, according to Li (1958, 1963a), it is rarely found in the wild and is quite variable, suggesting a possible hybrid origin. He says it was introduced to Japan and is widely planted there, mostly in temple grounds. It is traditionally believed that it was brought to Japan from China by a Buddhist monk about 1190 AD.

TOONA Meliaceae

T. sinensis 香椿 (Xiangchun) *Fragrant Chun*

T. sinensis is a handsome deciduous tree up to 20 m or more tall. The leaves are up to 60 cm long, pinnate with 10–24 entire leaflets, bronze on emergence and turning yellow before they fall. The tree bears a superficial resemblance to the Tree of Heaven (*Ailanthus altissima*), but is easily distinguished because the leaflets of the latter bear glandular teeth towards their bases. The small cream flowers are borne in lacy panicles and are fragrant. It has been cultivated for its wood since before the middle of the 3rd century BC (Wang, 1988) and has long been admired as a shade and ornamental tree. The character *chun* incorporates the character for 'spring' and part of that for 'tree', so could be interpreted as meaning 'spring tree'. Because of the esteem in which this tree is held, *chun* has become a literary figure of speech for 'father', the most respected member of the Chinese family (Bartholemew, 1985a; Li, 1955).

The leaves, fruits, and bark have been used in medicine, and, although the leaves are listed as a famine food in the *Jiuhuang Bencao* (Bretschneider, 1881), the young leaves and shoots seem to have achieved a certain popularity as a food in normal times. Cibot (1777a), who described the tree as 'Hiang-tchun' or 'Frêne Odorant', recorded that in Beijing the spring shoots were boiled, mixed with vinegar, and eaten with rice. Other writers say that they have an oniony taste and are used as a green vegetable or are salted. The salted leaves and twigs are eaten cold, usually mixed with sesame oil and fresh bean curd (Li, 1955; Wilson, 1913; Williams, 1975; Zee & Hui, 1990). In Chinese gardens the tree was often kept low so that the young shoots could be easily picked (Li, 1955).

Outside China *T. sinensis* is chiefly valued for the colour of its leaves in spring and autumn. In western gardens it is usually seen as the cultivar 'Flamingo', the leaves of which are bright pink on emergence (fig. 19.109), turning cream and then green. Perhaps those who grow it might care to try a few of the spring shoots, treated in the Chinese manner, as an entrée.

Figure 19.109 (right)
Spring shoots of
Toona sinensis
'Flamingo'

Figure 19.110 (below)
Trachycarpus fortunei
in the He Yuan,
Yangzhou

TRACHYCARPUS Palmae

T. fortunei 棕榈 (Zonglü)

CHUSAN PALM

This palm with fan-shaped leaves and characteristic fibrous sheaths surrounding the leaf bases is widespread in central and southern China. However, it has been cultivated for so long for its fibre, and so readily becomes naturalised, that its original habitat is uncertain. *T. fortunei* is occasionally planted in traditional Chinese gardens (fig. 19.110), and is sometimes kept more or less dwarfed in a large pot to ornament courtyards and entrances. Also the young flower buds are eaten in China in much the same way as are bamboo shoots (Thrower, 1988). It is occasionally represented in Chinese art (e.g. She & Yuhas, 1973, no. 48) and advice about painting it is given in the *Jieziyuan Huazhuan* (1975).

Because of its cold-hardiness this palm is perhaps now cultivated in more parts of the world than any other. It found its way to Japan long ago and was first introduced to Europe from there by Siebold in 1830, but it was probably as a result of later introductions that it became popular in the West. Robert Fortune first met with it on the island of Zhoushan (Chusan) in 1843 and gave it the vernacular name Chusan Palm. He sent plants to Kew in 1849, one of which was planted in the open and has survived every winter since. According to Bean (1980) the first general distribution took place in 1860, when plants raised from seed sent by Fortune from the Ningbo region were auctioned by Glendinning's nursery.

ULMUS Ulmaceae

Elms are among the many Chinese trees that have traditionally been cultivated for timber, shade, and their ornamental appearance, although it is not clear from the ancient literature which species were involved. It was early discovered that they could be raised from seed (Needham, 1996). At the present time the species most likely to be encountered as garden and roadside trees are those described below.

U. parvifolia 榔榆 (Lang Yu)

CHINESE ELM

This small-leaved elm, native to China, Korea, and Japan, flowers and produces its winged fruits in late summer and autumn (fig. 19. 111). It is semi-evergreen, spreading, and has a most graceful habit, the branches tending to be pendulous. The trunks of mature trees are particularly handsome, the smooth bark flaking to reveal patches of orange and brown. Thus it is not surprising that it has come to be admired as a garden plant not only in China, where it has been planted since very ancient times (Wang, 1988), but in many parts of the world. It responds well to dwarfing and, with its small leaves, is one of the most popular plants for penjing. Clarke Abel (1819) recorded that almost all the

dwarf trees he saw in Guangzhou were elms, and there is an illustration of a plant trained to resemble an elephant in *Penjing* by Hu Yunhua (1982). According to Huxley et al. (1992) there are now several named cultivars.

U. pumila 榆树 (Yushu) *Yu Tree*

SIBERIAN ELM

This elm, a native of northern China and adjoining regions, has leaves which are a little larger than those of *U. parvifolia*, from which it is easily distinguished by its more upright habit, by being completely deciduous, and by flowering and producing its winged fruits in spring. It is widely planted in China as an avenue, shade, and timber tree, particularly in the north, where it often becomes very tall (fig. 19.112). The typical form was sent to the United States by Meyer and is now much planted in the Middle West, where it has proved hardy, drought resistant, and resistant also to Dutch Elm Disease (Bean, 1980; Cunningham, 1984). Meyer also found a weeping form in 1908 on a grave at Fengtai near Beijing and introduced it to the United States (Cunningham, 1984). In 1996 I saw several young trees of this type planted along a street in Chengde, grafted as standards on what is presumably the typical form (fig. 19.113). Li (1956a) says that the weeping form is also favoured in China for pot culture, as it

Figure 19.111 (left) *Ulmus parvifolia* in fruit
Figure 19.112 (opposite page) *Ulmus pumila* at the Imperial Summer Villa, Chengde
Figure 19.113 (far left) *Ulmus pumila* 'Pendula' in Chengde

is amenable to being trained into picturesque shapes.

Tun (1965), writing of the 3rd lunar month in Beijing in days gone by, said that the seeds were collected at that time and boiled with sugar and flour to make elm-seed cakes. Also, the *Jiuhuang Bencao* records that the leaves, bark, and fruits can be eaten in times of famine (Bretschneider, 1881).

VIBURNUM Caprifoliaceae

V. farreri 香荚迷 (Xiang Jiami) *Fragrant Viburnum*

This upright deciduous shrub is notable for producing its terminal clusters of fragrant, white, often pink-tinged flowers from November until early spring (fig. 19.114). Under favourable conditions these are followed in autumn by red berries, which Reginald Farrer (1916) described as being 'hardly less beautiful in their way than the blossom, as well as offering a favourite dish for dessert (but you must spit out the poisonous cloven stone)'. *V. farreri* is native to the rather bleak province of Gansu and is very tolerant of low temperatures. It is amongst the best loved and most universal of garden plants all over northern China, from the coast to as far west as Xinjiang.

V. farreri was first described as a *Lonicera* from a specimen collected near the Great Wall in the early part of the 18th century, at about the same time as d'Incarville collected it in Beijing. It was subsequently collected in Beijing during his visit of 1830–31 by Bunge, who described it as *V. fragrans* (*Curtis's Botanical Magazine*, 1938, t. 8887). Unfortunately this name had earlier been used for another species and, to conform to the International Code of Botanical Nomenclature, it was eventually given a new name, *V. farreri*, after Reginald Farrer, who found it growing wild in southern Gansu in 1914. Although botanical specimens had been collected in the 18th and 19th centuries, it seems that it was not introduced elsewhere until both Farrer and William Purdom sent it to Britain early this century (Coats, 1963). F. N. Meyer sent cuttings from Beijing to the United States at about the same time (Cunningham, 1984). Rather amusingly, Farrer

(1916) records that they would have obtained more seed than they did 'had it not been for a falling-out with the Prince of Jô-ni, who, to avenge himself, set to and sedulously ate up all the Viburnum fruits in his palace garden, and threw away the seed'. Following these introductions it has become a popular winter-flowering shrub in many parts of the world.

V. macrocephalum f. *macrocephalum* 中国绣球花 (Zhongguo Xiuqiuhua) *Chinese Embroidered-ball Flower*

CHINESE SNOWBALL TREE

This is the viburnum most likely to be noticed by visitors to China (fig. 19.115). With its large globose heads of sterile flowers which change slowly from green to white, it is immediately recognised as the snowball tree which is now so popular in other parts of the world. It appears frequently in paintings from the 16th century on, so it has been in existence for at least 400 years. It is semi-evergreen in mild climates and can reach a height of 6 m. It was introduced to England by Fortune in 1844.

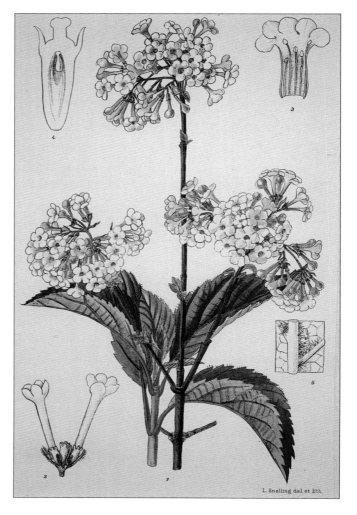

Figure 19.114 (right) *Viburnum farreri* from *Curtis's Botanical Magazine* 1938, t. 8887

Figure 19.115 (opposite page) *Viburnum macrocephalum* f. *macrocephalum* in the Liu Yuan, Suzhou

Figure 19.116 *Viburnum macrocephalum* f. *keteleeri* in Taiziwan Park, Hangzhou

As mentioned earlier, the name *Xiuqiuhua* (Embroidered-ball Flower) is also applied to hydrangeas, the inflorescences of which are superficially very similar. In view of all this it is not surprising that art experts have often mistaken this viburnum in paintings as being a hydrangea (e.g. Barnhart, 1983, cat. no. 26; Kao, 1988, no. 35.6). Other paintings of this plant are reproduced by Weidner et al. (1988, cat. nos 11.4, 24, 61).

V. macrocephalum f. *keteleeri* 八仙花 (Baxianhua)
Eight Immortals Flower

This is the wild type of *V. macrocephalum*, producing inflorescences similar to those of the lace-cap hydrangeas. However, as it was the sterile garden form which was first described, this is considered a form of it rather than the other way round. The name *Baxianhua* refers to the ring of large sterile flowers surrounding the fertile ones in the centre of the inflorescence (fig. 19.116). There is a beautiful painting of this flower by the northern Song emperor Huizong (Suzuki, 1982, vol. 3, JM 12–032), so it obviously attracted attention at least as early as the 12th century.

V. macrocephalum f. *keteleeri* seems to be grown in Chinese gardens at least as frequently as the sterile form, perhaps because it is considered to be the famous *Qionghua* (Jade Flower), about the identity of which there has been much

speculation. It seems that during the Tang and early Song the temple called Houtu Ci in Yangzhou was renowned for one or more shrubs or trees which were called *Qionghua*. These bore an abundance of white flowers and were considered one of the sights of the city. The fame of the Jade Flower was legendary and it was considered to be without peer amongst the ornamental plants. In fact, according to Li (1959), the gentry of Yangzhou built a pavilion beside it with the inscription *Wushuang* (Without Peer). Another story about it relates that Yang Di, second emperor of the Sui dynasty, had the Grand Canal dug so he could go from his capital to see it.

However, in 1129 A.D. the Jin Tartars captured Yangzhou, burned the temple, and dug up the plants and carried them away to the north. Subsequently fresh shoots producing something similar grew up in their place, but it was unclear whether this shrub was the same as the original. In view of the great fame of the plant, Du Yu, a horticulturalist, went there in 1191 to investigate and was told by an 80-year-old Daoist that the shoots had indeed come from the old roots about 30 years earlier. The results of his investigation were not published until 1234, but in the meantime another investigator, Zheng Xingyi, went to the temple and compared the flowers with those of hydrangeas, which some people said the Daoists had planted instead of the original Jade Flower. He concluded that the Jade Flower exhibited clearly recognisable differences from the commonly planted hydrangeas—for instance, the flowers were white, tinged with yellow, and scented, not blue and scentless. However, such was the fame of the plant that further monographs were written about it in the Ming and Qing, even though it was believed to have vanished and no one was sure of its identity (Needham, 1986).

Most authors have concluded that it was probably a white hydrangea but others have suggested that it may have been a pittosporum, gardenia, or viburnum. However, both Li (1959) and Needham (1986) quote a Ming author who said that he had seen a Song painting of it which looked very much like the wild hydrangea, except that most of the flower heads had 9 sterile florets each, while those of the hydrangea had 8. While this can hardly be considered a reliable difference, it seems most unlikely that it was a pittosporum or gardenia. There is no reason, however, why it should not have been a viburnum, particularly since viburnums and hydrangeas have been confused by Chinese gardeners in the past, the lace-cap forms of both being called *Baxianhua*. And, as Li points out, it seems clear that the Jade Flower was a good sized tree, whereas most species of *Hydrangea* are shrubby.

Whatever the truth of all this may be, there seems to be no doubt now in Yangzhou that it was and is *V. macrocephalum* f. *keteleeri*. It has been adopted as the floral emblem of the city and is widely planted there. Also the Houtu Ci, now known as the Qionghuaguan, has recently been restored and is open to the public. And in Hangzhou I bought a small vacuum flask, Jade Flower brand, made in Yangzhou and with a picture of *V. macrocephalum* f. *keteleeri* on the box. So the Jade Flower is certainly back in the limelight.

V. odoratissimum 珊瑚树 (Shanhushu) *Coral Tree*

V. odoratissimum is native to south-east Asia. It forms a large evergreen shrub or small tree with glossy, evergreen leaves and broadly conical panicles of small, white, scented flowers which appear in late spring or early summer (fig. 19.117). These are followed by red berries which eventually turn black. It is often cultivated in the Yangtze valley region and in southern China, being appreciated for its fragrance (Li, 1959; Thrower, 1988).

V. plicatum 雪球荚蒾 (Xueqiu Jiami) *Snowball Viburnum*

JAPANESE SNOWBALL BUSH

This well-known deciduous species, native to China and Japan, exhibits a characteristic horizontal branching habit and has leaves which appear pleated. Like *V. macrocephalum* it first became known in the West as the sterile garden form. Fortune (1850f) saw it growing in pots and in the ground at a nursery near Shanghai and sent it to the Horticultural Society of London. He is recorded as saying that it was 'a native of the northern parts of the Chinese Empire, and was found cultivated in the gardens of the rich, by whom it was much admired' (*Edwards' Ornamental flower Garden*, 1854). Li (1959) says that it is a popular

species in cultivation in China but I have not noticed it in any of the traditional gardens I have visited. Whether the sterile form arose in China or Japan I have not been able to determine.

WEIGELA Caprifoliaceae

W. florida 锦带花 (Jindaihua) *Brocade Ribbon Flower*

W. florida is a native of Japan, Korea, and northern China. It was mentioned in the first half of the 3rd century AD as a hedge plant (Li, 1959) and is commonly cultivated in China. It was originally described in the West by Bunge from specimens collected in Beijing gardens. However, it seems that

Figure 19.117 (right) *Viburnum odoratissimum*

Figure 19.118 (above right) *Weigela florida* in the Fragrant Hills Park, Beijing plants were not introduced elsewhere until Fortune, who had first encountered it in a mandarin's garden on Zhoushan, sent a plant from a Shanghai nursery to the Horticultural Society of London in 1844 (Bean, 1980). Later he remarked that it was one of the few frost-hardy plants stocked by the nurseries of Tianjin (Fortune, 1863). For half a century or more it was the commonest of the weigelas in Western gardens, but has now largely given way to its cultivars and hybrids (Bean, 1980). As seen in Chinese gardens (fig. 19.118) and in the wild in northern China, it is very variable both in the size of its flowers and their colour, which ranges from a very pale to a rich pink outside, the interior usually being paler, sometimes almost white. Some plants seemed to me to be quite the equals of the named cultivars of the West.

XANTHOCERAS Sapindaceae

X. sorbifolium 文冠果 (Wenguanguo) *Official's-hat Fruit*

X. sorbifolium has long been a feature of Beijing gardens, where the first European to have noticed it was d'Incarville

in the 18th century. It was collected in northern China in 1831 by Bunge, who subsequently named and described it (Lauener, 1996). It is a small deciduous tree which produces abundant racemes of white flowers at the same time as its pinnate leaves begin to expand. On opening, the 5 petals of each flower are stained with yellow at their bases but this soon darkens to a deep carmine (fig. 3.7). As seen in Beijing it is most striking when in bloom. There are splendid specimens in Behai Park (fig. 19.119) and it can still be seen wild in the mountains to the north of the city. Apparently the leaves, flowers, and fruits are edible as they are listed as famine foods in the *Jiuhuang Bencao* (Bretschneider, 1881). The plant gets its Chinese name because the open capsule resembles an official's hat.

This charming plant was introduced to Paris by Armand David who, while on an expedition to Mongolia in May 1866, wrote that he had found 'for the first time in its wild state the handsome *Xanthoceras sorbifolia*, cultivated in Pekin as an ornamental tree, of which I have just sent plants to the Paris Museum' (Fox, 1949). However, it is reported to have already been cultivated in the Crimea some 40 years before this

Figure 19.119 *Xanthoceras sorbifolium* in Behai Park, Beijing

Figure 19.120 *Yucca gloriosa* on either side of a doorway at the Liu Yong Yuan, Beijing, with a persimmon (foreground) and *Kerria japonica* (beyond)

(Lauener, 1996). Regrettably it seems rarely to have lived up to expectations in gardens in other parts of the world, apparently requiring the extremes of temperature of its homeland in order to grow well and bloom abundantly. It is said that the seeds are eaten by the Chinese (*Curtis's Botanical Magazine* 1887, t.6923).

YUCCA Agavaceae

Y. gloriosa 华丽丝兰 (Huali Silan) *Magnificent Silk Orchid,* 凤尾兰 (Fengweilan) *Phoenix-tail Orchid*

When or by what route this plant from eastern North America reached China is unclear. Nevertheless, unlike many foreign plants, it must have appealed to Chinese gardeners as it is present in many Chinese gardens, both traditional and modern. It is often encountered in the ground or potted and ornamenting entrances and courtyards (figs 13.18, 19.120). The bold rosettes of leaves do not rise above ground level and from their centres come the tall panicles of pendulous cream flowers. It may be that *Y. filamentosa*, which has a similar growth habit, is also grown, as perhaps are other species.

ZELKOVA Ulmaceae

Zelkova species are closely related to the elms but differ in their unwinged fruits and in the leaves, which are symmetrical at their bases. Identifying those encountered in China is not easy, as all three Asiatic species are said to occur there and it is not easy to distinguish between them.

The distinctions seem largely to be based on the shape and size of the leaves, on the number of lateral veins, and whether or not the midrib and veins are pubescent beneath.

Z. schneideriana 榉树 (Jushu) *Lifting Tree*

CHINESE ZELKOVA

Z. schneideriana is a tall deciduous tree which is native to central and eastern China and is sometimes planted in gardens and temple grounds. It has thick, ovate-elliptic leaves, 3–9.5 × (1–)3–5.5 cm, with 7–14 veins per side, and with the midrib and veins pubescent below (Andrews, 1994). The bark is dark grey and examples I have seen showed no sign of flaking to reveal orange patches as occurs in *Z. serrata* and *Z. sinica*. It is thought that its Chinese name refers to the use of the compact heavy wood to lift or support heavy objects. There are several huge trees at Tiger Hill, Suzhou, which are claimed to be 170 years old (figs. 19.121, 19.122). It is also used for penjing (Hu, 1982).

Z. serrata 光叶榉 (Guangye Ju) *Smooth-leaved Zelkova*

JAPANESE ZELKOVA

In this species the bark flakes to reveal orange patches, especially in winter, and the leaves are thin, ovate or ovate lanceolate, rarely broadly ovate, 3.5–11.2 × 1.3–7.2 cm, with 8–13(–16) veins per side, and are prominently toothed. The midrib and veins are mostly glabrous below. It occurs in Japan, South Korea, and Taiwan, but in mainland China it has been confused with *Z. schneideriana* and *Z. sinica* (Andrews, 1994). Thus if it occurs at all it may be only as planted specimens.

Li (1956a) says that *Z. serrata* is often used for penjing. While this may be so, he goes on to say that *Z. serrata* is a common tree in most parts of China. As this is not the case, perhaps it is *Z. schneideriana* which is generally used for this purpose. It would be necessary to carry out a survey of the zelkovas used for penjing in order to clarify the situation.

Z. sinica 大果榉 (Daguo Ju) *Large-fruited Zelkova*, 小叶榉 (Xiaoye Ju) *Small-leaved Zelkova*

Z. sinica has a grey trunk, smooth except towards the base, and peeling to reveal orange or brown patches. The leaves are thin,

Figure 19.121, 19.122 (left & below) *Zelkova schneideriana* **at Tiger Hill, Suzhou, and a close-up of its label giving its recorded number and age**

ovate or broadly ovate, 2.5–6.3(–9.3) × 1.5–5 cm, and have 8–10 veins per side (Andrews, 1994). It is native to central and eastern China but whether it is ever planted there I do not know.

ZIZIPHUS Rhamnaceae

Z. jujuba 枣 (Zao)

CHINESE DATE, JUJUBE

The Chinese Date is a small, spiny, deciduous tree which is native to temperate Asia and is widely cultivated there and elsewhere. It produces plum-like fruits, 2–3 cm long, each with a single stone. Zhu Xiao, the author of the *Jiuhuang Bencao*, may not have thought much of them as he lists both the fruits and the leaves as famine foods. However, the fruits are very popular in China, where many different varieties are grown. Cibot (1778b, 11) was aware of 60 kinds and reported that the plants were annually attacked by caterpillars. He said that, as the eggs are laid here and there on the branches, the Chinese run over them with a firebrand on a pole to destroy them. He also noted that the fruits were dried and used a great deal in medicine.

The Chinese Date was among the fruits excavated from a tomb of the 2nd century BC in Hunan. Together with the plum, apricot, peach, and chestnut it ranks as one of the 'five celebrated fruits of ancient China'. Because of its auspicious reddish colour, it was used as one of the offerings for official ceremonies as early as the Tang. Also, as the name *Zaozi* (Chinese Date Fruit) is homophonous for 'early son', it is among the fruits and nuts scattered on the wedding bed (Bartholemew, 1985a).

Figure 19.123 Fruits of *Ziziphus jujuba* bought from a street seller in Beijing

Tun (1965) records that, in his day, the fruits became available in Beijing towards the end of the 7th month, the sellers carrying them about on poles together with grapes. He said that their cries evoked a suggestion of chilliness and melancholy, and noted that there was no way to avoid feeling the emotions that go with the seasons of the year.

Meyer (1911, 1916) recorded that the plants were mostly propagated by suckers or root cuttings. He estimated that at the time of his visit there were at least 100 named varieties in China, possibly 300 or 400, different kinds being used for different purposes, eaten either fresh, dried, preserved in sugar, stewed, or smoked. Nowadays they usually seem to be dried, in which condition they resemble dates in texture and flavour, though they are rarely as sweet (fig. 19.123).

The Chinese Date is sometimes seen as a garden plant, particularly as the contorted form, 'Tortuosa'. This is known in China as *Longzhao Zao* (Dragon's-claw Chinese Date) and Meyer noted that it was highly prized by the rich as a horticultural curiosity, and propagated by being grafted onto the wild plant. It now appears to be rare.

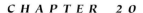

CHAPTER 20

OTHER HERBACEOUS PLANTS

While most of the plants in Chinese gardens
are trees, shrubs, or vines, there are also many
herbaceous plants to be seen. These may either
form part of the permanent planting or be grown
in containers and brought in at the appropriate
times to decorate courtyards and garden buildings.
Included here are annuals, perennials, bulbs,
and corms, many of which are as well known
as those herbaceous plants already dealt with
in earlier chapters.

John Barrow (1804), who accompanied Lord
Macartney's embassy of 1793–94, wrote of the
area along the Grand Canal near the Yellow River:
'In the small flower gardens, without which we
observed scarcely a single cottage, were balsams,
several kinds of beautiful asters, holy-hocks, two
species of *Malva*, an *Amaranthus*, and the showy
and handsome shrub the *Nerium Oleander*'. Similarly, when travelling from Beijing to the Great
Wall in the 1860s, Freeman-Mitford (1900) noted
that almost every cottage had a little flower garden

Figure 20.1 *Dicentra spectabilis* in a garden at Chengde

347

fenced in by a hedge of millet stalks. And Fortune (1863) saw large nursery gardens near Tianjin, where annuals and herbaceous plants were cultivated 'in enormous quantities'. Amongst these were balsams, cockscombs, African marigolds, China asters, tuberoses, and chrysanthemums. Other well-known Chinese herbaceous plants which spring to mind are the Japanese anemones, balloon flowers, bleeding hearts, Chinese lanterns, Chinese pinks, daylilies, hostas, and irises, most of which have been grown and developed for hundreds of years and have been the subjects of poems and paintings. Pre-eminent amongst the various bulbous plants is *Narcissus tazetta*, which, although not native to China, arrived early and has become invested with a mystique and symbolism which is scarcely matched by that of any other plant.

Potted evergreen herbaceous plants have long been popular for both indoor and outdoor decoration. As well as the *Cymbidium* species already dealt with in Chapter 5, plants frequently used in this manner are *Aspidistra elatior*, *Liriope* species, *Reineckia carnea*, and *Rohdea japonica*. Their evergreen habit has become a symbol of longevity and good for-

tune. Perhaps this is why in China at present the international range of indoor plants seems to be settling in readily.

According to Wang (1988), the use of evergreen foliage plants for interior decoration in winter became popular from the middle of the 10th century AD on. He quotes a poem from the Song dynasty which can be translated as follows:

Green foliage shines as protected with blue curtains. Nobody knows how cold it is at the frontier.

Apparently this literary effort was interpreted as a criticism of the administration, and so the official who wrote it lost his position.

A great many introduced annual and perennial flowers are now also widely planted in China, though not usually in traditional gardens. Hence in this chapter I have restricted myself to those herbaceous plants which are most likely to be encountered in such gardens or in temple grounds. And, as mentioned in the Introduction, I have taken the liberty of including one or two herbaceous food plants which are worthy of comment.

ABELMOSCHUS
Malvaceae

A. manihot 黄秋葵
(Huangqiu Kui) *Yellow Autumn Kui*

A well-known south-east Asian plant, *A. manihot* has narrowly-lobed leaves and large whitish to sulphur-yellow, hibiscus-like flowers, marked with maroon in their centres (fig. 20.2). It is native to tropical Asia and, although a perennial, can be grown as an annual. With its

Figure 20.2 *Hibiscus and Rock*, a Southern Song painting by Li Di, showing *Abelmoschus manihot* (centre, upper right), *Amaranthus tricolor* (lower left), and *Callistephus chinensis* and *Celosia cristata* (lower right)

autumn-flowering habit, subtle fragrance, and ornamental foliage it has a long history in China, embodying nuances of beauty and virtue particularly attractive to retired scholars (Barnhart, 1983). It is amongst the plants most frequently represented in art, at least from the Song dynasty on (e.g. figs 19.74, 21.2; Bartholemew, 1985a, cat. no. 41; Li & She, 1971, no. 21; Weidner et al., 1988, cat. nos 11, 34). John Barrow saw it in the nurseries outside Guangzhou in December 1793 (Barrow, 1804). The name *Kui*, which forms part of the name of this plant, is applied to various large flowers, particularly other members of the Malvaceae.

ACONITUM Ranunculaceae

A. carmichaelii 乌头 (Wutou) Crow's Head

MONKSHOOD

I have not seen this monkshood in cultivation in China but it is included in Volume 3 of the *Jieziyuan Huazhuan* (Sirén, 1949; Sze, 1977), and appears from time to time in paintings (e.g. Barnhart, 1983, cat. no. 33; Weidner et al., 1988, cat. no. 34). While this does not prove that this Chinese species was or is cultivated in China, it does suggest that this may be so, hence I have included it. It is a perennial with tuberous roots and has become well-known as a garden plant in other countries, where several cultivars have been named. It gets its Chinese name from the shape of the flowers.

ALCEA Malvaceae

A. rosea 蜀葵 (Shu Kui) Sichuan Kui

HOLLYHOCK

It is probably impossible now to know for certain the original habitat of the hollyhock. Perhaps it was carried both east and west from somewhere in western Asia. Whatever its origin it has featured as a garden plant in China for a very long time, becoming by the 9th century a symbol of passing time (Goody, 1993). It eventually reached Japan where it also became popular, a representation of its leaf being adopted by the Tokugawa shoguns as their family crest. A biennial or perennial, it has been grown in Europe for at least 500 years, and Coats (1968) records that a new phase in its development arose in the 18th century, when plants with variegated flowers were raised in his London garden by Lord Burlington from seeds which came from China. It remains a common garden plant in China, whether planted or self-sown, and

Figure 20.3
*Alcea
rosea*, the
Hollyhock

both single (fig. 20.3) and double forms occur. Like *Abelmoschus manihot* it has often been painted, and appears in the *Song Hundred Flowers* in the collection of the Palace Museum, Beijing, and in other paintings (figs 19.41, 19.74). Coats (1968) says that the flowers are a valued ingredient in Chinese cookery, though she does not divulge the source of this information. However, Wang and Ma (1995) record that the flowers and seeds are used as ingredients of medicines and that the bark can be used to make into textiles.

ALPINIA Zingiberaceae

The two species listed below are included by Li (1956a) in a list of plants grown in classical Chinese gardens, hence I have included them here. They are perennial plants, the rhizomes of which are used in cooking and, no doubt, in medicine, but whether they are ever planted for ornamental purposes nowadays I do not know. They produce terminal inflorescences of small, shell-like flowers and, like those of ginger, the stems bear leaves in two ranks.

Figure 20.4 *Amaranthus tricolor*

AMARANTHUS
Amaranthaceae

A. tricolor 三色苋 (Sanse Xian)
Three-coloured Amaranth, 雁来红
(Yanlaihong) *Red At The Time When The Wild Geese Come*

JOSEPH'S COAT

Cultivars of this annual plant (fig. 20.4) have been grown for ornament in China for hundreds of years, usually as pot plants. Its curious appearance has undoubtedly appealed to Chinese taste and it has frequently been painted. It can be seen, for instance, on the bottom left of the Song dynasty *Hibiscus and Rock* (fig. 20.2) and in many later works (e.g. Barnhart, 1983, cat. no. 31; Weidner et al., 1988, cat. nos 11, 46; Sirén, 1949, facing p. 62). The plant also appeals to Chinese taste in the culinary sense, and cultivars of simpler appearance are grown and cooked in the manner of spinach. It has been grown as a vegetable in south-east Asia from ancient times, and visitors to China in spring are likely to be presented with a dish of *Xiancai* (Amaranth Vegetable) and to regard with some curiosity the carmine juice exuding from the pile of stems and leaves, which are green when cooked. In the West this type is known as Chinese Spinach and in flavour it is very similar to true spinach, though perhaps with a little bit more of a bite to it.

A. globosa 脆果山姜 (Cuiguo Shanjiang)
Brittle-fruited Mountain Ginger

GLOBOSA GALANGAL

As mentioned above, this species is recorded by Li (1956a) as a garden plant in China, and he also records that it was listed by Zhang Qiande in his *Pinghua Pu* (Treatise on Vase Flowers) of 1595. As a result I have included it here, though it is grown principally for its seeds, which are used as a condiment, and its rhizomes, which are employed medicinally (Usher, 1974).

A. officinarum 高良姜 (Gaoliang Jiang) *Gaoliang Ginger*

LESSER GALANGAL

It seems likely that this is the galangal that Marco Polo saw in cultivation during his visit to China in the 13th century (Waugh, 1984). It is apparently native, as it was found wild last century in southern China (*Curtis's Botanical Magazine* 1888, t. 6995). The rhizome is used as a condiment and in medicine (Usher, 1974). Along with *A. globosa* it is listed as a garden plant by Li (1956a). Its flowers are borne singly and are white with a lip veined with red.

ANEMONE Ranunculaceae

A. hupehensis 秋牡丹 (Qiumudan) *Autumn Tree Peony*

JAPANESE ANEMONE

While it occurs in central China as a wild plant with single pink or white flowers, it first became known to the West as a cultivated form with multi-petalled, mauve-pink flowers (fig. 20.5), having been collected in Japan by Thunberg. It is, however, a Chinese plant, now known as *A. hupehensis* var. *japonica*, which probably was taken to Japan long before

Thunberg's residence there during 1775–78. It was introduced to cultivation in England from China by Fortune who sent it to the Horticultural Society of London in 1844 (Lindley, 1846). Fortune (1847a) recorded: 'When I first discovered the *Anemone Japonica*, it was in full flower amongst the graves of the natives, which are round the ramparts of Shanghae'. This aberrant form must have been selected and propagated by Chinese gardeners long ago.

Li (1959) reports that *A. hupehensis* was formerly used in medicine and has been called *Baitouweng* (White-haired Old Man), being a familiar plant in the autumn garden. There is a representation of a single form (fig. 20.6) in the *Song Hundred Flowers* in the collection of the Palace Museum, Beijing, so it seems likely that this has been cultivated too. At the present time, however, no form of the species seems to be commonly cultivated in China. In Europe it has given rise to several cultivars and hybrids, which have become popular autumn-flowering plants in many parts of the world.

AQUILEGIA Ranunculaceae

A. flabellata 扇形漏斗菜 (Shanxing Loudoucai)
Fan-shaped Funnel Vegetable

This columbine, which is native to eastern Asia, bears soft blue-purple flowers, 1–2 per stem, with white or cream petal tips and incurved spurs. It is occasionally grown in Chinese gardens. Like all the aquilegias, it is a perennial herb.

ARACHIS Leguminosae

A. hypogaea 落花生 (Luohuasheng)
Drop Flower Born

PEANUT

Like chillies, the peanut is such a regular feature of Chinese meals that one tends to forget that this annual plant was introduced from America. It has been claimed that it was introduced to China around 1608, but there is a description dating from the early 16th century of what appears to be this plant by a writer from Suzhou (Ho, 1955). Merrill (1954) believes it was brought from Brazil to various parts of the country from about 1500 until the middle of the

next century. Ho suggests that it may have been brought by sea to Fujian early in this period. Its Chinese name refers to the manner in which the seeds are produced after the fertilised flowers are buried.

Meyer (1911) noted the use of peanuts all over China for oil and food. He recorded that the seeds, which had been steamed and kept in weak brine, were everywhere eaten cold as an appetiser. This custom prevails, and every visitor to China today is sure to find them on the table in this and many other guises. Their absorption into Chinese culture is such that they are regarded as yet another symbol of fecundity (Bartholemew, 1985a, b).

Figure 20.5 (below)
Anemone hupehensis var. *japonica*
Figure 20.6 (right)
Anemone hupehensis, detail from the *Song Hundred Flowers*

ASPIDISTRA Convallariaceae

A. elatior 蜘蛛抱蛋 (Zhizhubaodan)
Spider Embracing Eggs

ASPIDISTRA

It would not seem unreasonable to claim that *A. elatior* is the cultivated ornamental plant most frequently seen in China. It is rarely planted in the ground but as a pot plant it reigns supreme, bordering pathways, framing entrances, lining stairways, forming part of mixed displays (fig. 19.10), or simply placed in a porcelain jardiniere on the floor or on one of the characteristic plant stands which form part of the furnishings of almost every room (fig. 20.7).

It seems to have arrived in Britain from Guangzhou in 1824 (*Edwards's Botanical Register* 1826, t. 977), and one is tempted to suggest that its extraordinary tolerance of indoor conditions has been a major factor in establishing the Western obsession with indoor plants which has prevailed ever since. In spite of its widespread success having provoked derision from certain quarters, it remains one of the most successful indoor plants of all time. The mature reader may well remember, as testimony of its popularity, the voice of Gracie Fields singing 'The Biggest Aspidistra in the World'.

In China it has certainly retained its pre-eminent position though, oddly, I have not succeeded in finding out anything of its history. Any depiction in paintings or mention in the literature must be rare. And, as in the West, the form with leaves striped with creamy white is rare, perhaps because, unless checked, its pronounced tendency to revert to the plain green type triumphs. The Chinese name refers perhaps to the somewhat sinister-looking flowers which appear at ground level.

ASTRAGALUS Leguminosae

A. sinicus 紫云英 (Ziyunying) *Purple Cloud Petals*

A low herb with pinnate leaves above which rise stalked umbels of pink flowers (fig. 20.8), *A. sinicus* is native to China and Japan, where, according to Thunberg it abounds by the wayside (*Curtis's Botanical Magazine* 1811, t. 1350). This is certainly so in eastern China, where it is planted as a green manure crop. Presumably its Chinese name refers to the way its flowers carpet many of the dry ricefields in spring, and which are the source of a favourite kind of honey. It is for this reason that I have included it here, though it does not seem to be grown in gardens. However, it occasionally appears in paintings, at least from the early 18th century on (e.g. Capon and Pang, 1981, cat. no. 65; Weidner et al., 1988, cat. no. 45.5).

Figure 20.7 (far left) *Aspidistra elatior* in a pavilion at Tiger Hill, Suzhou

Figure 20.8 (left) *Astragalus sinicus*, from *Curtis's Botanical Magazine* 1811, t. 1350

BEGONIA Begoniaceae

B. grandis subsp. *evansiana* 秋海棠 (Qiuhaitang)
Autumn Crabapple

This begonia has, in the past at least, been an esteemed garden plant in China. Li (1959) says it is frequently planted on rockwork, that it is a good groundcover for small areas in shade, and that it is also grown as a pot plant. Goody (1993) records that it was the feminine flower *par excellence* because it prefers cool shady places, and that it features in many stories of forlorn love. For instance an ancient legend tells how, on a spot watered by the tears of a woman deserted by her lover, this lovely flower first arose when it sprang up to console her (Bartholemew, 1985a; Li, 1959). It is also a symbol of modest beauty and virtue, celebrated by the poets (Cibot, 1778b, 3) and represented in art, at least from the Ming dynasty on (e.g. Bartholemew, 1985a, cat. nos 42, 43, 44; Kao, 1988, no. 35.1; Suzuki, 1982, vol.3, JM1–075, JM1–133-5, JM3–197; Weidner et al., 1988, cat. nos 28, 29, 49).

In his detailed description of it Cibot (1778b, 3) called it 'Le Tsieou-hai-tang' and said that its flowers formed 'jolies bouquets' at the ends of the branches. He said that, because it could not stand the sun, it was planted against north-facing walls, where it was scarcely in flower before the end of August. He wrote that the scent of the flowers answered to their beauty, resembling that of violets and roses, but that you could only smell it when close. He also gave instructions for its cultivation.

In the Zhuozheng Yuan in Suzhou there is a courtyard called the Spring Home of Begonia. There has been no sign of any begonias there at the times of my visits but perhaps potted plants are put there in their flowering season. Or maybe the name is a reference to some beautiful but rejected woman whose residence this was.

B. grandis subsp. *evansiana* is a tuberous species native to eastern Asia which produces semi-pendulous clusters of pink flowers in early summer on plants which may be up to 60 cm tall (fig. 20.9). In addition to the usual pink form there is a variant which is almost white. While not amongst the most showy of the begonias, it has a quiet charm which no doubt appealed to Chinese gardeners. It reached Europe from China in 1804 (*Curtis's Botanical Magazine* 1812, t. 1473) and proved to be remarkably tolerant of low temperatures in winter when, of course, it is dormant.

Figure 20.9 (above)
Begonia grandis
subsp. *evansiana*
Figure 20.10 (right)
Belamcanda
chinensis

BELAMCANDA Iridaceae

B. chinensis 射干 (Shegan) *Shooting Stem*

LEOPARD LILY, BLACKBERRY LILY

B. chinensis is an iris-like plant, usually about 1 m tall, which produces small pink, yellow, or orange flowers heavily spotted with maroon (fig. 20.10). Whilst it cannot be said to be a showy plant, the individual flowers have an interesting, barbaric air about them. It was introduced to England from China in 1823 (Chittenden, 1956) and has found its way to many of the warmer parts of the world. It is a deciduous perennial, the capsules of which open to reveal clusters of shiny black seeds, resembling elongated blackberries.

In China it is a drug plant of long standing and is mentioned under the name *Shegan* in a poem written in 120 BC (Li, 1959). Li says that it was first cultivated as a medicinal plant but is now used primarily as an ornamental. It is commonly planted and still grows wild in many parts of the country. It is said to have been given its Chinese name because of the long, thin stems which shoot up and produce the flowers. Krahl (1987) records its use as a decorative motif on porcelain and it is depicted in an 18th century *Hundred Flowers* reproduced by Capon and Pang (1981, cat. no. 65).

CALENDULA Compositae

C. officinalis 金盏花 (Jinzhanhua) *Golden Cup flower*

MARIGOLD

While the origin of this plant is unknown, it seems unlikely that it is native to China, probably having been introduced somewhere along the way. It was mentioned in the middle of the 12th century AD by Chen Shan, the writer of *Menshi Xinhua*, who marvelled at the fact that the flowers of central and southern China were so different from those of the north (Needham, 1986). It has become popular as a flowering pot plant and is widely used for massed displays (fig. 2.16). Its bright orange flowers suggest prosperity and longevity in the way other flowers and fruits of this colour do (Goody, 1993), and the leaves were recommended as a famine food in the *Jiuhuang Bencao* (Bretschneider, 1881). It appears in paintings of relatively late date (e.g. Capon & Pang, 1981, cat. no. 65; Weidner et al., 1988, cat nos 34, 44).

CALLISTEPHUS Compositae

C. chinensis 翠菊 (Cuiju) *Kingfisher Chrysanthemum*

CHINA ASTER

C. chinensis is native to China and is the only species of its genus. Needham (1986) thinks that it is the plant mentioned in two folk songs from the 8th century BC recorded in the *Shi Jing* (Book of Odes). He says it has been cultivated for a couple of millennia in China and has long been called *Cuiju* or *Lanju* (Blue Chrysanthemum) on account of its colour (fig. 3.2). It has often appeared in Chinese paintings, for instance in the Song dynasty *Hibiscus and Rock* reproduced here as fig. 20.2 and in various works reproduced by others

(e.g. Barnhart, 1983, cat. no. 26; Weidner et al., 1988, cat. nos 16, 34, 49). It has also been used as a decorative motif on porcelain (e.g. Bartholemew, 1985a, cat. no. 18; Fourcade, 1965, ceramic no. 22).

In China it seems to have been grown mostly as a pot plant, set out in its season in the same manner as other annual flowers. Dorothy Graham (1938), for instance, noted the 'misty purple blue of China asters' amongst the potted plants brought round by the itinerant plant sellers in Beijing, and welcomed their 'profuse feathery blossoms' in October for filling the spaces in her garden where the plants had thinned. It has also been grown as a cut flower. Fortune (1863) observed fields planted with it near Tianjin and said the flowers were in demand in that city.

Seeds of *C. chinensis* were sent to Europe in 1728 (Gorer, 1970; Needham, 1986) and subsequently pink and white forms arose, along with doubles of various types, all of which became popular garden plants (W[atson], 1898). However, the China asters are not often planted in Western gardens these days, which is a pity, as their various colours and forms have considerable charm. Nevertheless tall varieties of various types have been developed for the cut flower trade and are widely grown.

CANNA Cannaceae

C. indica 美人蕉 (Meiren Jiao) *Beautiful Woman Banana*

INDIAN SHOT

This plant from tropical America (fig. 20.11) was amongst the plants collected by Peter Osbeck in the neighbourhood of Guangzhou in 1751–52, where it is frequently cultivated (Bretschneider, 1880). It is also occasionally seen in paintings of the 17th and 18th centuries (e.g. Weidner et al., 1988, cat. nos 34, 39).

CAPSICUM Solanaceae

C. annuum Cerasiforme Group 小米椒 (Xiaomi Jiao) *Millet Chilli*

While ornamental cultivars of *C. annuum* have become popular in many parts of the world, it seems that those with very small fruits have appealed most to the Chinese. Loureiro (1790) records the cultivation of this type in China, and Livingstone (1822), writing of the Chinese method of

Figure 20.11
(right)
Canna indica
Figure 20.13
(far right)
Ceratostigma plumbaginoides, from *Curtis' Botanical Magazine* 1850, t. 4487

dwarfing plants, said that 'The Capsicum Cerasiforme, and some other species of the same genus, are planted in pots not more than two or three inches in diameter. The plants are reduced thereby to the most diminutive size, with fruit scarcely larger than our Currant'. The Chinese name given above no doubt refers to cultivars which produce fruits not or little larger than millet grains.

CELOSIA Amaranthaceae

C. cristata (C. argentea *var.* cristata) 鸡冠花 (Jiguanhua)
Cockscomb Flower

COCKSCOMB

The cockscomb, with its distorted inflorescences, is an annual with a long history in China. Sirén (1949) refers to a painting of the plant from the 11th century AD and it appears in the Song dynasty *Hibiscus and Rock* (fig. 20.2) and the *Song Hundred Flowers* in the collection of the Palace Museum, Beijing, so it is clear that this mutant form has been known for at least 900 years. As well as being represented in many paintings, it is also recorded as a decorative motif on porcelain (Krahl, 1987). It is known only in cultivation and is believed to have been derived from *C. argentea* (Mabberley, 1997a).

As with many other annuals, the cockscomb is usually grown in pots and set out for display, the prevailing colours being red and yellow (fig. 20.12, see over). Fortune (1847a) listed it among the favourite plants cultivated by the people living on boats in Guangzhou, and also saw it amongst the plants in bloom brought down to that city in boats at the Chinese New Year. Cantoniensis (1867) reported that yellow cockscombs were frequently grafted onto red stocks to produce plants bearing flowers of both colours. Judging from the appearance of the plants, he concluded that the grafting was done by the wedge and cleft method.

CERATOSTIGMA Plumbaginaceae

C. plumbaginoides 蓝雪花 (Lanxuehua) *Blue Snow Flower*

This perennial herb bears deep blue, 5-petalled flowers (fig. 20.13). What appears to be this species is included in many paintings of flowers of relatively late date (e.g. Barnhart, 1983, cat. no. 33; Capon & Pang, 1981, cat. no. 65). The '*Plumbago*'

Figure 20.12 *Celosia cristata* in the Daguan Yuan, Beijing

which Clarke Abel (1819) saw in 1816 grown in pots at Tong Xian, near Beijing, but could not determine, was possibly this plant. It was subsequently collected in the same region by Bunge, who described and named it in 1833 (Lauener, 1996). If it is still cultivated in China it is not common. Its Chinese name refers to the cool effect of its blue colour in summer.

COMMELINA Commelinaceae

C. communis 鸭舌草 (Yashecao) *Duck's Tongue Plant*

DAY FLOWER

A native of eastern Asia, this plant is closely related to the well-known tradescantias of America. However, its bright-blue flowers, each of which usually lasts less than a day, are distinctive, since the three petals are not equal in size, the lower one being much reduced. This gives the flowers a but-terfly-like appearance. The leaves of *C. communis* are listed as a famine food in the *Jiuhuang Bencao* (Bretschneider, 1881), and the Chinese name for the plant refers to the shape of these leaves.

I have included *C. communis* here as it appears in paint-ings (fig. 20.14, see over; Capon & Pang, 1981, cat. no. 65; Li, 1959, pl. 7B; Weidner et al., 1988, cat. nos 16, 30, 34) and Liu (1982) lists it amongst the plants grown in Suzhou gardens.

DELPHINIUM Ranunculaceae

Delphinium grandiflorum 大花飞燕草

(Dahua Feiyancao) *(Large-flowered Flying-swallow Herb)*,

翠雀 (Cuique) *Kingfisher*

Li (1956a, 1959) lists this north-east Asian species amongst the garden plants of China (fig. 20.15). Whether it is still grown there I do not know, but it has occasionally been de-picted in paintings, for instance in an 18th century *Hundred Flowers* reproduced by Capon & Pang (1981, cat. no. 65). Its Chinese names graphically express the shape of the flowers. It is best known in the West as its cultivar 'Blue Butterfly'.

DIANTHUS Caryophyllaceae

D. chinensis 石竹 (Shizhu) *Stone Bamboo*

CHINESE PINK

D. chinensis is a short-lived perennial usually grown as an annual. As seen in China it is a bushy plant bearing single or

Figure 20.15 *Delphinium grandiflorum*, from *Edwards's Botanical Register* 1820, t.472

double flowers of various colours (figs 17.7, 20.16). Li (1959) says it has many common names but it is generally known as *Shizhu* (Stone Bamboo) because of the conspicuous nodes on its stems and its slender leaves, and because it is often planted amongst rocks. Also it is mentioned as being edible in the *Jiuhuang Bencao* (Li, 1959). Bretschneider (1880) says it was introduced to Europe about 1702, and many strains and hybrids have been developed in the West.

The Chinese Pink appears frequently in paintings but, although Li (1959) says that several species of *Dianthus* have been cultivated 'from time immemorial', I have not identified *D. chinensis* with certainty in any early works. It appears commonly in Ming and Qing dynasty paintings, which show white, pink, red, and purple forms (e.g. figs 19.74, 20.14; Barnhart, 1983, cat. no. 33; Weidner et al., 1988, cat. no. 14), and a double variety appears in an 18th century *Hundred Flowers* reproduced by Capon and Pang (1981, cat. no. 65).

Figure 20.14 (opposite page)
A fan painted by Lo Guifen,
active late 19th century,
showing (left to right)
Saxifraga stolonifera,
Dianthus chinensis, and
Commelina communis
Figure 20.16 (left) *Dianthus
chinensis*

It is also often depicted on porcelain (e.g. Bartholemew, 1985a, cat. no. 18; Fourcade, 1965, ceramic no. 22).

DICENTRA Fumariaceae

D. spectabilis 荷包牡丹 (Hebaomudan)

Carrying-bag Tree Peony

BLEEDING HEART, DUTCHMAN'S BREECHES

This herbaceous perennial with characteristically shaped flowers is a native of north-east Asia and seems to have been a favourite garden plant for centuries. It appears in the *Jieziyuan Huazhuan* of 1701 (Sirén, 1949) and in paintings at least from that time on (e.g. Suzuki, 1982, vol. 4, JP7–064). Li (1959) states that it is frequently cultivated in small shady areas and amongst rockwork. Both pink and white forms are seen in China today, usually planted in the ground. The Chinese name given above, needless to say, refers to the shape of the flower (fig. 20.1). The *Jieziyuan Huazhuan* refers to it as *Yuermudan* (Fish's Son Tree Peony), because to Chinese eyes the flower resembles a fish with a smaller fish issuing from its mouth (Sze, 1977). Fortune (1847b) recorded yet another name, 'Hong-pak-Moutan Wha' or 'Red and White Moutan Flower', and pointed out that its leaves were like those of the tree peony. He first found it in a mandarin's garden on Zhoushan amongst the artificial rockwork, and said that the Chinese increased it by dividing the roots in spring when it began to grow. Later he saw it growing in a bed at a nursery near Shanghai (Fortune, 1850f). It was amongst the plants he sent to the Horticultural Society of London.

DUCHESNIA Rosaceae

D. indica 蛇莓 (Shemei) *Snake Berry*

MOCK STRAWBERRY

D. indica, which is native to Asia, closely resembles a strawberry but has yellow flowers. In China is called *Shimei* (Snake Berry) because of its creeping growth habit and because it is used for snake bite in Chinese medicine. It is depicted in a handscroll *Flowers and Insects after Song Masters* dated 1723 (Weidner et al., 1988, cat. no. 45) and perhaps is represented elsewhere in Chinese art. Mabberley (1997a) indicates that the genus *Duchesnia* has been merged with *Potentilla*, but for the purposes of this book I have retained the generic name by which it is still known to most gardeners.

GOMPHRENA Amaranthaceae

G. globosa 千日红 (Qianrihong) *Thousand Days Red*

GLOBE AMARANTH

This Central American annual has at times been a popular garden plant in China. It is not clear when it arrived but, according to Bretschneider (1880), Loureiro's *Flora Cochinchinensis* of 1790 mentioned it being cultivated in China. It was introduced to Britain from the East Indies as long ago as 1714 (*Curtis's Botanical Magazine* 1828, t. 2815), so it seems likely that it was amongst the early arrivals in Asia from America. It gets its Chinese name because the papery, reddish-purple bracts in the flower heads are very long-lasting.

Figure 20.17 A bed of *Hemerocallis* at the Tanzhe Si, Hebei

HEMEROCALLIS Hemerocallidaceae

The *Hemerocallis* species, the daylilies, are native to eastern Asia. In China they are widely grown in gardens and are usually seen planted in beds (fig. 20.17). They have been noticed and used by the Chinese since ancient times, being mentioned in the *Shi Jing* (Book of Odes) and in the ancient pharmacopoeias as herbs useful for helping one to forget sorrows. Thus a daylily is an ideal motif for pillow ends, and embroidery patterns depicting it were always prominent at fairs (Koehn, 1952). Also daylilies were believed to favour the conception of sons when worn in women's girdles (Li, 1959). Plants in bloom appear in many paintings and a picture of a daylily together with a pomegranate, splitting to reveal its many seeds, represents the saying 'May there be many sons among the many children' (Koehn, 1952).

As well as being grown as ornamentals, daylilies are also cultivated for their flower buds, which are dried for medicinal use, and as a vegetable known as *Jinzhencai* (Golden Needle Vegetable) or *Huanghuacai* (Yellow Flower Vegetable)

(Baker, 1937; Li, 1959; Stout, 1933). The dried buds are dark-brownish yellow, more or less translucent, and when cooked have an agreeable odour, a sweetish taste, and a mucilaginous texture. According to Li (1959), they can be found at all food stores throughout China. They have also been exported (Stout, 1933), and perhaps still are. I had a dish which included them in a restaurant in Shanghai and found the flavour pleasant, if unexciting. The most interesting part of the experience, if I may be excused for having, like Lot's wife, looked back, was that I observed that they passed through more or less intact.

It is difficult to tell for certain which species are grown for this purpose, but records include forms of both *H. lilio-asphodelus* and *H. fulva* (Baker, 1937; Li, 1959; Stout, 1933). However, both these species were originally described by Linnaeus from cultivated plants which had early reached Europe and which now probably cannot be identified with any wild or cultivated plants in the East. Since a detailed taxonomic study would be necessary to clarify the situation, I have decided here to follow the usual practice of referring the daylilies cultivated in China, whether for food, ornament, or medicine, to *H. fulva*, *H. lilio-asphodelus*, and *H. minor*.

Daylilies are amongst the earliest Chinese plants to have been brought to the West, having reached the Mediterranean by the 1st century AD. They are referred to at this time by Pliny the Elder and by Dioscorides, who gave a description and noted the medicinal properties of the leaves and roots (Li, 1959). However, in spite of these early mentions, they do not seem to have been introduced to the rest of Europe until the 16th century. They are now favourite garden plants all over the world, having given rise to numerous hybrids in a wide range of colours.

H. fulva 黄花菜 (Huanghuacai) *Yellow Flower Vegetable,* 萱草 (Xuancao) *Xuan Plant*

ORANGE DAYLILY, TAWNY DAYLILY

The origin of the orange or tawny daylily, *H. fulva*, is uncertain, as the typical form has not been found in the wild (Li, 1959). It probably arose as a mutant and has been propagated vegetatively, as it is a triploid and rarely, if ever, sets seed. However, various other types referred to this species occur in China (fig. 20.18), where more than 100 cultivars are known (Wang & Ma, 1995). Some with pink flowers were sent to the New York Botanical Garden in 1924, and in 1930 a clone grown for its edible buds in Sichuan was also sent to this garden, where it was given the name 'Chengtu'. It had 'attractive scarlet colouring in a banded pattern'. Another clone grown for its buds was sent from Hankou in 1933. This had large orange flowers with a bold mid-band of scarlet in the petals (Stout, 1942). These variants have played an important part in the development of the modern hybrid daylilies in America.

A daylily of the *H. fulva* type, very like that shown in figure 20.18, is depicted (fig. 2.14) in the *Song Hundred Flowers* in the collection of the Palace Museum, Beijing, and similar flowers appear in many subsequent paintings. The double form, which is obviously also an old cultivar, is represented on an album leaf of 1540 (Barnhart, 1983, cat. no.26) and in the *Jieziyuan Huazhuan* (1975). The orange daylily also features as a decorative motif on porcelain (e.g. Bartholemew, 1985a, cat. no. 18). The word *xuan* has two associations in China, firstly with forgetting sadness, and secondly with 'mother'. Thus the orange daylily is regarded as the 'forgetting sadness herb' and is also used to honour one's mother (Bartholemew, 1985a).

H. lilio-asphodelus 北黄花菜 (Bei Huanghuacai)
Northern Yellow Flower Vegetable

YELLOW DAYLILY

H. lilio-asphodelus has clear yellow flowers and, not surprisingly, is generally known as the Yellow Daylily (fig. 20.19). However, it seems that none of the yellow daylilies observed in China can definitely be assigned to this species, which was originally described by Linnaeus from cultivated plants in Europe (Stout, 1935). Nevertheless the tall yellow daylilies found in Chinese gardens are usually given this name. Like their orange counterparts, their buds are eaten and they have also been considered appropriate subjects for painting, for instance in a work dated 1737 reproduced by

Capon and Pang (1981, cat. no. 76). No doubt a search would reveal other and perhaps earlier depictions.

H. minor 小黄花菜 (Xiao Huanghuacai)
Small Yellow Flower Vegetable

This species is shorter and has narrow, grass-like leaves and yellow flowers. It is native to north-eastern Asia and is said to be cultivated in China (Li, 1959). Li (1979) says that this is the species used since ancient times for 'boy-getting'. The leaves are listed as a famine food in the *Jiuhuang Bencao* (Bretschneider, 1881).

Figure 20.18 (above right)

Hemerocallis fulva

Figure 20.19 (right) ***Hemerocallis lileo-asphodelus***

HOSTA Hostaceae

The hostas, which have become so popular in Western gardens, are found principally in north-east Asia, particularly Japan. Like the daylilies discussed above, they have also long been cultivated in Chinese gardens. The species usually planted there are *H. plantaginea* and *H. ventricosa*, both of which are native. On rare occasions I have encountered one or two others. However, as they were not in bloom, I was unable to identify them.

H. plantaginea 玉簪 (Yuzan) *Jade Hairpin*

FRAGRANT PLANTAIN LILY

H. plantaginea became popular as long ago as the Han dynasty and remains the species most admired and most frequently planted. As an ornamental it is found in courtyards, flower beds, and potted landscapes, and every part of the plant is useful for making ointments for relieving inflammation and fever (Wang & Ma, 1995). Its large, conspicuously veined leaves are held on firm upright petioles, giving the plants a robust appearance (fig. 20.20). The flowers are produced in summer in short, dense racemes and, unlike those of other species, are white, narrowly tubular, and 6 cm or more long (fig. 20.21). They are also strongly fragrant, especially at night. No doubt the Chinese name for the flower arose from its resemblance to a jade hairpin, and probably to its use as a hair ornament as well. A traditional woman's hairpins were long, made of precious materials, and valuable. Selling them was a last resort (Pratt & Chiang in Shen Fu, 1983). It is easy, then, to understand the esteem in which this plant is held in China. *H. plantaginea* has also been considered a suitable subject for painting. It can be seen, for example, in the *Song Hundred Flowers* in the collection of the Palace Museum, Beijing. There are many subsequent examples (e.g. Li, 1959, pl. 7B; Wang & Barnhart, 1990; Weidner et al., 1988, cat. nos 34, 48), and it is included in the *Jieziyuan Huazhuan* of 1701 (Sirén, 1949).

In the West it was first described in 1789 as *Hemerocallis plantaginea* by the French botanist Lamarck, who stated that it had already been cultivated for a few years in the French king's garden. Apparently it had been sent to France from China by M. de Guines (Lauener, 1996). So it is clear that it has been known in Europe for over 200 years.

In China *H. plantaginea* has been mistaken by various Western writers for the tuberose, *Polianthes tuberosa*. Since both plants have similarly shaped, scented, white

Figure 20.20 (left) *Hosta plantaginea* at the Biyun Si, Western Hills, Beijing
Figure 20.21 (above left) *Hosta plantaginea* in bloom

Figure 20.22 *Impatiens balsamina*

flowers and have given rise to double cultivars, there is some basis for this confusion.
In the West there is a double form known as
'Aphrodite', the flowers of which are said
hardly ever to open (Brickell, 1996). However, this defect is not exhibited by the double form shown in a colour photograph
reproduced by Yu & Wu (1993).

H. ventricosa 紫萼玉簪 (Zie Yuzan)
Purple-calyx Jade Hairpin

Although not appearing in Chinese gardens
as frequently as *H. plantaginea*, nor attracting the same attention from artists, *H.
ventricosa* is often planted. In foliage it resembles *H. plantaginea*, but it differs in that
the leaves tend to be a little narrower, are
thinner textured, and are borne in greater
numbers on individual shoots. It also forms
large, dense clumps, whereas *H. plantaginea*
usually forms small, more open clumps.
When it is in bloom there can be no confusing the two, as *H. ventricosa* produces tall,
loose racemes of deep mauve flowers.

HYLOTELEPHIUM—see *Sedum*

HYPERICUM Guttiferae

H. ascyron 黄海棠 (Huanghaitang)
Yellow Crabapple

H. ascyron is a herbaceous perennial with stems 0.5–1.5 m tall
and yellow, star-shaped flowers. Lancaster (1989) described
its petals as resembling the blades of a propeller. It is native
to China and I have included it as Sirén (1949) mentions it
amongst the garden flowers. He identifies the *Jinsitao* (Gold
Silk Peach) of the *Jieziyuan Huazhuan* as this species but,
as the name is also applied to the shrubby *H. monogynum*, it
is not clear whether *H. ascyron* has ever been grown in
Chinese gardens. That I have not seen it may mean little.

IMPATIENS Balsaminaceae

I. balsamina 凤仙花 (Fengxianhua)
Phoenix Immortal Flower

BALSAM

This sparsely branched annual, which bears pink, red, purple or white flowers in the axils of the leaves in summer (fig.
20.22), has a long history of cultivation all over China. It may
be native to the country or perhaps have been introduced
from India in very early times. Box (1905) recorded that it
was the floral symbol of the 7th month in Shanghai, and

Bredon and Mitrophanov (1927) noted that the same applied in Beijing.

During hundreds of years of cultivation many forms have arisen with much variation in colour, size, and doubling of the flowers (Li, 1959). The resemblance of the spurs of the flowers to the feet of birds led to the name *Fengxianhua* (Phoenix Immortal Flower), and the bursting open of the pods at a touch gave the synonym *Jixingzi* (Impatient Seed). During the Song and perhaps earlier the flowers were used for dyeing fingernails, a process involving garlic juice and fixation with alum (Li, 1959; Needham, 1986). The plant is thus also called *Zhijiacao* (Fingernail Plant) and has been confused with henna (*Lawsonia inermis*), which also has this name. According to Tun (1965), the colour 'penetrates to the bones' and will last a year before disappearing. As a result of this another name for it is *Tougucao* (Penetrating to the Bones Plant). As well as all this, it is recorded in the *Jiuhuang Bencao* that its leaves can be eaten when times are hard (Bretschneider, 1881).

It appeared in paintings quite early, for example in the *Song Hundred Flowers* in the collection of the Palace Museum, Beijing, and subsequently has been frequently depicted (e.g. Barnhart, 1983, cat. no. 26; Suzuki, 1982, vol. 3, JM3–235; Weidner et al., 1988, cat. no. 57). However, unlike many popular flowers, it had to wait until 1785 before a monograph, the *Fengxian Pu* by Zhao Xuemin, was devoted to it (Needham, 1986).

INULA Compositae

I. brittanica 旋复花 (Xuanfuhua) *Return Replicate Flower,* 六月菊 (Liuyueju) *Sixth Moon Chrysanthemum*

Li (1959) includes amongst the garden flowers of China '*I. chinensis*' and gives as one of its Chinese names *Xuanfuhua*, a name applied to what in China is now referred to I. *japonica*. *I. japonica*, in turn, is considered elsewhere to be a synonym of *I. brittanica*, a species widely distributed throughout Eurasia, so this is the name I have chosen to give this plant. It is a perennial with yellow, daisy-like flowers.

Li merely lists it and I have not succeeded in finding any other mention of its cultivation in China. However, the name *Liuyueju* suggests that it is a familiar plant, and *Xuanfuhua* refers to the fact that it regenerates easily from pieces of rhizome left in the soil and is hard to kill. This, too, suggests that it must be well known to gardeners, even if only as a weed. Its leaves are listed as a famine food in the *Jiuhuang Bencao* (Bretschneider, 1881).

IRIS Iridaceae

Although irises are commonly cultivated in Chinese gardens and appear occasionally in paintings, comparatively little seems to have been written about them in China. Also, the symbolical and ritual uses attributed to them by Eberhard (1958) and Goody (1993) would seem to apply more correctly to the Sweet Flag (*Acorus calamus*). Both *I. japonica* and *I. tectorum* are widely cultivated, and these species, together with *I. lactea*, are the only irises I have seen in traditional gardens.

I. japonica 蝴蝶花 (Hudiehua) *Butterfly Flower*

Apparently native to both China and Japan, this decorative species is usually seen in Chinese gardens growing in shady places, sometimes, as in the Jichang Yuan at Wuxi, planted as an extensive ground cover. As in the West, there appear to be two forms in Chinese gardens, a pale blue (fig. 20.23) and a white, in both of which the flowers exhibit orange and purple markings. As well as being grown as an ornamental, *I. japonica* is used medicinally for the treatment of fever, poisoning, tuberculosis, and other conditions (Waddick & Zhao, 1992), and is depicted in paintings (e.g. Capon & Pang, 1981, cat. no. 80; Weidner et al., 1988, cat. no. 11).

I. lactea 白花马蔺 (Baihua Malin) *White-flowered Horse Rush*

I. lactea is a clump-forming species from northern China, Mongolia, and Siberia. It has slender, grassy leaves amongst which it produces masses of narrow-petalled flowers in late spring. While the plant originally described had white flowers and was given the name *Baihua Malin*, it is the blue form, known simply as *Malin*, which is generally planted in northern China. It is present in many gardens in and around Beijing (fig. 20.24).

I. lactea is remarkably drought resistant and its foliage can be used for grazing. Its leaves have also been used for making paper and brushes and to make cord for tying bunches of vegetables and grapes (Dykes, 1974). The flowers and seeds are used medicinally (Waddick & Zhao, 1992).

Figure 20.23 *Iris japonica* in the Qiuxia Pu, Jiading

Figure 20.24 *Iris lactea* at Badachu, south-west of Beijing

I. tectorum 鸢尾 (Yuanwei) *Kite Tail*

ROOF IRIS

Because it is grown on the ridges of thatched roofs in Japan, from which it was first introduced to Europe, that this iris has been given the specific epithet *tectorum*, which means 'of the roofs of houses'. In China, where roofs are usually tiled, I have not seen it growing in this situation, nor has George Waters (1996), but it is commonly planted in the ground. According to Bretschneider (1895) its cultivation in China can be traced back to the 7th century AD and it is now found throughout the country. Whether its occurrence in neighbouring Burma and Japan is natural, or whether it has been introduced from China, it is impossible to say (Dykes, 1974).

In Chinese gardens this species is usually planted against rocks (fig. 20.25). It has broad, pale-green leaves and in late spring produces its somewhat flattened lilac-blue flowers, which are about 10 cm in diameter. There is a white-flowered variant and another in which the leaves are striped with white. It is depicted in Chinese paintings (e.g. Capon & Pang,

Figure 20.25 *Iris tectorum* at Gongwangfu, Beijing

1981, cat. no. 65; Suzuki, 1982, vol. 3, JM1–075) and on porcelain (Krahl, 1987), and guidance for painting it is included in the *Jieziyuan Huazhuan* (1975).

In climates which suit it *I. tectorum* is one of the easiest irises to grow, though it appears to require a sunny position to bloom prolifically. It may be propagated by division or from seed.

LACTUCA Compositae

L. sativa var. *asparagina* 莴笋 (Wosun) *Wo Bamboo-shoot*, 千金菜 (Qianjincai) *Thousand Pieces of Gold Vegetable*

ASPARAGUS LETTUCE, CELTUCE, STEM LETTUCE

Westerners are so used to eating only the leaves of lettuce and throwing away the stem that it comes as a surprise to find that the lettuce commonly grown in China is treated in exactly the opposite way. *L. sativa* var. *asparagina*, which perhaps should be regarded as a group of cultivars rather than a botanical variety, forms no heart but develops a stem 3–4 cm thick and about 30 cm tall. The cut stems are marketed with a topknot of leaves remaining. They are prepared by removing the leaves and peeling the stem, which is cut into slices before being lightly cooked, usually stir-fried with other ingredients. It has an excellent flavour and texture.

Its centre of origin is China, although it is also cultivated in many parts of south-east Asia. Herklots (1972) says that in northern China several distinct cultivars are known, including white, wrinkled, cleft-leaved, pointed-leaved, purple-leaved, and red-leaved types. Seed was sent to England by Fortune in 1845, along with the following instructions for preparing it: 'Mode of dressing the Hoo-Sung. —Pare off the outer skin, cut off the leaves, and take the stalk; either simply boil it with salt and eat it with pepper; or stew it with a few spoonfuls of soup, or with a little soy, salt, and pepper. The last is the preferable way of dressing this vegetable' (Lindley, 1846). Almost a century later, in 1938, seeds were sent from China to the W. Atlee Burpee Company in the United States and subsequently it was offered in their catalogue under the name of 'Celtuce' (Herklots, 1972). It seems, however, never to have become popular in the West.

It was given the name *Wosun* because it was said to have been brought from the country of Wo in the Sui dynasty and was regarded as a kind of bamboo shoot. *Qianjincai* refers to the valuable gift believed to have been given to the diplomat from Wo who brought the seeds.

LILIUM Liliaceae

Though few *Lilium* species are planted for ornamental purposes in China, in many places several kinds, particularly the tiger lily (*L. lancifolium*) and *L. brownii*, are cultivated on a commercial scale for their edible bulbs. Likewise the bulbs of many wild species are collected for food or medicine, as they are believed to have a cooling, moistening, and calming effect. As well as this the flowers of many species are fragrant, containing essential oils which can be used in perfume (Haw, 1986b). According to Li (1959), the bulbs are consumed in large quantities, especially during the summer, the scales being peeled and cooked in sugar and water. Herklots (1972) says that the bulbs may be baked, or grated and added to soup to thicken it.

Lilies are known collectively as *Baihe* (100 United), an allusion to the numerous scales which make up the bulb. On account of these tightly overlapping scales, the bulbs are seen as symbols of harmony and friendship and are often depicted with other auspicious plants and objects, their presence also suggesting the multiplication of the good omens 100 times (fig. 2.12).

The history of lilies in China has been discussed in detail by Stephen Haw (1984, 1986b), and readers should refer to his accounts should they wish to know more than I have set out in what follows. It is probable that lilies were originally valued by the Chinese chiefly for their medicinal uses. Haw considers it likely that they have been used medicinally for at least 2000 years. The early writers describe them as *Baihe* and say that the bulbs are formed from earthworms which knot themselves together and are transformed.

A pharmacopoeia of 659 AD records two kinds, one with narrow leaves and red flowers and another with large leaves, a tall stem, a large bulb, and white flowers. A writer of the early 8th century says that those with red flowers are called *Shandan* (Mountain Red) and are not suitable for eating. In the early texts the name *Shandan* is thought to apply to either or both of *L. pumilum* and *L. concolor*, and perhaps *L. callosum* and other species as well, though the name is now used only for *L. pumilum*. In the 10th and 11th centuries other kinds were mentioned, including two with stem bulbils, one with yellowish-white flowers and the other with orange ones with black spots.

It is clear, then, that by the 12th century several kinds of *Baihe* were recognised by Chinese writers, some with white and some with red flowers, some with stem bulbils and some without, and differing in other characteristics. It also seems clear that at that time the chief source of *Baihe* bulbs for medicine and eating was the common and widespread *L. brownii*, and that the bulbiferous lily with the orange flowers and black spots must have been the tiger lily, which was not at that time considered suitable for medicinal use.

It seems probable that originally lily bulbs were only collected from the wild, but a 10th century reference describes the cultivation of *Baihe* from scales and the drying, pounding, and sieving of the bulbs to make a flour. This appears to be the earliest reference to the cultivation of lilies in China and it is interesting to note that the Chinese were already familiar with the method of propagating them from bulb scales. Perhaps they were already being cultivated for the beauty of their flowers as well, as poems in praise of them date from as early as the 6th century. An 11th century verse describes *Shandan* planted in front of a hall and, as the small bulbs were not rated so highly for medicinal use, it may well have been more popular in gardens than the white-flowered *L. brownii*. That lily bulbs were not rated highly as food is suggested by the inclusion of *Baihe* as a famine food in the *Jiuhuang Bencao*, which notes that the *Shandan* is said not to be suitable for consumption.

By the latter half of the 16th century it seems likely that many of the lily bulbs used medicinally came from cultivated plants, as in Li Shizhen's *Bencao Gangmu* of 1596 the entry for *Baihe* appears in the 'Vegetables' section. He distinguishes three kinds of lily: the broad-leaved *Baihe* with white, nodding flowers, the *Shandan* with narrow leaves and red flowers which do not nod, and the *Juandan* (Rolled-up Red) with bulbils and orange-red flowers spotted with black. These are clearly *L. brownii*, *L. concolor*, and *L. lancifolium* respectively. Li Shizhen wrote that the *Shandan* was not as useful medicinally as the *Baihe*, but that the people of Shandong and Hebei dried the unopened flower buds and sold them as 'Red Flower Vegetable'. He also expressed his disbelief of the transformation of worms into the bulbs.

The *Qunfang Pu* of 1630, which is largely concerned with plants of ornamental value, describes several species including a plant called *Wodan*, also called *Shandan*, which was probably *L. concolor*. This is also described in the *Hua Jing* of 1688, the author of which, Chen Haozi, lists several other species including a 'foreign' *Shandan*, *Tianxiang* (Heavenly Fragrance) and *Shexianghua* (Musk-Scented Flower),

which Haw (1986b) suggests were probably *L. leitchlinii* var. *maximowiczii*, *L. auratum*, and *L. longiflorum* respectively, and which presumably had arrived from Japan. Whatever the truth may be, it is clear that by this time there were at least seven or eight species cultivated in Chinese gardens and that, although lily bulbs were still consumed, several of these species were grown solely for their flowers.

Other species have no doubt been cultivated locally. For instance in 1914 Reginald Farrer found *L. leucanthemum* var. *centifolium* in the gardens of two cottages in southern Gansu, and George Forrest, also in the early decades of this century, saw *L. davidii* grown as a field crop in western Yunnan. It is still cultivated in various areas in the west of the country (Haw, 1986b). And, according to Li (1959), *L. formosanum* is said to be grown locally in Taiwan. However, although there are many species of *Lilium* in China, few of them seem to have achieved any real prominence as garden plants. The fact that *L. brownii* and *L. concolor* are the only species I have so far seen depicted in paintings tends to confirm this. According to Krahl (1987), *L. lancifolium* is represented on porcelain but, not having seen the evidence for myself, I cannot be certain that her identification is accurate. On the whole it seems that *L. brownii*, *L. concolor*, *L. lancifolium*, and perhaps *L. pumilum* are the principal Chinese species which have been grown, and it is these which I have included below.

L. brownii 野百合 (Ye Baihe) *Wild Lily*

The name *L. brownii* was given to a form of uncertain origin in which the outside of the flowers is heavily stained with purplish-brown. Another type often seen in China differs in having oblanceolate to obovate leaves and usually less colour on the outside of the flowers. It has been described as var. *viridulum*, which Wilson (1925) says is known to have reached England along with the tiger lily from the Fa Tee Gardens, being sent by William Kerr to Kew on the East Indiaman *Henry Addington* in 1804. Both varieties occur widely in China and are still quite common in some areas in spite of having been dug for centuries for their edible bulbs (Haw, 1986b). Wilson (1925) wrote that *L. brownii* is 'the common trumpet-flowered lily of China and probably grows wild in every one of the eighteen provinces of the Empire from sea level to 5000 feet according to climate. It is the "Peh ho" of the Chinese, and is often cultivated for its bulb which is esteemed as a table delicacy'. The bulbs, as mentioned above, are also used in medicine and can be used to make wine (Haw, 1986b). *L. brownii* has been cultivated from at least as early as the 10th century and continues to be grown as a crop over a wide area.

Herklots (1932b) says that the bulbs are slightly bitter when boiled in water and that the water is subsequently used for medicinal purposes. They are sometimes also used in soup. As well, Herklots (1972) quotes a Chinese author as saying that a kind of arrowroot is made of its starch.

L. brownii is the lily most frequently depicted in Chinese art (e.g. Bartholemew, 1985a, cat. no. 25; Sirén, 1973, plates 251, 324; Suzuki, 1982, vol. 3, JM1–075, JM1–133-5; Weidner et al., 1988, cat. nos 11, 39). While this fragrant, yellow-throated lily is very variable, particularly with regard to the amount of purplish colouration on the outside of the flowers, most of the representations in Chinese paintings show them to be white. However, the flower which appears in the *Song Hundred Flowers* in the collection of the Palace Museum, Beijing, is strongly pigmented on the outside (fig. 20.26), so there is probably some variation in the forms which are cultivated.

L. concolor 渥丹 (Wodan) *Wo Red*

MORNING STAR LILY

L. concolor is a dainty species from central and northern China with glossy, upward-facing flowers. In the typical form

these are scarlet and unspotted, hence the specific epithet *concolor* (all one colour), although variants with spotted flowers occur. Haw's (1984, 1986b) evidence suggests that it has probably been cultivated in Chinese gardens for hundreds of years. It also occasionally appears in Chinese paintings, for example the *Song Hundred Flowers* in the collection of the Palace Museum, Beijing (fig. 20.27), and later examples reproduced by Weidner et al. (1988, cat. nos 15, 34, 40, 44, 50.3). The bulbs can be eaten, or employed in medicine, or used to make wine (Haw, 1986b). It seems from Li Zhizhen's description mentioned above that it was the flower buds of this species which the people of Hebei and Shandong dried and sold as 'Red Flower Vegetable' in the 16th century.

L. lancifolium 卷丹 (Juandan) *Rolled-up Red*

TIGER LILY

L. lancifolium is now considered to be the correct name for the tiger lily, formerly called *L. tigrinum*. Perhaps this is just as well, as orange-red tigers with purple-black spots, as noted by Woodcock and Coutts (1935), are not often seen, even in far Cathay. Its Chinese name, *Juandan*, presumably refers to the shape of the flowers (fig. 20.28). Swaying in the breeze over a green carpet of creepers, tiger lily flowers have been described in China as bringing a painted dragon to life (Wang & Ma, 1995). Even so, it does not appear to have appealed to the painters, as so far I have not come across an example of its representation.

This lily occurs wild in many provinces of China as well as in Korea and Japan. As noted above, it was introduced to England along with *L. brownii* var. *viridulum* by William Kerr in 1804. It is perhaps the best-known of all cultivated lilies and several variants are known, including a somewhat grotesque double form. Although various authors have stated that it has been raised for over 2000 years by the Chinese, principally for its edible bulbs, Haw (1986b) has produced evidence to the contrary. While it has certainly been grown as an ornamental for perhaps 1000 years, he thinks that it has been

Figures 20.26, 20.27 *Lilium brownii* (below), and *Lilium concolor* (left), details from the *Song Hundred Flowers*

Figure 20.28 (opposite page) *Lilium lancifolium*, the Tiger Lily

cultivated for its bulbs only relatively recently, probably no earlier than the first half of the 19th century. Anyway it is now the lily most commonly grown on a large scale in China. The main centres of its cultivation are in Jiangsu, but it is quite commonly grown in other provinces. It is also cultivated by the Japanese and Koreans (Wilson, 1925). Wilson (1913) recorded that in his time the white bulbs of this lily were more expensive in China than they were in Britain. He said that when properly cooked they were 'not at all bad eating', somewhat resembling parsnip in flavour. So perhaps they would be worth trying.

L. pumilum 山丹 (Shandan) *Mountain Red,* 细叶百合 (Xiye Baihe) *Small-leaved Lily*

CORAL LILY

This diminutive lily with small, scarlet, turk's cap flowers is found in northern China and adjoining regions. The bulb is edible and can be used in medicine. Haw (1986b) says that it is 'the most common lily of north China, occurring in every province and autonomous region through which the Yellow River flows and northward to beyond the borders'. The bulbs are said to have been eaten by the 'wandering Tartars' (*Edwards' Ornamental Flower Garden*, 1854) and

Figure 20.29 *Liriope muscari* in the Guyi Yuan, Jiading

perhaps they still are. According to Li (1959), *L. pumilum* is cultivated in northern China, and Woodcock and Coutts (1935) quote Reginald Farrer as saying that nothing could yield more brilliant results than this species 'as the Thibetans grow it, for they ram a handful of bulbs at haphazard into the hard mud of their flat roofs and there above the eave sprouts a living pyramid of fire from year to year'. Like *L. lancifolium*, apparently it has not appealed to Chinese artists, as I have not recognised it in any of the paintings I have seen.

LIRIOPE Convallariaceae

The species of *Liriope* are natives of eastern Asia and, along with *Ophiopogon japonicus*, are known generally as *Maidong* (Winter Wheat), on account of their grass-like leaves which are evergreen. As well as being grown in Chinese gardens for ground cover or edging, they are often grown in pots for outdoor display or for bringing indoors. The identification of *Liriope* species is not easy. In particular *L. graminifolia* and *L. muscari* have been much confused.

L. graminifolia 禾叶山麦冬 (Heye Shanmaidong) *Cereal-leaved Mountain Winter Wheat*

L. graminifolia is described as an evergreen, clump-forming perennial with strap-shaped leaves up to 30 cm long, which produces spike-like racemes of small, purple to white flowers borne on black-purple scapes. Although it is rare in Western gardens, Li (1956a) says it is the most commonly planted species in China. He records that it is often planted in artistically decorated pots, sometimes with rocks, to be placed on the desk of a scholar. Not only does the plant serve as a restful diversion for the eyes, but the flat leaves are useful as bookmarks. Hence it is also known as the Book-tape Herb.

L. muscari 阔叶麦冬 (Kuoye Maidong) *Broad-leaved Winter Wheat*

LILY TURF

L. muscari is similar to the above but taller. The dark green, strap-shaped leaves are about 1.5 cm broad and up to 45 cm long and have conspicuous parallel veins. In late summer and early autumn it produces spikes, shorter than the leaves, of deep mauve flowers. It is much cultivated in gardens, usually being placed amongst rocks or grown in pots, and several cultivars have arisen exhibiting variations in flower colour,

Figure 20.30 (right) Pots of *Liriope spicata* arranged around a bronze pagoda at the Baiyunguan, Beijing

Figure 20.31 (far right) *Lychnis senno* from *Curtis's Botanical Magazine* 1837, t.3594 (as *L. bungeana*)

leaf variegation, and so on. It is native to China, including Taiwan, and to Japan.

Due to my unfamiliarity with *L. graminifolia*, I have chosen to refer the larger liriopes I have seen in Chinese gardens to *L. muscari* (fig. 20.29), but as they were not in bloom I may well be mistaken.

L. spicata 土麦冬 (Tu Maidong) *Native Winter Wheat*
DWARF LILY TURF, CREEPING LILY TURF

This species occurs in China, Japan, and Vietnam. It is very similar to the other species but the leaves are usually no more than 35 cm long, and only 5–8 mm broad (fig. 20.30). It too has produced a range of cultivars. However, as with *L. muscari*, usually the typical form is cultivated in China. It is frequently seen as a component of an arrangement of potted plants or, less commonly, as a ground cover. Needham (1986) suggests that the '*mên tung*' mentioned in the literature of the 1st millenium BC was either this plant or *Asparagus lucidus*.

LYCHNIS Caryophyllaceae

Lychnis is now included in *Silene* (Greuter, 1995). However, since this change has not been generally adopted by gardeners at the time of my writing, I have retained the name *Lychnis* here.

L. coronata 剪夏萝 (Jianxialuo) *Cut Summer Plant*

The Chinese lychnis with orange-red petals toothed at their ends is usually referred to this species. Its Chinese name refers to the 'cut' margins of the petals. It is depicted on the bottom left of *The Jade Hall Peony* (fig. 2.7), painted in the 10th century. Thus it has attracted attention and perhaps been cultivated for at least 1000 years. Also it occasionally appears in later paintings of the 'flowers of the seasons' or 'hundred flowers' genres (e.g. Weidner et al., 1988, cat. nos 16, 34). Loureiro (1790) recorded it as a cultivated plant in Guangzhou in the 18th century and Clarke Abel (1819) saw it growing in pots at Tong Xian, near Beijing, in 1816 while accompanying Lord Amherst's embassy. More recently Li (1959) says that it is cultivated in Chinese gardens. On the other hand, Steward (1958), in his *Manual of Vascular Plants of the Lower Yangtze Valley China*, records it, along with *L. senno* (fig. 20.31), as a plant of grassland but makes no mention of either species as a cultivated ornamental. However, it is recorded that *L. senno* (*Curtis's Botanical Magazine* 1837, t. 3594, as *L. bungeana*) was collected by Bunge in a Beijing garden, so it seems that this plant is or was cultivated as well. It differs from *L. coronata* in having deep scarlet flowers with petals which are both bifid and laciniate. It blooms later than *L. coronata* and is known as *Jianqiuluo* (Cut Autumn Plant). However, David Mabberley (in press) has come to the conclusion that the east-Asian plants currently called *L. coronata*, *L. fulgens*, *L. senno*, and *L. sieboldii* are all forms of a variable species or hybrid group for which the valid name is *Silene banksia*.

LYCOPERSICON Solanaceae

L. esculentum (Solanum lycopersicum) 番茄 (Fanqie)
Foreign Eggplant, 西红柿 (Xihongshi) *Western Red Persimmon*

TOMATO

Originally from South America, the tomato has been taken round the world and is to be met with at meal times almost everywhere one goes. In China, however, it did not fit comfortably into the cuisine and does not appear to have become popular until recent times. Along with the melons, cucumbers, and so on grown in the traditional greenhouses of the north, it can now be made to ripen in the spring, in which season it is served as a cold entree, sliced and sprinkled with sugar. It also turns up frequently in the large bowls of soup which are served at the termination of almost every meal.

LYCOPODIUM Lycopodiaceae

Lycopodium spp. 石松 (Shisong) *Stone Pine*

CLUB MOSSES

The genus *Lycopodium* belongs to a family of primitive plants known as club mosses, often described as being allies of the ferns. In China one or more species known as *Shisong* have found favour with gardeners in the south. Clarke Abel (1819), for instance, wrote that 'The *Lycopodium*, which I had met with on barren hilltops in the province of Canton, growing to a height of four or five inches, and which might perhaps best be compared to a fir-tree in miniature, was to my surprise cultivated in pots kept in a tub filled with water; for it had always been found in very arid situations'. But then his travels took place in winter in an area where the rains fall chiefly in summer. Robert Fortune (1852) encountered what sounds from his description to be the same plant in cultiva-

tion in Messrs Dent & Co.'s garden in Hong Kong, saying it was the 'man-neen-chung', highly prized by the Chinese. He also said (Fortune, 1857) that parts of a garden at Ningbo 'were carpeted with the pretty little *Lycopodium caesium* I introduced to England some years ago'. Fortune (1846) records having sent this species and *L. wildenovii* to England from Hong Kong, but what these species really were I do not know. More recently Zhu (1992) mentions the growing of dwarf club mosses in Chinese gardens, and Hu (1982) records the use of *L. cernuum* for penjing, so it may well be that this is the species commonly grown. The name *Shisong* refers to the natural habitat and conifer-like appearance of the club mosses.

LYCORIS Amaryllidaceae

Usually known as spider lilies in the West, the members of the genus *Lycoris*, named for a mistress of Mark Antony, are the east-Asian equivalent of the South African genus *Nerine*. These bulbous perennials bear their flowers on naked scapes in late summer and early autumn before the leaves appear. According to Li (1959) the bulbs are deadly poisonous and hence the plants are known as Zhongkui flowers. Zhongkui is a fierce-looking immortal who fights demons.

I have not been in China at the flowering time of *Lycoris* species and have noticed plants only once in a garden, that of the Dajue Si west of Beijing. As this was in May and they were in leaf, I was not able to identify them with certainty. However, the species included below have been recorded as being cultivated in China (Li, 1959; Steward, 1958).

L. aurea 忽地笑 (Hudixiao) *Suddenly The Soil Smiles*

YELLOW SPIDER LILY

This species produces scapes up to 60 cm tall bearing umbels of comparatively large, yellow flowers with exserted stamens (fig. 20.32). Its Chinese name refers to its habit of quickly producing its flowers before the leaves. The spider lily shown in the *Song Hundred Flowers* in the collection of the Palace Museum, Beijing, looks very like this species, but as it has been drawn only with black ink it is difficult to be certain. However, its presence in this work suggests that it may have been cultivated at the time.

It was introduced to Britain in 1777 (Chittenden, 1956) and has become a popular garden plant in many parts of the world. It is also grown as a cut flower.

Figure 20.32 (right) *Lycoris aurea*

Figure 20.33 (below right) *Lycoris radiata*

L. radiata 石蒜 (Shisuan) *Rock Garlic*

RED SPIDER LILY

Smaller than the above in all its parts, *L. radiata* has bright red flowers with stamens which are very strongly exserted (fig. 20.33). It is known as *Shisuan* (Rock Garlic) as it grows amongst rocks and looks like garlic. The 'brilliant purple *Lycoris*' which Fortune (1847a) saw flowering in autumn amongst tombs at Shanghai was probably this species. Wang and Ma (1995) say the bulbs contain an alkaloid used in medicine and that, as well as growing the plants in gardens, people also arrange the cut flowers in vases.

L. sanguinea 铁色箭 (Tiesejian) *Rust-coloured Arrow*

L. sanguinea has dull red or reddish orange flowers, the stamens of which are not exserted. Whether it is ever grown in Chinese gardens nowadays I do not know.

L. squamigera 鹿葱 (Lucong) *Deer Onion*

With its large, pinkish-mauve flowers, the stamens of which are not exserted, this is an exceptionally beautiful species. It was first sent to Europe from Japan, and Li (1959) says it was introduced to American gardens around 1898 from China. Dr G. Hall of Rhode Island, who had it in his garden in Shanghai before 1860, is reported as saying that the Chinese often used to decorate their cemeteries with it (*Curtis's Botanical Magazine* 1897, t. 7547). It is also known as *Xiashui* (Summer Narcissus).

MALVA Malvaceae

M. sylvestris 欧锦葵 (Ou Jinkui) *European Splendid Kui*

TALL MALLOW

This erect biennial or perennial, native to Europe, North Africa, and south-west Asia, has upright stems to 1.2 m, rounded leaves with 3, 5, or 7 lobes, and axillary clusters of pinkish purple flowers up to 6 cm across, with notched petals and darker veins. It is cultivated as an ornamental in Chinese gardens and, according to Li (1959), it is called just *Jin Kui* (Brocade or Splendid Kui).

Figure 20.34 *Narcissus* bulbs on sale in Beijing

M. verticillata 野葵 (Ye Kui) *Wild Kui*

An annual or biennial, native to Europe and Asia, with stems to 1.8 m, rounded leaves with 5 or 7 shallow lobes, *M. verticillata* bears axillary clusters of small white or mauve flowers. It was an important vegetable in ancient times, being recorded with the name *Kui* in most of the important classics, including the *Shi Jing* (Book of Odes). After the Tang and Song dynasties it became less and less popular as a vegetable, and is probably now cooked only on rare occasions in outlying areas. It is, however, a common wild plant throughout the country and is occasionally cultivated for its flowers (Li, 1959). Its name, *Kui*, gradually became a general term for malvaceous plants.

MIRABILIS Nyctaginaceae

M. jalapa 紫茉莉 (Zimoli) *Purple Jasmine,* 胭脂花
(Yanzhihua) *Rouge flower*

FOUR O'CLOCK, MARVEL OF PERU

M. jalapa, an erect perennial from Peru, is recorded as a garden plant in the Yangtze delta region (Liu, 1982; Steward,

1958). The fragrant flowers open in the late afternoon and may be purple, crimson, yellow, or white, often striped or mottled. It is now grown in many parts of the world.

NARCISSUS Amaryllidaceae

N. tazetta subsp. *tazetta* 水仙 (Shuixian) *Water Immortal*

POLYANTHUS NARCISSUS

Of all the plants which are associated with the Chinese New Year, *N. tazetta* seems to lead the field. The plants are carefully forced into bloom in time for this festival, as it is believed that they will ensure good fortune for the ensuing 12 months. The bulbs are placed in special bowls and dishes, usually with pebbles or pieces of stone, and given nothing but water. No doubt this practice is responsible for the plant's Chinese name, *Shuixian*, which is usually translated as 'Water Fairy', though *xian* is better rendered as 'immortal', since it does not denote 'fairy' in the usual Western sense. Wherever the narcissus is depicted in Chinese arts and crafts, whether alone or with other emblems, it always suggests the deities of the Daoist pantheon, bringers of great happiness (Koehn, 1952).

As the end of the year approaches large quantities of bulbs, pebbles, and special dishes appear on the market (fig. 20.34), as it takes only a few weeks for the flowers to be produced. Nevertheless bringing them into bloom at the appropriate time requires considerable skill. And since only a single and a double kind are known, additional variations are induced by carving away part of the bulb so that the leaves and flower stems grow in a distorted manner. A large part of the bulb is cut away on one side before it is placed in a horizontal position over water with the excised area uppermost. The leaves then develop in a curved and crooked manner, and the flower stems remain short and are trained so that the mass of flowers occupies a central position amongst the weirdly shaped foliage. These plants are called 'crab's claw narcissi' (fig. 20.35). Sometimes the plants are trained to resemble a phoenix, peacock, elephant, or some other creature or object (fig. 20.36), or grown amongst stones and jagged rocks in elaborate penjing arrangements. Detailed descriptions of the cultural methods are given by Dorsett (1928), McClure (1932), and Xu and Li (1992).

Apparently the bulbs sold annually all over the country come from a small area in Fujian, where one small village once monopolised the industry. The bulbs are grown in raised beds, well spaced, and mulched with straw. They are 3 years old when marketed and at this stage consist of a very large central bulb, usually with two or more offsets (fig. 21.34).

Although *N. tazetta* appears to be growing wild in parts of Fujian and Zhejiang, and also in Japan, it is thought that it has probably been introduced from the West (Blanchard, 1990). This theory is supported by the fact that its cultivation in China is so local and that its old name, *Naiji*, must surely have been derived from the Persian *nargis* or the Arabic *narjis*, from which *Narcissus* probably also arose (Li, 1959). It seems likely that it was brought during the Song by Arab traders from western Asia to Fujian, where the chief trading port of the time was situated, since it first appeared in paintings and in the literature at this time. The first poet to have made it well-known was Huang Tingjian (1045–1105). It became associated poetically with the two goddesses of the Xiang River, and this association in turn linked it with Qu Yuan (343–377 BC), the loyal minister of the state of Chu who, having failed to win the ear of his prince, drowned himself in the Miluo River, a tributary of the Xiang (Fong & Fu, 1973). And amongst the most remarkable of the early paintings of it is an extraordinary Song handscroll by Zhao Mengjian (1199–c.1267), which depicts a sea of flowering plants extending horizontally for 3.731 m (Fong & Fu, 1973, no. 12). Since then it has been depicted innumerable times, invariably it seems as the single form (fig. 2.12). It may be that the double cultivar was a later introduction, perhaps having arrived only relatively recently.

Figure 20.35 (above right) Double *Narcissus tazetta*, **trained in the 'crab's claw' manner; from Xu & Li (1992)**

Figure 20.36 (right) *Narcissus tazetta* **trained to resemble a teapot; from Xu & Li (1992)**

Figure 20.37 (far right) *Narcissus tazetta*, **double form, at the Taihua Si, Western Hills, Kunming**

Figure 20.38 (below left) *Orychophragmus violaceus* in a hotel garden at Jiaxing

Figure 20.39 (left) *Orychophragmus violaceus* beneath old junipers and thujas at the Temple of Heaven, Beijing

OPHIOPOGON
Convallariaceae

O. japonicus 麦冬 (Maidong)
Winter Wheat

MONDO GRASS, DWARF LILY TURF

O. japonicus forms dense clumps of dark green, grass-like leaves up to 30 cm long and 3–4 mm broad, and in summer it bears short racemes of inconspicuous, white or pale lilac flowers. It is sometimes called *Yanjiecao* (Bordering-the-steps Grass), and might well be called 'Bordering-everything Grass', as in almost every garden it appears fulfilling this role (figs 12.9, 13.18, 19.8, 19.65). It can also be seen planted in this manner in an 18th century painting (She & Yuhas, 1973, no. 48). Although indigenous to north-east Asia, it is now grown in similar situations in many parts of the world. In China it also features in tray gardens and landscapes. In addition its tuberous roots are among the famine foods listed in the *Jiuhuang Bencao* (Bretschneider, 1881) and, preserved in sugar, are used in medicine (*Curtis's Botanical Magazine* 1807, t. 1063).

ORYCHOPHRAGMUS Cruciferae

O. violaceus 二月兰 (Eryuelan) *Second Moon Orchid*

Although rarely mentioned by anybody elsewhere, in Chinese gardens *O. violaceus* (fig. 20.38) shares the ubiquity of bamboo, mondo grass, osmanthus, and many other favourite plants, at least from the Yangtze valley north. It is an annual or biennial herb which in bloom bears a general resemblance to honesty (*Lunaria annua*) and sweet rocket (*Hesperis matronalis*). Like them, more often than not it is self-sown and has, for instance, established itself over considerable areas in the gardens of Beijing, transforming the ground beneath the old junipers and thujas into a carpet of mauve in spring (fig. 20.39). It is particularly impressive in

As it occurs in China, this narcissus has been described as a separate variety of *N. tazetta* or even as a separate species, but there seems to be no good botanical reason for regarding it as distinct from subsp. *tazetta* of this variable plant. The single variety occurring in China is a broad-petalled white with a yellow cup (fig. 20.36), which Edouard André (1892) thought very close to the European cultivar 'Grand Monarque'. The double cultivar (figs 20.35, 20.37) appears identical with the old double form widely distributed in other parts of the world.

the area surrounding the Temple of Heaven, as it must once have been on the site of the nearby Temple of Agriculture. Juliet Bredon (1931), who knew this area before it was 'tidied up', regretted the disappearance of 'a wild waste of grassland carpeted in summer with mauve wild turnip flowers'.

O. violaceus was also admired in Beijing by Dorothy Graham (1938), one of the few other people to mention it. She recorded that it was called 'second month orchid' and that in Behai Park it completely covered the hillside, rippling like silk in the breeze, as it still does today. She also noted that the Manchus were said to have brought this flower from the steppes when they became rulers in Beijing and that, undulating in the wind, it evoked wide spaces—plains extending to the far horizons. Huxley et al. (1992) record that the leaves are eaten as a vegetable, but it seems unlikely that this is common nowadays. Although it became known in Europe from dried specimens collected by Bunge, it appears not to have been seen alive in Britain until it flowered in 1876 at Kew, having been raised from seeds sent from Beijing (*Curtis's Botanical Magazine* 1876, t. 6243).

OXALIS Oxalidaceae

O. articulata 紫花酢浆草 (Zihua Zuojiangcao) *Purple-flowered Vinegar-juice Plant*

For those of us used to regarding *Oxalis* species with the appearance of *O. articulata* with alarm, it comes as a surprise to find it planted as a neat border or used as a ground cover in Chinese gardens. However, it is not invasive, as it is a perennial with a semi-woody, tuberous rootstock. It produces its bright mauve-pink flowers in late spring and I have seen it used extensively in gardens in Nanjing and Yangzhou (figs 12.9, 20.40). When this South American plant was introduced is unclear.

Figure 20.40 *Oxalis articulata* in the Xi Yuan, Nanjing

O. corymbosa 铜锤草 (Tongchuicao) *Copper Hammer Plant*, 三叶醋 (Sanyecu) *Three-leaved Vinegar*

Although Steward (1958) records *O. corymbosa* as a cultivated ornamental in the lower Yangtze valley, I have not seen it in Chinese gardens. It may perhaps be that Steward wrongly identified what was actually the superficially similar *O. articulata*, as *O. corymbosa* is also a South American species with mauve-pink flowers. However, as this bulbous species produces numerous small bulbils around the parent bulb, it has become a troublesome weed in many parts of the world. Its Chinese name, *Tongchuicao*, appears to have something to do with the shape of the leaves and the spots on their undersurfaces. *O. corymbosa* is now included in *O. debilis* (Mabberley, 1997a), but I have retained the name here as it is with this designation that it is to be found generally in the literature.

PANAX Araliaceae

P. ginseng 人参 (Renshen) *Human Tonic*

GINSENG

While ginseng is not found in Chinese gardens, no visitor to the country can fail to be unaware of this plant, the roots of which are regarded as the ultimate panacea. These roots are displayed in every traditional pharmacy, and pills and potions purporting to contain it are on sale everywhere.

Ginseng is a herbaceous perennial of the ivy family which has a carrot-shaped root, branching when mature. The root sometimes resembles a human being, hence the Chinese name. It is native to north-eastern China and Korea, which are the chief sources of supply, though this is augmented by importations of *P. quinquefolius*, American Ginseng.

PAPAVER Papaveraceae

P. rhoeas 虞美人 (Yumeiren) *Beauty Yu*, 丽春花 (Lichunhua) *Beautiful Spring Flower*

CORN POPPY, FLANDERS POPPY

No doubt this annual poppy from the Old World has become popular in China on account of its bright red colour (fig. 20.41). When it was introduced is not clear, but it is likely that it arrived as a weed seed long ago. It has probably been cultivated in China for over 300 years, as it appears in paintings at least as early as the 17th century (e.g. Bartholemew, 1985a, cat. no. 7; Weidner et al., 1988, cat. no. 34) and is included in the *Jieziyuan Huazhuan* of 1701 (Sirén, 1949). The young leaves and the seeds can be eaten in times of famine according to the *Jiuhuang Bencao* (Bretschneider, 1881).

The name *Yumeiren* (Beauty Yu) was given to *P. rhoeas* in honour of the favourite concubine of the famous warrior Xiang Yu (233-202 B.C.). She went into battle with him and committed suicide after his death. From the ground saturated with her blood the corn poppy is said to have sprung up (Bartholemew, 1985b). *P. rhoeas* is sometimes also called *Wucao* (Dancing Plant) on account of the way the petals flutter in the wind.

P. somniferum 罂粟 (Yingsu) *Jar and Millet*, 大烟花 (Dayanhua) *Big-tobacco Flower*, 鸦片花 (Yapianhua) *Opium Flower*

OPIUM POPPY

It is unclear when and where the double, ornamental varieties of the opium poppy first appeared, but it seems probable that they arose somewhere in Asia and spread from there (Coats, 1968). There seems to be little mention of them in the Chinese literature, but Williams (1975) says that *P. somniferum* was first grown in China for ornamental purposes and that its medicinal properties were made known subsequently by Mohammedan merchants who came from central Asia. Typical double forms are depicted in a painting

of the 13th century reproduced in Sirén (1949, facing p.30), and in the *Song Hundred Flowers* from the same period in the collection of the Palace Museum, Beijing. Mauve, pink, red, and white doubles appear in many subsequent paintings (e.g. Barnhart, 1983, cat. no. 33; Bartholemew, 1985a, cat. no. 8k; Weidner et al., 1988, cat. nos 34, 39). Hence it seems that the ornamental kinds have been cultivated in China for at least several centuries. It seems to have been regarded chiefly as a food and ornamental plant, opium being largely imported into the country from India from the 18th century on (Grey-Wilson, 1993).

Ying is a small-mouthed jar and *su* is a millet seed, hence the name *Yingsu* refers to the shape of the capsule and the small seeds it contains. *Dayanhua* (Big-tobacco Flower) refers to the fact that the drug is more powerful than tobacco, and *Yapianhua* is a name based on the pronunciation of the word 'opium'.

PERSICARIA Polygonaceae

P. orientalis (*Polygonum orientale*) 水红花子 (Shuihonghuazi) *Water [Knotweed] with Red Flower and Fruit*, 红蓼 (Hong Liao) *Red Knotweed*

PRINCE'S FEATHER

This and many other plants long regarded as being species of *Polygonum* are these days considered to be placed more appropriately in the genus *Persicaria*. *P. orientalis*, native to east Asia and Australia, is an annual herb to 2 m tall with broadly ovate leaves and dense, drooping spikes of pink flowers. In Chinese gardens it is often planted beside ponds. It sometimes appears in paintings, at least from the 17th century on, often as part of a composition in which the principal feature is the lotus (e.g. Barnhart, 1983, cat. no. 28; She & Yuhas, 1973, no. 44), and is included in the *Jieziyuan Huazhuan* (Sze, 1977). The leaves can be eaten as a famine food according to the *Jiuhuang Bencao* (Bretschneider, 1881).

PHYSALIS Solanaceae

P. alkekengi 酸浆 (Suanjiang) *Sour Paste*, 灯笼花 (Denglonghua) *Lantern Flower*, 红姑娘 (Hong Guniang) *Red Girl*

CHINESE LANTERN

This perennial occurs in central and southern Europe and across Asia to Japan. From the flowers produced in the axils of the leaves, round red fruits like small tomatoes develop,

Figure 20.41 *Papaver rhoeas* at Tiger Hill, Suzhou

each enclosed in an inflated calyx which becomes orange or bright red in late summer and autumn. The resemblance of this calyx to a Chinese lantern accounts for the plant's English common name. Together with its close relatives, the ornamental-fruited species of *Capsicum* and *Solanum*, it is, or at least has been, a popular plant in China, no doubt due to the colour of its fruits and their calices. These fruits, along with the leaves, were listed as famine foods in the *Jiuhuang Bencao* (Bretschneider, 1881).

PLATYCODON Campanulaceae

P. grandiflorus 桔梗 (Jiegeng)

BALLOON FLOWER, CHINESE BELLFLOWER

Closely resembling its near relatives the campanulas, this perennial from north-east Asia produces balloon-shaped buds which open to become widely bell-shaped blue flowers. It has been cultivated since early times in China and Japan, where it has often been depicted in paintings. According to Wilson (1913), an extract made by boiling the rootstock of *P. grandiflorus*, 'a campanulaceous plant known locally as

"Chieh K'eng", is a cure for chill in the stomach'. Like many blue flowers it has given rise to both white (fig. 20.42, see over) and pink variants, as well as double forms. Robert Fortune, for instance, found a white form in nursery gardens near Shanghai and sent it to Britain in 1845 (Lindley, 1846). This had hose-in-hose flowers which appeared like a star with ten points. He sent the ordinary blue form as well (Lindley & Paxton, 1853), although this had reached Europe much earlier (Coats, 1968). According to the *Jiuhuang Bencao* the leaves can be eaten in emergencies (Bretschneider, 1881).

POLIANTHES Agavaceae

P. tuberosa 晚香玉 (Wanxiangyu) *Evening Fragrant Jade*

TUBEROSE

The tuberose was introduced to the rest of the world from Mexico, where it is known only as a cultivated plant. It was used in pre-Columbian times as a flavouring for chocolate (Huxley et al., 1992). How and when it got to China is unclear, but it became appreciated along with the jasmines and other white-flowered, strongly scented plants, for similar

Figure 20.42 A white form of *Platycodon grandiflorum*

reasons. John Barrow (1804) noted it in Guangzhou in late 1793, and Robert Fortune (1847a, 1852) mentioned seeing it worn by the women of Fuzhou for the decoration of their hair. Steward (1958) records the tuberose as a cultivated ornamental in the lower Yangtze valley, though it is doubtful whether it is grown there much nowadays. A painting of what appears to be the single form can be seen in an 18th century *Hundred Flowers* reproduced by Capon and Pang (1981, cat. no. 65). There is also a double form. Since the tuberose has been confused by Western observers with *Hosta plantaginea*, as mentioned earlier, records of its occurrence in China should be treated with caution.

POLYGONUM—see *Persicaria*

PRIMULA Primulaceae

As with *Rhododendron* and various other genera with numerous, highly ornamental, indigenous species, the primulas seem to have attracted little attention from Chinese gardeners. *P. sinensis* seems to have been the only species brought into cultivation in earlier times, though these days *P. malacoides* is more commonly seen, apparently having become popular after its introduction to the West early this century. Both species flower in winter and early spring and, although perennial, they are usually grown as annuals.

P. malacoides 报春花 (Baochunhua)
Announcing Spring Flower

In China *P. malacoides* is usually grown in pots for display with various other flowering and foliage plants. The original, narrow-petalled, mauve-pink form was the type most commonly grown at the time of my visits. *Baochunhua* is now also the general name for members of the genus *Primula*.

P. sinensis 藏报春 (Zang Baochunhua)
Superb Announcing Spring Flower

CHINESE PRIMROSE

This species is grown as a pot plant in China and is apparently no longer known in the wild. The form usually seen has pale mauve flowers. Attention was drawn to this plant in Britain in 1819 when a painting of it was sent to the Horticultural Society of London by John Reeves, and consequently it was introduced from Guangzhou (Synge, 1953; Whitehead

Figure 20.43 *Primula sinensis* at the Baoguang Si, Chengdu

& Edwards, 1974). Subsequently many colour variants were developed which became popular plants all over the world, particularly for displays under glass. Like *P. malacoides* it flowers in winter and early spring in climates which suit it. Morrison (1822) listed it as flowering in Guangzhou in the 11th moon (December). I saw the typical mauve form blooming out of doors in Chengdu, Sichuan, in January (figs 7.3, 20.43). According to Li (1959) it is also known simply as *Baochun* (Announcing Spring).

REINECKEA Convallariaceae

R. carnea 吉祥草 (Jixiangcao) *Good Luck Plant*

R. carnea is native to China and Japan and when not in flower closely resembles *Liriope muscari*. However, the flowers are produced in early summer, are a dusty, pale pink and, unlike those of the liriopes, open widely. They are followed by round scarlet berries. As its name suggests it is one of the many evergreen plants reputed to bring good fortune and long life. According to Li (1956a) the flowers appear so infrequently that the blooming of the plant is seen as a good omen. As with many plants of this general appearance, it is either planted in the ground or used as a pot plant for bringing indoors. There is a variegated form with leaves that are striped with cream.

ROHDEA Convallariaceae

R. japonica 万年青 (Wannianqing) *Ten Thousand Years Green*

In general appearance *R. japonica* is similar to *Aspidistra elatior*. The thick leaves are up to 45 cm long and 7–8 cm broad, and the curious, greenish-white flowers are borne in short dense spikes in late spring (fig. 20.44). These flowers are pollinated by slugs (Hattori & Grounds, 1994) and are succeeded by bright red berries, the fruiting spikes resembling those of *Arum italicum*. It is native to both China and Japan.

R. japonica is of interest as it was the national flower of the Manchus (Steward, 1958), and is yet another of the evergreen plants associated with good fortune. It is almost always grown in pots and is a traditional gift for housewarmings, birthdays, and the openings of new businesses. On account of its Chinese name it becomes involved in many

Figure 20.44 *Rohdea japonica* at the Lanting, Shaoxing

rebuses, having a multiplying effect on good wishes, as pointed out in Chapter 2. New restaurants and shops are always full of *Wannianqing* or its modern substitute, the philodendron, which has been given the same Chinese name, as they suggest 'ten thousand years of flourishing business' (Bartholemew, 1985a). How long *R. japonica* has had this significance is uncertain, but it appears in paintings and other works of art, at least from the 18th century on (e.g. Bartholemew, 1985a, cat. no. 7; Fourcade, 1965, painting no. 49). In the Nanjing Museum there is a Qing dynasty replica with leaves made of green jade and berries of coral.

R. japonica has prompted extraordinary enthusiasm in Japan, where it has been cultivated for over 500 years and where approximately 1500 varieties and cultivars have been named, including forms with twisted, variegated, or curled leaves (Hattori & Grounds, 1994).

SALVIA Labiatae

S. splendens 一串红 (Yichuanhong) *A String of Red*

SCARLET SALVIA

Although Brazilian, the scarlet salvia is now so commonly seen in China that I have felt obliged to include it. Presumably it is on account of the colour of its flowers that it is so widely planted. Although it is a perennial sub-shrub, it is usually cultivated as a herbaceous annual. Like so many herbaceous plants, it is generally grown in pots and included in massed displays (fig. 19.10).

SAXIFRAGA Saxifragaceae

S. stolonifera 虎耳草 (Huercao) *Tiger's Ear Plant*

MOTHER OF THOUSANDS, STRAWBERRY GERANIUM

S. stolonifera is a perennial Chinese plant with round or kidney-shaped leaves which are reddish beneath and have silvery veins above. It spreads by means of long, thin stolons and in summer produces large panicles of tiny white flowers, spotted with yellow or red. It seems to have attracted attention during the Qing dynasty and appears in works of art of relatively late date (e.g. fig. 20.14; Weidner et al., 1988, cat. nos 45.3, 50).

It was introduced to Britain from China in 1770 and its variegated form, 'Tricolor', was probably the saxifrage so highly prized by Robert Fortune that he brought it back under his personal care in 1862 (Coats, 1968). I have

Figure 20.45 *Sedum spectabile*

not come across it in Chinese gardens, so perhaps it is rarely cultivated there these days. It also occurs in Japan.

SEDUM

The two plants listed here are herbaceous perennials native to eastern Asia and grown as garden plants there. They are included by some in the genus *Hylotelephium*. Both species annually produce upright, unbranched stems, 30–70 cm tall, bearing opposite or whorled, fleshy, grey-green leaves, and flat heads of small, star-shaped flowers in summer. Presumably it is these plants which are included in the *Jiuhuang Bencao* with the advice that the leaves are edible (Bretschneider, 1881).

S. erythrostictum 八宝 (Babao)
Eight Treasures

Less well-known in the West than *S. spectabile*, this species tends to be the taller of the two and to produce larger flower-heads, up to 15 cm across. The flowers are greenish white with pink centres.

S. spectabile 长药八宝 (Changyao Babao)
Long-stamen Sedum

ICE PLANT

S. spectabile has become a popular garden plant in the West, where it is generally known as Ice Plant on account of its fleshy, bluish-green or grey-green leaves. The flowers are usually pink (fig. 20.45), but white forms are also known and several cultivars have been named.

Li (1959) notes that in China *S. spectabile* is much planted amongst rocks in courtyards. He also says it is grown in pots to be placed on roofs, as since ancient times it has been believed to offer protection against fire. For this reason it is also called *Shenhuo* (Guarding Against Fire) and other similar names.

SILENE—see *Lychnis*

SOLANUM Solanaceae

As well as the eggplant (*S. melongena*) and the potato (*S. tuberosum*), which are cultivated as food plants, several species of *Solanum* are grown in China as ornamentals. Those with bright red fruits have become popular, particularly as symbols of good fortune for the New Year (Metcalf, 1942). Although they are really small shrubs they are usually grown in pots as annuals, hence I have included them here.

S. aculeatissimum 牛茄子 (Niu Qiezi) *Cow Eggplant*

S. aculeatissimum is an American species introduced early into the Old World tropics (Bailey, 1976). It is a soft shrub to about 1 m tall but is usually treated as an annual in China, a procedure which allows it to be grown out of doors at least as far north as Beijing. It has small, white flowers which are usually borne singly in the axils of the leaves, which are lobed

Figure 20.46 *Solanum aculeatissimum* in Behai Park, Beijing

and bear spines on the veins. The flowers are followed by bright red fruits up to 5 cm in diameter. These are ridged and lobed in the manner of some tomato cultivars. This species is occasionally seen planted in garden beds (fig. 20.46).

S. capsicastrum 珊瑚豆 (Shanhudou) *Coral Bean*

FALSE JERUSALEM CHERRY

Although it forms a small, much-branched shrub, this species from Brazil is also grown as an annual, usually as a pot plant for winter decoration. It does not bear spines, the leaves are ovate to lanceolate, unlobed but with undulate margins, and it bears abundant, small, orange-scarlet fruits. These are ovoid, usually somewhat pointed, and about 2.5 cm long.

S. pseudocapsicum 珊瑚樱 (Shanhuying) *Coral Cherry*

JERUSALEM CHERRY

The Jerusalem Cherry is very similar to *S. capsicastrum* but its long-lasting, orange-red fruits are spherical and about

1.5–2 cm in diameter. Apparently an Old World species, it is now widely naturalised in the warmer regions. This is the ornamental solanum most frequently seen in China, usually as a pot plant.

TAGETES Compositae

Here belong the French and African marigolds, *T. patula* and *T. erecta*, which gained their misleading names as a result of routes by which they reached Britain, even though they must first have been brought to Europe from their native Mexico by the Spanish. It seems that *T. erecta* was early taken to India by the Portuguese who, you may remember, had established a trade route from eastern Brazil to Goa as early as 1500 and from whom it was taken over by the Hindus (Merrill, 1954). It may have reached China from there or perhaps it arrived via Manila, after being brought there by the Spanish direct from Mexico. Whatever the truth may be, both species, or at least their hybrids and derivatives, are now very popular in China, no doubt for their auspicious yellow and orange colours. Potted plants are often seen in outdoor displays, either alone or with other plants, and low growing kinds are used to plant up the large dragons and other creatures made of wire and mud which are such striking features of many public spaces.

T. erecta 万寿菊 (Wanshouju)
Ten Thousand Birthdays Chrysanthemum

AFRICAN MARIGOLD

T. erecta is a tall, fast-growing annual which has angular green stems and yellow or orange flowers (figs 6.15, 17.7). As its Chinese name suggests, it has become endowed with auspicious symbolism. This is due in part to the association of the colour yellow with the emperor in days gone by. It occasionally appears in paintings, for example a Qing handscroll reproduced by Suzuki (1982, vol.3, JM1–075).

T. patula 孔雀草 (Kongquecao) *Peacock Plant*

FRENCH MARIGOLD

T. patula is usually smaller in all its parts than *T. erecta*, with cylindrical, purplish stems and often with some red-brown colouration in the flowers. Like *T. erecta*, it occasionally appears in paintings. The earliest I have seen is in a 17th century *Hundred Flowers* illustrated by Weidner et al. (1988, cat. no. 34).

TROLLIUS Ranunculaceae

T. chinensis 金莲花 (Jinlianhua) *Gold Lotus Flower*

T. chinensis is a species from northern China and adjoining regions. It can reach 1 m in height and has the much-divided leaves and bowl-shaped, golden flowers typical of the genus. It is included amongst the garden flowers of China by Li (1959). I have not come across it, so it seems likely that it is rarely planted now.

TROPAEOLUM Tropaeolaceae

T. majus 旱金莲花 (Han Jinlianhua) *Dryland Golden Lotus Flower*

NASTURTIUM

When and by what route the nasturtium reached China from South America I do not know. As its Chinese name suggests, the yellow-flowered types are the most popular. It is an annual and at present in China one usually encounters it in displays of potted plants of a dwarf form (fig. 17.7). It gets its Chinese name as, despite its lotus-like leaves, it grows on dry land and has yellow flowers.

VIOLA Violaceae

Viola hybrids 三色堇 (Sansejin) *Three-coloured Violet,* 蝴蝶花 (Hudiehua) *Butterfly Flower*

PANSY

Owing to the present popularity of pansies in China, I have included them here. The modern pansies are derived from complex hybrids of *V. tricolor* and other species made during the 19th century. Hence the pansies one sees in China today must be relatively recent arrivals. They are treated as annuals or biennials, usually grown in pots for spring display.

ZINGIBER Zingiberaceae

Z. officinale 姜 (Jiang)

GINGER

Z. officinale is a plant characteristic of its family, the leaves being borne in two parallel ranks on unbranched reed-like stems about 1 m tall. These arise from the knobbly, branching rhizomes which provide the ginger of commerce. It rarely flowers and is always propagated vegetatively (Larkcom, 1991). Being a tropical plant it is grown principally in the south, particularly in Guangdong. Meyer (1916) says that several varieties are known.

The exact origin of ginger, which is widely cultivated in tropical Asia, is unknown. I have included it here because, like the lychee, it has so long been associated with China by Westerners. This is due principally to the export from the south of the country of pieces of the rhizome preserved in syrup and packed in the well-known ginger jars, very often dark blue and decorated with white plum blossom (fig. 6.7). In China itself ginger is offered for sale all over the country and is one of the most common flavourings, being incorporated in many dishes.

In addition to the rhizomes, the young shoots are eaten as a delicacy, being cut when they are about 7.5 cm high. They are sometimes marinated in vinegar, sugar, and sesame oil (Larkcom, 1991).

GLOSSARY

anther: pollen-producing part of the stamen

axil: angle between the upper surface of a leaf and the stem bearing it

bifid: divided at one end into two parts by a cleft

bipinnate: of leaves where both primary and secondary divisions are pinnate

bract: modified protective leaf associated with the inflorescence, with buds, and with newly emerging shoots

calyx: outer whorl of flower parts made up of sepals, which are often green and protect the flower in bud

clone: population of genetically identical individuals produced by propagation from a single parent

column: feature of orchids—style and stamens fused together in a single structure

cordate: heart-shaped

corolla: petals of a flower (collective term)

corymb: racemose inflorescence with the lower flower stalks longer than those above, so that all the flowers are at the same level

culm: aerial stem of a grass

cultivar: assemblage of cultivated plants, retaining distinguishing features when reproduced

digitate: diverging from a central point, like the fingers on a hand

elliptic: oval, widest in the middle, with two planes of symmetry

exserted: projecting beyond the surrounding parts

filament: stalk of a stamen

glabrous: without hairs

inarch: approach graft

inflorescence: group of flowers borne on one stem

internode: portion of a stem between two successive nodes

laciniate: margin irregularly cut into narrow slender teeth

lanceolate: lance-shaped, narrowly ovate, 3–6 times as long as broad and tapering to a point

leaflet: separate portion of a compound leaf

node: portion of a stem from which one or more leaves, bracts, branches, or flowers arise laterally

obconical: cone-shaped but attached at the narrower end

orbicular: with a circular or almost circular outline

ovary: central portion of a flower containing the ovules, developing after fertilisation into the fruit

ovate: shaped like a longitudinal section of an egg, widest at one end

ovoid: egg-shaped, and attached by the broader end

ovule: body inside the ovary which becomes the seed after fertilisation

palmate: having several lobes or leaflets radiating from a central point, like the fingers of a hand

panicle: branched racemose inflorescence

pedicel: stalk of an individual flower in an inflorescence

peduncle: stalk of an inflorescence or of a solitary flower

pentamerous: with parts in groups of five

petal: one of the parts forming the corolla of the flower, usually brightly coloured and conspicuous

petiole: stalk of leaf

pinnate leaf: compound leaf, with leaflets arranged on opposite sides of a common rachis, as in *Wisteria*

pinnatisect: describes a leaf blade cut into lobes on both sides down to or almost to the midrib

pollard: to cut the branches of a tree back to the trunk

pollen: mass, usually powdery, of reproductive bodies produced by the stamens

pome: fruit of the type exemplified by the apple and pear

pseudobulb: thickened bulb-like stem found in many orchids, for example *Cymbidium*

pubescent: covered with short soft hairs

raceme: inflorescence of stalked flowers, the youngest at the apex, borne on unbranched, elongated axis, as in *Cymbidium* and *Wisteria*

racemose: of flowers borne in a raceme

rhizome: underground stem

rugose: wrinkled

scape: leafless peduncle arising from near the ground

sepal: one of the parts or lobes of the calyx

serrate: describes a margin with apically directed teeth resembling those of a saw

sessile: lacking a stalk, or apparently so, and thus close to the body from which it arises

sinus: depression or notch between two adjacent lobes

stamen: flower structure producing the pollen; consists of an anther, usually borne on a stalk, the filament

stigma: structure borne by the style adapted to receive the pollen grains

stolon: horizontally growing stem producing roots at the nodes

stoloniferous: producing stolons

style: stalk rising from the ovary, bearing the stigma

tomentose: covered with closely matted short hairs

trifoliate: describes a compound leaf consisting of three leaflets

truncate: terminating abruptly as if cut off transversely

umbel: racemose inflorescence of stalked flowers in which all the pedicels arise at the tip of the peduncle and the flowers lie at the same level

umbellate: with the characters of an umbel

undulate: wavy

Chrysanthemums at the Temple of Heaven, Beijing

REFERENCES

Abel, C. 1819. *Narrative of a Journey in the Interior of China and of a Voyage to and from that Country in the Years 1816 and 1817*. Longman, Hurst, Rees, Orme and Brown, London.

Aleksinas, D.A. 1995. Chinese Tree Peonies. *Green Scene* May/June 1995: 9–12.

Andoh, Y. 1972. Camellia reticulata as illustrated in ancient pictorial books in Japan. *New Zealand Camellia Bulletin* 7(4): 8–20.

André, E. 1892. Le Narcise Sacré De Chine. *Revue Horticole* 1892: 198–9.

Andrews, S. 1994. Tree of the Year: *Zelkova*. *International Dendrology Society Yearbook* 1993: 11–30.

—— 1996. Tree of the Year: *Ginkgo biloba*. *Ibid.* 1995: 14–31.

Arlington, L.C., and Lewisohn, W. 1935. *In Search of Old Peking*. Paragon Book Reprint Corp., New York, 1967.

Bailey, L.H., and Bailey, E.Z. 1976. *Hortus Third*. Revised and expanded by staff of Liberty Hyde Bailey Hortorium. Macmillan, New York.

Baker, G.P. 1937. Hemerocallis: The Day Lily. *Journal of the Royal Horticultural Society* 62(9): 399–411.

Barnhart, R. 1972. *Wintry Forests, Old Trees: Some Landscape Themes in Chinese Painting*. China Institute in America, New York.

Barnhart, R.M. 1983. *Peach Blossom Spring: Gardens and Flowers in Chinese Painting*. Metropolitan Museum of Art, New York.

Barrow, J. 1804. *Travels in China*. T. Cadell and W. Davies, London.

Bartholemew, B. 1982. *Tiannan Chahua Xiaozhi* and the old *Camellia reticulata* cultivars from Yunnan, China. *American Camellia Yearbook* 1982: 147–55.

Bartholemew, T.T. 1980. Examples of Botanical Motifs in Chinese Art. *Apollo* 112(221): 48–54.

—— 1985a. *The Hundred Flowers: Botanical Motifs in Chinese Art*. Asian Art Museum of San Francisco.

—— 1985b. Botanical Puns in Chinese Art from the Collection of the Asian Art Museum of San Francisco. *Orientations* 16(9): 18–34.

Bean, W.J. 1970–1988. *Trees and Shrubs Hardy in the British Isles*. 8th edn. Vol.1, 1970; Vol.2, 1973; Vol.3, 1976; Vol. 4, 1980; Supplt, 1988. John Murray, London.

Bickford, M. 1985. *Bones of Jade, Soul of Ice: The Flowering Plum in Chinese Art*. Yale University Art Gallery, New Haven, Connecticut.

—— 1996. *Ink Plum: The Making of a Chinese Scholar Painting Genre*. Cambridge University Press.

Blanchard, J.W. 1990. *Narcissus*. Alpine Garden Society, Woking, Surrey.

Blofield, J. 1961. *City of Lingering Splendour*. Hutchinson & Co., London.

Bodde, D. 1991. *Chinese Thought, Society, and Science*. University of Hawaii Press, Honolulu.

Booth, W.B. 1830. History and Description of the Species of Camellia and Thea; and of the Varieties of the Camellia Japonica that have been imported from China. *Transactions of the Horticultural Society of London* 7: 519–62.

Borrell, O.W. 1996. *Flora of the Shanghai Area*. Vols I &2. O.W. Borrell, Bulleen, Victoria.

Box, E. 1902. Shanghai Folk-Lore. *Journal of the China Branch of the Royal Asiatic Society* 34: 101–35.

—— 1905. Shanghai Folk-Lore. *Ibid.* 36: 130–56

Bredon, J. 1931. *Peking*. 3rd edn. Kelly and Walsh Ltd, Shanghai.

—— and Mitrophanov, I. 1927. *The Moon Year*. Paragon Book Reprint Corp., New York, 1966.

Bretschneider, E. 1880. Early European Researches into the Flora of China. *Journal of the North China Branch of the Royal Asiatic Society* n.s. 15: 1–194.

—— 1881. *Botanicon Sinicum*. Part 1. *Ibid.* n.s.16: 18–230.

—— 1893. *Botanicon Sinicum*. [Part 2] *Ibid.* 25: 1–468.

—— 1895. *Botanicon Sinicum*. Part 3. *Ibid.* 29: 1–623.

—— 1898. *History of European Botanical Discoveries in China*. Sampson Low, Marston and Co., London.

Brickell, C. (Editor-in-Chief). 1996. *The Royal Horticultural Society A–Z Encyclopedia of Garden Plants*. Dorling Kindersley, London.

—— and Mathew, B. 1981. *Daphne, the Genus in the Wild and in Cultivation*. Alpine Garden Society, Woking, Surrey.

Bridge, A. 1932. *Peking Picnic*. Little, Brown, and Co., Boston.

Brummitt, R.K. 1997. *Chrysanthemum* once again. *The Garden* 122(9): 662–3.

Buckley, M., Samagalski, A., Storey, R., Taylor, C., and Lindenmeyer, C. 1994. *China*. 4th edn. Lonely Planet Publications, Hawthorn, Victoria.

Burkill, I.H. 1946. On the Dispersal of the Plants most intimate to Buddhism. *Journal of the Arnold Arboretum* 27(4): 327–39.

But, P.P.-h., Chia, L.-c. and Fung, H.-l. 1985. *Hong Kong Bamboos*. The Urban Council, Hong Kong.

Cahill, J. 1978. *Parting at the Shore*. Weatherhill, New York and Tokyo.

Callaway, D.J. 1994. *The World of Magnolias*. Timber Press, Portland.

—— (ed.). 1998. Magnolia Society Awards its First Research Grant. *Magnolia Magazine* 5(1): 1.

Cantoniensis. 1867. Grafting. *Notes and Queries on China and Japan* 1(11): 157–8.

Capon, E., and Pang, M.A. 1981. *Chinese Paintings of the Ming and Qing Dynasties*. International Cultural Corporation of Australia Ltd.

Carrière, E.-A. 1879. *Armeniaca davidiana*. *Revue Horticole* 1879: 236–7.

Chamberlain, D.F., and Rae, S.J. 1990. A revision of *Rhododendron*. IV Subgenus *Tsutsusi*. *Edinburgh Journal of Botany* 47(4): 89–200.

Chang, H. T., and Bartholemew, B. 1984. *Camellias*. B.T. Batsford, London.

Chen, S.-c., and Tang, T. 1982. A General Review of the Orchid Flora of China. In Arditti, J. (ed.) *Orchid Biology, Reviews and Perspectives* 2: 39–81. Cornell University Press, Ithaca and London.

Chen, S.-l., and Chia, L.-c. 1988. *Chinese Bamboos*. Science Press, Beijing, and Dioscorides Press, Portland.

Chen, Y. 1953. Illustrated Manual of Chinese Trees and Shrubs (*Zhonguo Shumu Fenleixue*). 2nd edn. Nanjing.

Cheng, N. 1987. *Life and Death in Shanghai*. Grove Press, New York.

Cheng, T.-k. 1969. Jade Flowers and Floral Patterns in Chinese Decorative Art. *Journal of the Institute of Chinese Studies* 2(2): 251–348.

Chin, C.-h. 1989. *Chung-kuo Shu Mu Hsueh* (Explicationes Nominem Scientificiorum Dendrophytorum Chinae). Peking.

Chittenden, F.J. (ed.). 1956. *The Royal Horticultural Society Dictionary of Gardening*. 4 vols. Oxford University Press.

Cibot, P.M. 1777a. Notice du Frêne du Chine Nommé *Hiang-Tchun*. *Mémoires concernant l'Histoire, les Sciences, les Arts, les Moeurs, les Usages, etc. des Chinois par les Missionaires de Pé-kin* 2: 598–601.

—— 1777b. Sur le Bambou. *Ibid.* 2: 623–42.

—— 1778a. Serres Chinoises. *Ibid.* 3: 423–37.

—— 1778b. Notices de quelques Plantes, Arbrisseaux, etc. de la Chine. *Ibid.* 3: 437–99. 1. *Nenuphar de Chine*, pp. 437–41; 2. *Le Yu-lan*, pp. 441–3; 3. *Le Tsieou-hai-tang*, pp. 443–6; 4. *Le Mo-li-hoa*, pp. 446–9; 5. *La Châtaigne d'eau*, pp. 449–51; 6. *Le Lien-kien ou Ki-teou*, pp. 451–5; 7. *Le Kiu-hoa ou la Matricaire de Chine*, pp. 455–61; 8. *Le Mou-tan ou Pivoine, arbrisseau de Chine*, pp. 461–78; 9. *Le Yê-hiang-hoa*, pp. 478–9; 10. *Le Pé-gé-hong*, pp. 480–2; 11. *Le Jujubier*, pp. 482–4; 12. *Le Chêne*, pp. 484–90; 13. *La Chataigner*, pp. 490–4; 14. *Les Oranges-Coings*, pp. 494–9.

—— 1780a. Notes sur Différens Objets. II. Raisins Secs de Ha-Mi. *Ibid.* 5: 481–6.

—— 1780b. Notes sur Différens Objets. VI. Abricotier. *Ibid.* 5: 505–13.

—— 1786a. Le Pêcher. *Ibid.* 11: 280–93.

—— 1786b. Notices *sur* le Mou-chou-kuo-tsée, *le* Chou-keou, & *le* Tsée-tsao. *Ibid.* 11: 294–7.

—— 1786c. Notice *sur* la Pivoine. *Ibid.* 11: 470–92.

—— 1786d. Notice *sur* le Tsao-kia, ou Fébier Chinois. *Ibid.* 11: 493–500.

Clough, R. 1996. Charles Fraser's Record of Plants in the Sydney Botanic Garden in 1827–28. *Australian Garden History* 8(3): 14–16.

Clunas, C. 1996. *Fruitful Sites: Garden Culture in Ming Dynasty China.* Reaktion Books, London.

—— 1997. *Art in China.* Oxford University Press.

Coats, A.M. 1963. *Garden Shrubs and their Histories.* Vista Books, London.

—— 1968. *Flowers and their Histories.* A. & C. Black, London.

Collas, L. 1786a. Observations sur les plantes, les fleurs & les arbres de Chine, qu'il est possible & utile de se procurer en France. *Mémoires concernant l'Histoire, les Sciences, les Arts, les Moeurs, les Usages, etc. des Chinois par les Missionaires de Pékin* 11: 183–268.

—— 1786b. Sur le Bambou. *Ibid.* 11: 353–4.

Cooper, W.C. 1989. *Odyssey of the Orange in China.* Dr William C. Cooper, Winter Park, Florida.

Couling, S. 1917. *The Encyclopaedia Sinica.* Kelly and Walsh, Shanghai.

Cox, E.H.M. 1944. The Hon'ble East India Company and China. *Proceedings of the Linnean Society of London* Sess. 156(1): 5–7.

—— 1986. *Plant-hunting in China.* Oxford University Press. First published 1945.

Crawford, M. 1997. *Bamboos.* 2nd edn. Agroforestry Research Trust, Totnes, Devon.

Crook, A.H. 1935. The Chinese Habit of Eating Melon-Seeds. *Hong Kong Naturalist* 6(1): 18–22.

Cunningham, I.S. 1984. *Frank N. Meyer, Plant Hunter in Asia.* Iowa State University Press, Ames.

Curtis's Botanical Magazine. 1787–. London.

Cusack, V. 1997. *Bamboo Rediscovered.* Earth Garden Books, Trentham, Victoria.

Davidian, H.H. 1995. *The Rhododendron Species.* Vol. IV. Timber Press, Portland.

den Boer, A.F. 1959. *Ornamental Crab Apples.* American Association of Nurserymen.

Der Ling, Princess. 1911. *Two Years in the Forbidden City.* Moffat, Yard and Co., New York.

Doolittle, J. 1866. *Social Life of the Chinese.* Sampson Low, Son, & Marston, London.

Dorsett, P.H. 1928. Chinese Sacred Lilies in Pekin. *National Horticultural Magazine* 7(4): 147–8.

—— 1931. Glimpses of the white-barked pine in Peiping. *National Horticultural Magazine* 10(4): 237–9.

Du Cane, F. 1908. *The Flowers and Gardens of Japan.* A. & C. Black, London.

Du Halde, J.B. 1736. *Description Géographique, Historique, Chronologique, Politique, et Physique de l'Empire de la Chine et de la Tartarie Chinoise.* 4 vols. The Hague.

Du Puy, D. and Cribb, P. 1988. *The Genus Cymbidium.* Christopher Helm, London.

Dykes, W.R. 1974. *The Genus Iris.* Dover Publications, Inc., New York.

Eberhard, W. 1958. *Chinese Festivals.* Abelard-Schuman Ltd, London and New York.

Edwards' Ornamental Flower Garden (The Ornamental flower Garden and Shrubbery). 1854. 4 vols. G. Willis, London.

Edwards's Botanical Register. 1815–47. London.

Engel, D.H. 1986. *Creating a Chinese Garden.* Timber Press, Portland.

Evreinoff, V.A. 1934. Les Pêches 'Peen-Too'. *Revue Horticole* 1934 (1): 11–13.

Faber, E., and Macgregor, D. 1907. Contribution to the Nomenclature of Chinese Plants. *Journal of the North China Branch of the Royal Asiatic Society* 38: 97–164.

Farrer, R. 1916. Report of Work in 1914 and 1915 in Kansu and Tibet. *Journal of the Royal Horticultural Society* 42(1): 47–114.

—— 1917. *On the Eaves of the World.* 2 vols. Reprint by Theophrastus Publishers, Little Compton, Rhode Island, 1977.

Feng, G., Xia, L., and Zhu, X. 1986. *Yunnan Camellias of China.* Science Press, Beijing.

Ferguson, R. 1997. Yellow Camellias. *Rhododendrons with Camellias and Magnolias* 1996–97: 37-9, fig. 11.

Ferry, E.S. 1949. Flower Compositions for the Home in Eastern China. I. The Chinese Manner of Flower Arrangement. *Journal of the New York Botanical Garden* 50(589): 8–15.

Fiala, J.L. 1988. *Lilacs.* Timber Press, Portland.

—— 1994. *Flowering Crabapples.* Timber Press, Portland.

Fong, W., and Fu, M. 1973. *Sung and Yuan Paintings.* Metropolitan Museum of Art, New York.

Forrest, T. 1995. Two Thousand Years of Eating Bark: *Magnolia officinalis* var. *biloba* and *Eucommia ulmoides* in Traditional Chinese Medicine. *Arnoldia* 55(2): 12–18.

Fortune, R. 1844. Letter from Hong Kong. *Gardeners' Chronicle* 1844: 590.

—— 1846. Sketch of a Visit to China, in search of New Plants. *Journal of the Horticultural Society of London* 1: 208–24.

—— 1847a. *Three Years' Wanderings in the Northern Provinces of China.* John Murray, London.

—— 1847b. *Dielytra spectabilis*, a new Plant introduced from China. *Journal of the Horticultural Society of London* 2: 178–9.

—— 1848. New Plants, etc., from the Society's Garden. 17. *Citrus japonica. Journal of the Horticultural Society of London* 3: 239–40.

—— 1849a. Notes of a Traveller. III. *Gardeners' Chronicle* 1849: 197–8.

—— 1849b. Notes of a Traveller. IV. *Ibid.* 1849: 214.

—— 1849c. 'Sacred Bamboo'. *Ibid.* 1849: 277.

—— 1849d. Notes of a Traveller. VI. *Ibid.* 1849: 484–5.

—— 1850a. Notes of a Traveller. VII. *Ibid.* 1850: 70.

—— 1850b. Notes of a Traveller. VIII. *Ibid.* 1850: 84.

—— 1850c. Notes of a Traveller. X. *Ibid.* 1850: 212.

—— 1850d. Notes of a Traveller. XI. *Ibid.* 1850: 228.

—— 1850e. Notes of a Traveller. XII. *Ibid.* 1850: 372.

—— 1850f. Notes of a Traveller. XIII. *Ibid.* 1850: 757.

—— 1850g. Notes of a Traveller. XIV. *Ibid.* 1850: 821.

———— 1851. Notes of a Traveller. XVI. *Ibid.* 1851: 228–9.

———— 1852. *A Journey to the Tea Countries of China.* John Murray, London.

———— 1853a. Leaves from my Chinese Note Book. II. *Gardeners' Chronicle* 1853: 741–2.

———— 1853b. Leaves from my Chinese Note Book. III. *Ibid.* 1853: 822.

———— 1854. Leaves from my Chinese Note Book. IV. *Ibid.* 1854: 54.

———— 1855a. Leaves from my Chinese Note Book. VIII. *Ibid.* 1855: 242.

———— 1855b. Leaves from my Chinese Note Book. X. *Ibid.* 1855: 502–3.

———— 1857. A *Residence among the Chinese.* John Murray, London.

———— 1863. *Yedo and Peking.* John Murray, London.

———— 1864. Note on a New Forsythia from Peking. *Gardeners' Chronicle* 1864: 412.

———— 1880. The Chinese Tree Peony. *Ibid.* 1880: 179–80.

Fourcade, F. 1965. *Peking Museum Painting and Ceramics.* Thames and Hudson, London.

Fox, H.M. 1949. *Abbe David's Diary.* Harvard University Press, Cambridge, Mass.

Freeman-Mitford, A.B. 1900. *The Attache at Peking.* Macmillan, London.

Froncek, T. (ed.). 1969. *The Horizon Book of the Arts of China.* American Heritage Publishing Co., New York.

Fusek, L. 1982. *Among the Flowers. The* Hua-chienchi. Columbia University Press, New York.

Galle, F.C. 1985. *Azaleas.* Timber Press, Portland, Oregon.

Gardeners' Chronicle. 1856. Obituary of John Reeves. *Gardeners' Chronicle* 1856: 212.

Geil, W.E. 1911. *Eighteen Capitals of China.* J.B. Lippincott Co., Philadelphia & London.

Giles, H.A. 1901. *A History of Chinese Literature.* William Heinemann, London.

Glidden, H.W. 1937. The Lemon in Asia and Europe. *Journal of the American Oriental Society* 57: 381–96.

Goody, J. 1993. *The Culture of Flowers.* Cambridge University Press.

Gordon, G. 1846. Some Account of the Cryptomeria Japonica, or Japan Cedar. *Journal of the Horticultural Society of London* 1: 57–60.

Gorer, R. 1970. *The Development of Garden Flowers.* Eyre and Spottiswood, London.

Graham, D. 1938. *Chinese Gardens.* Dodd, Mead & Co., New York.

Green, P.S., and Chang, M.C. 1995. Some Taxonomic Changes in *Syringa* L. (Oleaceae), Including a Revision of Series *Pubescentes. Novon* 5(4): 329–33.

Greuter, W. 1995. *Silene* (Caryophyllaceae) in

Greece: a subgeneric and sectional classification. *Taxon* 44: 543–81.

Grey-Wilson, C. 1993. *Poppies.* Timber Press, Portland.

Gulland, W.G. 1928. *Chinese Porcelain.* Vol. 1. Chapman & Hall, London.

H.F.W.H. 1868. The Wax Insect. *Notes and Queries on China and Japan* 2(11): 161–2.

Hagerty, M.J. 1923. Han Yen-Chih's Chü Lu (Monograph on the Oranges of Wen-Chou, Chekiang). Introduction by Paul Pelliot. *Toung Pao* 22: 63–96.

Hargett, J.M. 1988-89. Huizong's Magic Marchmount: The Genyue Pleasure Park of Kaifeng. *Monumenta Serica* 38: 1–48.

Harkness, J. 1978. *Roses.* J.M. Dent & Sons, London, Toronto, Melbourne.

Harrist, R.E. Jr. 1987. Ch'ien Hsüan's *Pear Blossoms*: The Tradition of Flower Painting and Poetry from Sung to Yuan. *Metropolitan Museum Journal* 22: 53–70.

———— 1993. Site names and their meanings in the Garden of Solitary Enjoyment. *Journal of Garden History* 13(4): 199–212.

Hattori, K., and Grounds, R. 1994. Plant of Good Fortune. *The Garden* 119(7): 316–18.

Haw, S. 1984. Lilies in China. *The Garden* 109(9): 367–9.

———— 1985a. China teas. *Ibid.* 110(2): 84–7.

———— 1985b. Mudan: the king of flowers. *Ibid.* 110(4): 154–9.

———— 1985c. *Bletilla striata. Ibid.* 110(7): 328–9.

———— 1986a. The origins of the garden chrysanthemum. *Ibid.* 111(11): 525–8.

———— 1986b. *The Lilies of China.* B.T. Batsford, London.

———— 1987. Chinese flowering plums and cherries. *The Garden* 112(5): 224–8.

Haworth-Booth, M. 1984. *The Hydrangeas.* 5th edn. Constable, London.

Hearn, M.K. 1996. *Splendors of Imperial China.* Metropolitan Museum of Art, New York; National Palace Museum, Taipei; Rizzoli International Publications, New York.

Henry, A. 1888. Chinese Names of Plants. *Journal of the China Branch of the Royal Asiatic Society* 22: 233–83.

Henry, L. 1902. L'Armandier de David. *Revue Horticole* 1902: 290–2.

Herklots, G.A.C. 1932a. Flowering Shrubs and Trees of Hong Kong. *Hong Kong Naturalist* 3(1): 16–31.

———— 1932b. *Lilium brownii* var. *colchesteri,* Wilson. *Ibid.* 3(2): 82–6.

———— 1932c. Flowering Shrubs and Trees of Hong Kong. Part III. *Ibid.* 3(3–4): 194–206.

———— 1936. Flowering Shrubs etc. Part 7. *Ibid.* 7(3–4): 224–32.

———— 1937. Flowering Shrubs etc. Part 8. *Ibid.* 8(2): 94–105.

———— 1964. The Foxglove Tree, *Paulownia fortunei* Hemsley. *Journal of the Royal Horticultural Society* 89(7): 300–2.

———— 1972. *Vegetables in South-East Asia.* George Allen & Unwin, London.

Ho, P.-T. 1955. The Introduction of American Food Plants into China. *American Anthropologist* 57: 191–201.

Hoffman, D.I. 1849. Notes Relating to the History, Distribution and Cultivation of the Paeony in China and Japan. *Paxton's Magazine of Botany and Register of Flowering Plants* 16: 85–9, 109–14.

Holmes, W.B.F. 1996. *Ginkgo biloba,* the last of an illustrious line. *International Dendrology Society Yearbook* 1995: 38–43.

Hooker, W. 1858. *Camellia rosaeflora. Curtis's Botanical Magazine* 84: t. 5044.

Hookham, H. 1972. *A Short History of China.* The New American Library, Inc., New York.

Hou, M.-k. 1986. *Chung-kuo Ti Hua Hui.* Beijing.

Hu, H.H. 1938. Recent Progress in Botanical Exploration in China. *Journal of the Royal Horticultural Society* 63(8): 381–9.

Hu, S.-y. 1959. A Monograph of the Genus Paulownia. *Quarterly Journal of the Taiwan Museum* 12(1-2): 1–54.

———— 1971. Orchids in the Life and Culture of the Chinese People. *Chung Chi Journal* 10(1,2): 1–27.

Hu, Y. 1982. *Penjing.* Timber Press, Portland.

———— 1988. *Chinese Penjing.* Timber Press, Portland.

Hu, T.-W., and Chang, H. J. 1975. A New Species of Paulownia from Taiwan—*P. taiwaniana* Hu & Chang. *Taiwania* 20(2): 165–71.

Huang, H. T., and Pei, Y. 1987. The Ancient Cultured Citrus Ant. *Bioscience* 37(9): 665–71.

Hubbard, G.E. 1923. *The Temples of the Western Hills visited from Peking.* La Librairie Francaise, Peking and Tientsin.

Hulton, P., and Smith, L. 1979. *Flowers in Art from East and West.* British Museum Publications, London.

Hume, H.H. 1957. *Citrus Fruits.* Macmillan, New York.

Hurst, C.C. 1941. Notes on the Origin and Evolution of our Garden Roses. II. Modern Garden Roses (1800–1940). *Journal of the Royal Horticultural Society* 66: 242–50, 282–9.

Huxley, A., Griffiths, M., and Levy, M. (eds). 1992. *The New Royal Horticultural Society Dictionary of Gardening.* 4 vols. The Macmillan Press, London.

Ingram, C. 1948. *Ornamental Cherries.* Country Life, London.

Jieziyuan Huazhuan (Mustard Seed Garden Painting Manual). 1975. Zhonghua Shuju, Hong Kong. Originally published 1677–1701.

Johnstone, G.H. 1955. *Asiatic Magnolias in Cultivation*. The Royal Horticultural Society, London.

K. 1870. Grafting. *Notes and Queries on China and Japan* 4(1): 6.

Kaempfer, E. 1712. *Amoenitatum Exoticarum*.

Kao, M. (ed.). 1988. *Paintings of the Ming Dynasty from the Palace Museum*. The Art Gallery, Institute of Chinese Studies, The Chinese University of Hong Kong.

Kates, G.N. 1967. *The Years That Were Fat: the last of old China*. M.I.T. Press, Cambridge, Mass.

Keswick, M. 1978. *The Chinese Garden*. Academy Editions, London.

Knowlton, M.J. 1868. The Tallow Tree. *Notes and Queries on China and Japan* 2(7): 112.

Koehn, A. 1952. Chinese Flower Symbolism. *Monumenta Nipponica* 8(1,2): 121–46.

Krahl, R. 1987. Plant Motifs of Chinese Porcelain: Examples from the Topkapi Saray Identified through the Bencao Gangmu. Part 1. *Orientations* 18(5): 52–65; Part 2. *Ibid*. 18(6):24–37.

L. 1870. The Grease Tree. *Notes and Queries on China and Japan* 4(4): 64.

Lai, T.C. 1977. *Noble Fragrance*. Swindon Book Company, Kowloon, Hong Kong.

Lancaster, R. 1989. *Travels in China*. Antique Collectors' Club, Woodbridge, Suffolk.

Larkcom, J. 1991. *Oriental Vegetables*. John Murray, London.

Lauener, L.A. 1996. *The Introduction of Chinese Plants into Europe*. D.K. Ferguson (ed.). SPB Academic Publishing, Amsterdam.

Laufer, B. 1934. The Lemon in China and Elsewhere. *Journal of the American Oriental Society* 54: 143–60.

—————— 1967. *Sino-Iranica*. Ch'eng-Wen Publishing Company, Taipei.

Lay, G.T. 1846. Outlines of a Natural History Calendar at Foo-chow-foo, the capital of the Chinese province of Fokien. *Journal of the Horticultural Society of London* 1:119–26.

Le Rougetel, H. 1982. The Fa Tee Nurseries of South China. *Garden History* 10(1): 70–73.

—————— 1988. *A Heritage of Roses*. Stenmer House Publishers, Owings Mills, Maryland.

Lee, F.P. 1978. *The Azalea Book*. Theophrastus, Little Compton, Rhode Island. Reprint of 2nd edn, 1965.

Li, H.-L. 1955. Cedrela sinensis. *Morris Arboretum Bulletin* 6(4): 42–4.

—————— 1956a. *Chinese Flower Arrangement*. Hedera House, Philadelphia.

—————— 1956b. A Horticultural and Botanical History of Ginkgo. *Morris Arboretum Bulletin* 7(1): 3–12.

—————— 1956c. The Story of the Cultivated Horse-chestnuts. *Ibid*. 7(3): 35–9.

—————— 1957. The Discovery and Cultivation of Metasequoia. *Ibid*. 8(4): 49–53.

—————— 1958. The Cultivated Lindens. *Ibid*. 9(3): 39–44.

—————— 1959. *The Garden Flowers of China*. Ronald Press, New York.

—————— 1963a. *The Origin and Cultivation of Shade and Ornamental Trees*. University of Pennsylvania Press, Philadelphia.

—————— 1963b. The Cultivated Mahonias. *Morris Arboretum Bulletin* 14(3): 43–50.

—————— 1968a. The Lace-bark Pine. *Ibid*. 19(1): 3–7.

—————— 1968b. The Golden Larch, *Pseudolarix amabilis*. *Ibid*. 19(2): 19–25.

—————— 1979. *Nan-fang Ts'ao-mu Chuang*. The Chinese University Press, Hong Kong.

—————— and Wright, J.W. 1954. The Chinese Tulip Tree (*Liriodendron chinense*) in Northeastern United States. *Morris Arboretum Bulletin* 5(3): 34–5.

Li, J.-c. 1965. *Flowers in the Mirror*. Translated from the Chinese by Ching Hua Yuan. Peter Owen, London.

Li, L.-t., and She, C. (eds). 1971. *Masterpieces of Chinese Album Painting in the National Palace Museum*. National Palace Museum, Taipei.

Lian, W., and Hu, T. (Editors-in-Chief). 1988. *Zhongguo Bencao Tulu* (Illustrated Chinese Herbal). 10 vols. Commercial Press (Hong Kong), Hong Kong.

Lin, J.-x., Hu, Y.-s., and Wang, X.-p. 1996. Old *Ginkgo* trees in China. *International Dendrology Society Yearbook* 1995: 32–7.

Lindley, J. 1827. Camellia reticulata, Captain Rawes's Camellia. *Edwards's Botanical Register* 13: t. 1078.

—————— 1828. *Renanthera coccinea*—Scarlet Air-Plant. *Ibid*. 14: t. 1131.

—————— 1846. New Plants, Etc., From the Society's Garden. *Journal of the Horticultural Society of London* 1: 61–77, 146–59, 226–40, 298–308.

—————— 1857. *Camellia reticulata flore pleno*. *Curtis's Botanical Magazine* 83: t. 4976.

Lindley, Professor, and Paxton, Sir Joseph. 1853. *Paxton's Flower Garden*. 3 vols. Bradbury and Evans, London.

Little, A.J. 1888. *Through the Yang-tse Gorges or Trade and Travel in Western China*. Sampson Low, Marston, Searle, & Rivington, London.

Little, Mrs A. 1901. *Intimate China*. Hutchinson & Co., London.

Liu, D. 1982. *The Traditional Gardens of Suzhou*. Abridged trans. Frances Wood. *Garden History* 10(2): 108–41.

Livingstone, J. 1820a. Account of a Method of ripening Seeds in a wet Season; with some Notices of the Cultivation of certain Vegetables and Plants in China. *Transactions of the Horticultural Society of London* 3: 183–5.

—————— 1820b. Observations on the Difficulties which have existed in the Transportation of Plants from China to England, and Suggestions for obviating them. *Ibid*. 3: 421–9.

—————— 1822. Account of the Method of Dwarfing Trees and Shrubs, as practised by the Chinese, including their Plan of Propagation from Branches. *Ibid*. 4: 224–31.

—————— 1824. On the State of Chinese Horticulture and Agriculture; with an Account of several Esculent Vegetables used in China. *Ibid*. 5: 49–56.

Loureiro, J. de. 1790. *Flora Cochinchinensis*. 2 vols. Lisbon.

Mabberley, D.J. 1997a. *The Plant Book*. 2nd edn. Cambridge University Press.

—————— 1997b. A classification for edible *Citrus* (Rutaceae). *Telopea* 7(2): 167–72.

—————— 1998. Australian Citreae with notes on other Aurantioideae (Rutaceae). *Telopea* 7(4): 333-44.

McClure, F.A. 1932. The Chinese Sacred Lily. *Hong Kong Naturalist* 3(3,4): 186–93.

—————— 1966. *The Bamboos*. Harvard University Press, Cambridge, Mass.

Macmillan, H.F. 1991. *Tropical Planting and Gardening*. 6th edn, rev. H.S. Barlow, I. Enoch and R.A. Russell. Malayan Nature Society, Kuala Lumpur.

McMullen, D.L. 1987. Review of *Chinese Ornament: The Lotus and the Dragon* by Jessica Rawson. *Modern Asian Studies* 21(1): 198–200.

Major, R.H. 1854. In J. Gonzalez de Mendoza, *The History of the Great and Mighty Kingdom of China*, as publ. Hakluyt Society 1854. Republ. Lenox Hill, New York, 1970.

Matsumura, J. 1915. *Shokubutsu-Mei-I*. Part 1. Maruzen Company, Tokyo, Osaka, Kyoto & Fukuoka.

Mendoza, J. Gonzalez de. 1588. *The History of the Great and Mighty Kingdom of China*. Originally publ. Rome 1585. Original trans. R. Parke. Reprinted by Hakluyt Society 1854, with intro. by R.H. Major. Republ. Lenox Hill, New York, 1970.

Merrill, E.D. 1954. *The Botany of Cook's Voyages*. Chronica Botanica Company, Waltham, Mass.

Metcalf, F.P. 1942. Flowers of the Chinese New Year. *Arnoldia* 32(1): 1–8.

Meyer, F.N. 1911. *Agricultural Explorations in the Fruit and Nut Orchards of China.* US Dept Agriculture Bureau of Plant Industry Bulletin no. 204. Government Printing Office, Washington, D.C.

——— 1916. China a Fruitful Field for Plant Exploration. *US Dept Agriculture Yearbook* 1915: 205–24.

Millais, J.G. 1917. *Rhododendron Species and the Various Hybrids.* Vol. 1. Longmans, Green, London.

Mirov, N.T. 1967. *The Genus Pinus.* Ronald Press, New York.

Morris, E.T. 1983. *The Gardens of China.* Charles Scribner's Sons, New York.

Morrison, R. 1822. Names of Plants which flower or blossom in each month of the year at Canton, in China. *A Dictionary of the Chinese Language.* The Honorable East India Company's Press, Macao. Vol. 3: 172–4.

Na, C.-l., Shih, L.-y., and Hu, S.-l. 1969. *Masterpieces of Chinese Painting in the National Palace Museum.* National Palace Museum, Taipei.

Needham, J. 1954. *Science and Civilisation in China.* Vol. I. *Introductory Orientations.* With research assistance by Wang Ling. Cambridge University Press.

——— 1956. *Ibid.* Vol. 2. *History of Scientific Thought.* With research assistance by Wang Ling. Cambridge University Press.

——— 1971. *Ibid.* Vol. 4. *Physics and Physical Technology.* Pt. 3. *Civil Engineering and Nautics.* With collaboration of Wang Ling and Lu Gwei-djen. Cambridge University Press.

——— 1974. *Ibid.* Vol. 5. *Chemistry and Chemical Technology.* Pt. 2. *Spagyrical Discovery and Invention: Magisteries of Gold and Immortality.* With collaboration of Lu Gwei-djen. Cambridge University Press.

——— 1984. *Ibid.* 1984. Vol. 6. *Biology and Biological Technology.* Pt. 2. *Agriculture,* by Francesca Bray. Cambridge University Press.

——— 1986. *Ibid.* Vol. 6. *Biology and Biological Technology.* Pt. 1. *Botany.* With the collaboration of Lu Gwei-djen, special contribution by Huang Hsing-tsung. Cambridge University Press.

——— 1988. *Ibid.* Vol. 5. *Chemistry and Chemical Technology.* Pt. 9. *Textile Technology: Spinning and Reeling,* by Dieter Kuhn. Cambridge University Press.

——— 1996. Ibid. Vol. 6. *Biology and Biological Technology.* Pt. 3. *Agro-industries and Forestry,* by Christian Daniels and Nicholas K. Menzies. Cambridge University Press.

——— and Lu Gwei-Djen. 1968. The Esculentist Movement in Mediaeval Chinese Botany; Studies on Wild (Emergency) Food Plants. *Archives Internationales d'Histoire des Sciences* 21(84,85): 225–48.

Nelmes, E. 1944. Robert Fortune, Pioneer Collector, the Centenary of Whose Departure for China has fallen this Year. *Proceedings of the Linnean Society of London* Sess. 156(1): 8–14.

Ogisu, M. 1996. Some thoughts on the history of China roses. *The New Plantsman* 3(3): 152–7.

Ohga, I. 1923. On the Longevity of Seeds of Nelumbo nucifera. *Botanical Magazine (Tokyo)* 37: 87–95.

——— 1926. On the Structure of Some Ancient but still Viable Fruits of Indian Lotus, with Special Reference to Their Prolonged Dormancy. *Japanese Journal of Botany* 3: 1–20.

Osti, G.L. 1994. Tree peonies revisited. *The New Plantsman* 1(4): 195–205, 247.

Page, M. 1997. *The Gardener's Guide to Growing Peonies.* Florilegium, Sydney.

Pang, J., Feng, Z., Zhu, B., and Guo, S. 1995. *Camellias of China.* China Esperanto Press, Beijing.

Parks, C.R., Kondo, K., and Swain, T. 1981. Phytochemical evidence for the genetic contamination of *Camellia sasanqua* Thunberg. *Japanese Journal of Breeding* 31: 167–82.

Payne, C.H. 1890. A Brief History of the Chrysanthemum. *Journal of the Royal Horticultural Society* 12: 115–21.

Peer, R.S. 1951. New Varieties of Camellia reticulata Established in California. *Journal of the Royal Horticultural Society* 76: 301–7.

Peking: A Tourist Guide. 1960. Foreign Languages Press, Peking.

Phillips, R., and Rix, M. 1993. *The Quest for the Rose.* BBC Books, London.

Phipps, J.B., Robertson, K.R., Smith, P.G., and Rohrer, J.R. 1990. A checklist of the subfamily Maloideae (Rosaceae). *Canadian Journal of Botany* 68: 2209–69.

Pizzetti, I., and Cocker, H. 1975. *Flowers: A Guide for Your Garden.* 2 vols. Harry N. Abrams, New York.

Purseglove, J.W. 1968. *Tropical Crops.* Dicotyledons 2, p. 500. John Wiley and Sons, New York.

Read, B.E. 1936. *Chinese Medicinal Plants from the Pen Ts'ao Kang Mu.* 3rd edn. Peking Natural History Bulletin, Peking.

Recht, C., and Wetterwald, M.F. 1992. *Bamboos.* Timber Press, Portland, Oregon.

Reynolds, P.K. 1951. Earliest Evidence of Banana Culture. *Journal of the American Oriental Society.* Suppl. No. 12.

——— and Fang, C.Y. 1940. The Banana in Chinese Literature. *Harvard Journal of Asiatic Studies* 5: 165–81.

Richards, P. 1997. Builders of Bridges. *CAM* (Cambridge Alumni Magazine), Easter 1997: 15–18.

Rivoire, P. 1928. Le Chrysanthème en Chine au Dix-Septième Siècle. *Revue Horticole* 1928: 211–13.

Rushforth, K.D. 1987. *Conifers.* Facts on File Publications, New York and Oxford.

Sabine, J. 1817. An account of seven double herbaceous peonies now cultivated in England. *Transactions of the Horticultural Society of London* 2: 273–81.

——— 1822. Account and Description of the Varieties of Chinese Chrysanthemum which are at present cultivated in England. *Ibid.* 4: 326–54.

——— 1826. On the *Paeonia Moutan,* or *Tree Peony* and its varieties. *Ibid.* 6: 465–92.

Sampson, T. 1868. The China Pine. *Notes and Queries on China and Japan* 2(4): 52-53.

——— 1869a. The Fung Tree. *Ibid.* 3(1): 4–7.

——— 1869b. Grapes in China. *Ibid.* 3(4): 50–4.

——— 1869c. The P'u-T'i Tree. *Ibid.* 3(7): 100–5.

——— 1869d. Palm Trees. *Ibid.* 3: 115–17, 129–30, 147–50, 170–2.

Sargent, C.S. 1917. *Plantae Wilsonianae.* Vol. 3. Cambridge University Press. Repr. 1988, Dioscorides Press, Portland.

Saunt, J. 1990. *Citrus Varieties of the World.* Sinclair International, Norwich, England.

Savige, T.J. 1993. *The International Camellia Register.* 2 vols. The International Camellia Society.

Schafer, E.H. 1948. Notes on a Chinese Word for Jasmine. *Journal of the American Oriental Society* 68(1): 60–5.

——— 1963. *The Golden Peaches of Samarkand.* University of California Press, Berkeley and Los Angeles.

——— 1965. Li Te-Yü and the Azalea. *Asiatische Studien* 18–19: 105–13.

——— 1967. *The Vermilion Bird.* University of California Press, Berkeley and Los Angeles.

Schmidt, M., and Schmoll, H. 1994. *Ginkgo, Ur-baum und Arzneipflanze, Mythos, Dichtung und Kunst.* Wissenschaftliche Verlagsgesellschaft, Stuttgart.

Sealy, J.R. 1937. Species of Camellia in Cultivation. *Journal of the Royal Horticultural Society* 62(8): 352–69.

Sealy, J.R. 1954. Camellia oleifera. *Curtis's Botanical Magazine* 170: t. 221.

She, C., and Yuhas, L. 1973. *Masterpieces of Chinese Painting in the National Palace Museum.* Supplement. National Palace Museum, Taipei.

Shen, F. 1983. *Six Records of a Floating Life.* Trans. from orig. 1809, intro. and notes by Leonard Pratt and Chiang Su-hui. Penguin Books.

Siebold, P.F. von, and Zuccarini, J.G. 1835–41. *Flora Japonica.* Vol. 1. Lugduni Batavorum.

Sims, J. 1819. *Camellia sasanqua,* Palmer's double sasanqua. *Curtis's Botanical Magazine* 46: t. 2080.

Sirén, O. 1949. *Gardens of China.* Ronald Press, New York.

—— 1973. *Chinese Painting.* 7 vols. Hacker Art Books, New York.

Sitwell, O. 1935. *Penny Foolish.* Macmillan, London.

—— 1949. *Escape with Me!* Macmillan, London.

—— 1974. *Queen Mary and Others.* Michael Joseph, London.

Smith, F.P. 1869. The Wax Tree. *Notes and Queries on China and Japan* 3(6): 88.

South China Institute of Botany. 1997. *The International Symposium on the Family Magnoliaceae, 18–22 May, 1998. First Circular.*

Spence, J. 1996. *God's Chinese Son.* Harper Collins, London.

Spongberg, S.A. 1990. *A Reunion of Trees.* Harvard University Press, Cambridge, Mass., and London, England.

—— 1993. Exploration and Introduction of Ornamental and Landscape Plants from Eastern Asia. In: Janick, J. and Simon, J.E. (eds). *New Crops.* Wiley, New York.

Steward, A.N. 1958. *Manual of the Vascular Plants of the Lower Yangtze Valley China.* Oregon State College, Corvallis.

Stout, A.B. 1933. Gum-Jum or Gum-Tsoy: a food from the flowers of daylilies. *Journal of the New York Botanical Garden* 34: 97–100.

—— 1935. The Lemon Daylily (*Hemerocallis flava* L.): its origin and status. *Ibid.* 36: 61–8.

—— 1942. Daylilies of Chinese Origin. *Ibid.* 43: 237–43.

Sun, W., Kong, F., and Yue, Z. 1998. Magnoliaceae in Kunming Botanical Garden, Yunnan, China. In Hunt, D. (ed.). *Magnolias and their allies.* The International Dendrology Society and The Magnolia Society, 1998, pp. 284–8.

Suzuki, K. 1982. *Comprehensive Illustrated Catalog of Chinese Paintings.* 5 vols. University of Tokyo Press.

Swingle, W.T. 1915. A new genus, *Fortunella,* comprising four species of kumquat oranges. *Journal of the Washington Academy of Science* 5: 165–76.

—— and Reece, P.C. 1967. The Botany of Citrus and its Wild Relatives. In Reuter, W, Webber, H.J., and Batchelor, L.D. (eds). *The Citrus Industry.* Rev. edn. University of California, Division of Agricultural Sciences. Vol. I: 190–430.

Synge, P.M. 1953. Chinese Flower Paintings. *Journal of the Royal Horticultural Society* 78: 209–13.

—— (ed.). 1969. *Royal Horticultural Society Dictionary of Gardening. Supplement.* Clarendon Press, Oxford.

Sze, M.-m. 1977. *The Mustard Seed Garden Manual of Painting.* Princeton University Press, Princeton, New Jersey.

Thrower, S.L. 1988. *Hong Kong Trees.* Urban Council, Hong Kong.

Tolkowsky, S. 1937. *Hesperides.* Staples and Staples, London.

Trehane, P. 1995. Proposal to conserve *Chrysanthemum* L. with a conserved type (Compositae). *Taxon* 44(3): 439–41.

——, Brickell, C.D., Baum, B.R., Hetterscheid, W.L.A., Leslie, A.C., McNeill, J., Spongberg, S.A., and Vrugtman, F. (eds). 1995. *International Code of Nomenclature for Cultivated Plants—1995.* Quarterjack Publishing, Wimborne, UK.

Tschichold, J. 1952. In *Chinese Color-Prints from the Painting Manual of the Mustard Seed Garden.* George Allen and Unwin, London; Holbein, Basel.

Tun, L.-c. 1965. *Annual Customs and Festivals in Peking.* As recorded in the *Yen-ching Sui-shih-chi* by Tun Li-Ch'en, c.1901, trans. and annotated Derek Bodde. Hong Kong University Press, Hong Kong.

Uehara, K. 1961. *Jumoku Dai-Zusetsu* (Illustrated Dictionary of Trees). Ariake Shobo.

Urban Services Department. 1971. *Hong Kong Shrubs.* Government Printer, Hong Kong.

Usher, G. 1974. *A Dictionary of Plants Used by Man.* Constable, London.

Valder, P. 1995. *Wisterias.* Florilegium, Sydney.

Van Braam, A.E. 1798. *An authentic account of the Embassy of the Dutch East India Company to the court of the Emperor of China in the years 1794 and 1795.* 2 vols. London.

van Gelderen, D.M., de Jong, P.C., and Oterdoom, H.J. 1994. *Maples of the World.* Timber Press, Portland.

Waddick, J.W., and Zhao, Y.-t. 1992. *Iris of China.* Timber Press, Portland.

Waley, A. 1938. *The Analects of Confucius.* George Allen & Unwin, London.

—— 1969. *The Book of Songs.* 2nd edn. Allen & Unwin, London.

Walker, E.H. 1944. The plants of China and their Usefulness to Man. *Annual Report of the Smithsonian Institution* 1943: 325–61.

Wang, D. 1988. The History of ornamental Plants in China. *Camellia News* 107: 14–16.

—— and Shen, S.-J. 1987. *Bamboos of China.* Christopher Helm, London.

Wang, F., and Barnhart, R.M. 1990. *Master of the Lotus Garden. The Life and Art of Bada Shanren (1626–1705).* Yale University Press, New Haven and London.

Wang, J., and Ma, Y. 1995. *China's Rare Flowers.* Trans. Deng Xin. Morning Glory Press, Beijing.

Watanabe, S. 1990. *The Fascinating World of Lotus.* Parks and Open Space Assoc. of Japan.

Waters, G. 1996. Yet More Wild Irises. *Pacific Horticulture* 57(2): 23–6.

Watling, H. 1928. Researches into Chinese Orchid History. *The Orchid Review* 36: 235–40, 295–304.

W[atson], W. 1898. Callistephus hortensis (The China Aster). *The Garden* 53: 258–9.

Waugh, T. 1984. The Travels of Marco Polo. Sidgwick & Jackson, London.

Wei, T.-p. 1986. *Ho-nan Ming Hua.* Cheng-chou Shih.

Weidner, M., Laing, E. J., Lo, I.Y., Chu, C., and Robinson, J. 1988. *Views from Jade Terrace: Chinese Women Artists 1300–1912.* Indianapolis Museum of Art and Rizzoli, New York.

Weng, W.-g. 1982. *The Palace Museum: Peking.* Harry N. Abrams, New York.

Westland, A.B. 1894. Chinese Orchids. *Garden and Forest* 7: 76–7.

Whitehead, P.J.P., and Edwards, P.I. 1974. *Chinese Natural History Drawings.* Pub. no. 721, British Museum (Nat. Hist.), London.

Williams, C.A.S. 1975. *Outlines of Chinese Symbolism and Art Motives.* 3rd rev. edn. Charles E. Tuttle, Rutland, Vermont, and Tokyo, Japan.

Wilson, E.H. 1906. *Jasminum primulinum* : its history and culture. *Gardeners' Chronicle* 1906: 44.

—— 1913. *A Naturalist in Western China.* 2 vols. Methuen & Co., London.

—— 1916. The Wistarias of China and Japan. *Gardeners' Chronicle* 1545: 61–2.

—— 1925. *The Lilies of Eastern Asia.* Dulau & Co., London.

—— and Rehder, A. 1921. *A Monograph of Azaleas.* University Press, Cambridge, Mass.

Wister, J.C. (ed.). 1962. *The Peonies.* American Horticultural Society, Washington, D.C.

Wolters, O.W. 1960. The 'Po-Ssu' Pine Trees. *Bulletin of the School of Oriental and African Studies, University of London* 23: 323–50.

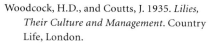

Woodcock, H.D., and Coutts, J. 1935. *Lilies, Their Culture and Management*. Country Life, London.

Wyman, D. 1949. The Wisterias. *Arnoldia* 9 (5–6): 17–28.

Xu, R., and Li, Y. 1992. *Zhongguo Shuixian* (Chinese Narcissus). Fujian Meisu Shubanse.

Yamaguchi, K., and Barry D. 1974. The Culture of Rhapis in Japan. *Principes* 18(3): 75–83.

Young, J., and Chong, L.-s. 1980. *Rhododendrons of China*. Binford & Mort, Portland.

Yu, S.-h., and Wu, Y.-h. 1993. *Hua Hui Tz'u Tien* (A Dictionary of Ornamental Plants). Peking.

Yu, T.-t. 1950. *Camellia reticulata* and its garden varieties. In P.M. Synge (ed.). *Camellias and Magnolias*. Royal Horticultural Society, London. Pp. 13–26.

———— and Bartholemew, B. 1980. The Origin and Classification of the Garden Varieties of Camellia reticulata. *American Camellia Yearbook* 1980: 1–29.

Zee, S.Y., and Hui, L.H. 1990. *Hong Kong Food Plants*. Urban Council, Hong Kong.

Zhu, J. 1992. *Chinese Landscape Gardening*. Foreign Languages Press, Beijing.

Zou, X., Zhao, X., and Jin, X. 1997. *Flowering Lotus of China*. Jindun Publishing House, Beijing.

CHINESE DYNASTIES

XIA	c. 2000 BC	to	c. 1520 BC
SHANG	c. 1500 BC	to	c. 1030 BC
ZHOU	c. 1030 BC	to	220 BC
QIN	221 BC	to	207 BC
HAN	202 BC	to	220 AD
THREE KINGDOMS	221 AD	to	280 AD
SIX DYNASTIES	265 AD	to	589 AD
SUI	581 AD	to	618 AD
TANG	618 AD	to	906 AD
FIVE DYNASTIES	907 AD	to	960 AD
NORTHERN SONG	960 AD	to	1126 AD
SOUTHERN SONG	1127 AD	to	1279 AD
YUAN	1279 AD	to	1368 AD
MING	1368 AD	to	1644 AD
QING	1644 AD	to	1911 AD

INDEX

Entries consist of common and botanical names of plants, general topics
covered in the early chapters, and sites of horticultural interest mentioned
in the text. Page numbers in **bold** refer to illustrations.